Study Guide and Solutions Book

ORGANIC CHEMISTRY: A SHORT COURSE

Seventh Edition

Harold Hart
Michigan State University

HOUGHTON MIFFLIN COMPANY   BOSTON

Dallas   Geneva, Illinois
Lawrenceville, New Jersey   Palo Alto

Printed in the U.S.A.

Library of Congress Catalog Card Number: 86-82495

ISBN: 0-395-42322-8

ABCDEFGHIJ-H-8987

CONTENTS

CONTENTS

CONTENTS

INTRODUCTION FOR THE STUDENT

This study guide and solutions book was written to help you learn organic chemistry. The principles and facts of this subject are not easily learned by simply reading them, even repeatedly. Formulas, equations, and molecular structures are best mastered by *written* practice. To help you become thoroughly familiar with the material, I have included many problems within and at the end of each chapter in the text.

It is my experience that such questions are not put to their best use unless correct answers are also available. Indeed, answers alone are not enough. If you know how to work a problem and find that your answer agrees with the correct one, fine. But what if you work conscientiously, yet cannot solve the problem? You then give in to temptation, look up the answer, and encounter yet another dilemma--how in the world did the author get that answer? This solutions book has been written with this difficulty in mind. For many of the problems, all the reasoning involved in getting the correct answer is spelled out in detail. Many of the answers also include cross-references to the text. If you cannot solve a particular problem, these references will guide you to parts of the text that you should review.

Each chapter in the text is summarized briefly here. Wherever pertinent, the chapter summary is followed by a list of all the new reactions and mechanisms encountered in that chapter. These lists should be especially helpful to you as you review for examinations.

When you study a new subject, it is always useful to know what is expected. To help you, I have included in this study guide a list of learning objectives for each chapter, that is, a list of what you should be able to do after you have read and studied that chapter. Your instructor may want to delete items from these lists of objectives or add to them. However, I believe that if you have mastered these objectives--and the problems should help you to do this--you should have no difficulty with examinations. Furthermore, you should be very well prepared for future courses that require this course as a prerequisite.

Near the end of this study guide you will find some additional sections that may help you to study for the final examination in the course. The SUMMARY OF SYNTHETIC METHODS lists the important ways to synthesize each class of compounds discussed in the text. This is followed by the SUMMARY OF REACTION MECHANISMS. Both of these sections have references back to appropriate portions of the text, in case you feel that further review is necessary. Finally, you will find two lists of sample test questions. The first deals with synthesis, and the second is a list of multiple-choice questions. Both of these sets should help you prepare for examinations.

Finally, I offer you a brief word of advice about how to learn the many reactions you will study during this course. First, learn the nomenclature systems thoroughly for each new class of compounds that is introduced. Then, rather than memorizing the particular examples of reactions given in the text, study reactions as being typical of a class of compounds. For example, if you are asked how compound A will react with compound B, proceed in the following way. First ask yourself, to what class of compounds does A belong? How does this class of compounds react with B (or with compounds of the general class to which B belongs)? Then proceed from the general reaction to the specific case at hand. This approach will probably help you to eliminate some of the memory work often associated with organic chemistry courses.

I urge you to study regularly, and I hope that this study guide and solutions book will make it easier for you to do so.

Great effort has been expended to ensure the accuracy of the answers in this book. It is easy for errors to creep in, however, and I will be particularly grateful to anyone who will call them to my attention. Suggestions for improving the book will also be welcome. Send them to:

Harold Hart
Department of Chemistry
Michigan State University
East Lansing, Michigan 48824

CHAPTER ONE   BONDING AND ISOMERISM

CHAPTER SUMMARY*

An **atom** consists of a nucleus surrounded by **electrons** arranged in **orbitals**. The electrons in the outer shell, or the **valence electrons**, are involved in bonding. **Ionic bonds** are formed by electron transfer from an **electropositive** atom to an **electronegative** atom. Atoms with similar electronegativities form **covalent bonds** by sharing electrons. A **single bond** is the sharing of one electron pair between two atoms.

Carbon, with four valence electrons, mainly forms covalent bonds. It usually forms four such bonds, and these may be with itself or with other atoms such as hydrogen, oxygen, chlorine, and sulfur. In pure covalent bonds, electrons are shared equally, but in **polar covalent bonds**, the electrons are displaced toward the more electronegative element. **Multiple bonds** consist of two or three electron pairs shared between atoms.

Structural (or **constitutional**) **isomers** are compounds with the same **molecular formulas** but different **structural formulas** (that is, different arrangements of the atoms in the molecule). **Isomerism** is especially important in organic chemistry because of the capacity of carbon atoms to be arranged in so many different ways: continuous chains, branched chains, and rings.

Structural formulas can be written so that every bond is shown, or in various abbreviated forms. For example, the formula for n-pentane (n stands for normal) can be written as

$$
\begin{array}{ccccc}
\text{H} & \text{H} & \text{H} & \text{H} & \text{H} \\
| & | & | & | & | \\
\text{H}-\text{C}-\text{C}-\text{C}-\text{C}-\text{C}-\text{H} \\
| & | & | & | & | \\
\text{H} & \text{H} & \text{H} & \text{H} & \text{H}
\end{array}
\quad \text{or} \quad CH_3CH_2CH_2CH_2CH_3 \quad \text{or} \quad \diagdown\diagup\diagdown\diagup
$$

Some atoms, even in covalent compounds, carry a **formal charge**, defined as the number of valence electrons in the neutral atom minus the sum of the number of unshared electrons and half the number of shared electrons. **Resonance** occurs

*In the chapter summaries, terms whose meanings you should know appear in boldface type.

1

when we can write two or more structures for a molecule or ion with the same arrangement of atoms but different arrangements of the electrons. The correct structure of the molecule or ion is a **resonance hybrid** of the **contributing structures**, which are usually drawn with a double-headed arrow (◄────►) between them. Organic chemists use a curved arrow (⌒) to show the movement of an electron pair.

A **sigma (σ) bond** is formed between atoms by overlap of two atomic orbitals along the line that connects the atoms. Carbon uses sp^3**-hybridized orbitals** to form four such bonds. These bonds are directed from the carbon nucleus toward the corners of a tetrahedron. In **methane**, for example, the carbon is at the center and the four hydrogens are at the corners of a regular tetrahedron with H—C—H bond angles of 109.5°.

Carbon compounds can be classified according to their molecular framework as **acyclic** (*not* cyclic), **carbocyclic** (containing rings of carbon atoms), or **heterocyclic** (containing at least one ring atom that is not carbon). They may also be classified according to **functional group** (Table 1.5).

REACTION SUMMARY

$$2 \text{ ROH} \quad + \quad 2 \text{ Na} \quad \longrightarrow \quad 2 \text{ RO}^-\text{Na}^+ \quad + \quad \text{H}_2$$

alcohol   sodium   sodium alkoxide   hydrogen

LEARNING OBJECTIVES*

1. Know the meaning of: nucleus, electrons, protons, neutrons, atomic number, atomic weight, shells, orbitals, valence electrons, valence, kernel.

2. Know the meaning of: electropositive, electronegative, ionic and covalent bonds, radical, catenation, polar covalent bond, single and multiple bonds, nonbonding or unshared electron pair.

3. Know the meaning of: molecular formula, structural formula, structural (or constitutional) isomers, continuous and branched chain, formal charge, resonance, contributing structures, sigma (σ) bond, sp^3 hybrid orbitals, tetrahedral carbon.

4. Know the meaning of: acyclic, carbocyclic, heterocyclic, functional group.

*Although the objectives are often worded in the form of imperatives (i.e., determine . . . , write . . . , draw . . .), these verbs are all to be preceded by the phrase "be able to. . . ." This phrase has been omitted to avoid repetition and to save space.

5. Know the meaning of the symbols: $\delta+$, $\delta-$, $\longmapsto$, $\longleftarrow$, $\frown$.

6. Given a periodic table, determine the number of valence electrons of an element and write its electron-dot formula.

7. Given two elements and a periodic table, tell which element is more electropositive or electronegative.

8. Given the formula of a compound and a periodic table, classify the compound as ionic or covalent.

9. Given an abbreviated structural formula of a compound, write its electron-dot formula.

10. Given a covalent bond, tell whether it is polar. If it is, predict the direction of bond polarity from the electronegativities of the atoms.

11. Given a molecular formula, draw the structural formulas for all possible structural isomers.

12. Given a structural formula abbreviated on one line of type, write the complete structure and clearly show the arrangement of atoms in the molecule.

13. Given a line formula, such as $\diagup\!\!\diagdown\!\!\diagup\!\!\diagdown$ (pentane), write the complete structure and clearly show the arrangement of atoms in the molecule. Tell how many hydrogens are attached to each carbon, what the molecular formula is, and what the functional groups are.

14. Given a simple molecular formula, draw the electron-dot formula and determine whether each atom in the structure carries a formal charge.

15. Draw electron-dot formulas that show all important contributors to a resonance hybrid.

16. Predict the geometry of bonds around an atom, knowing the electron distribution in the orbitals.

17. Draw in three dimensions with solid, wedged, and dashed bonds the tetrahedral bonding around sp^3-hybridized carbon atoms.

18. Distinguish between acyclic, carbocyclic, and heterocyclic structures.

19. Given a series of structural formulas, recognize compounds that belong to the same class (same functional group).

20. Begin to recognize the important functional groups: alkene, alkyne, alcohol, ether, aldehyde, ketone, carboxylic acid, ester, amine, nitrile, amide, thiol, and thioether.

3

ANSWERS TO PROBLEMS

Problems Within the Chapter

1.1   Elements with fewer than four valence electrons tend to give them up and
        form positive ions; those with more than four valence electrons tend to
        gain electrons to complete the valence shell, becoming negative ions:
        Al^{3+}, F^-, Li^+, Mg^{2+}, S^{2-}, H^+, or H^-.

1.2   Within any horizontal row in the periodic table, the most electropositive
        element appears farthest to the left. Na is more electropositive than
        Al and B is more electropositive than C. In a given column in the peri-
        odic table, the lower the element, the more electropositive it is. Al
        is more electropositive than B.

1.3   Within any horizontal row in the periodic table, the most electronegative
        element appears farthest to the right. F is more electronegative than O,
        O more than N. In a given column, the higher the element, the more
        electronegative it is. F is more electronegative than Cl.

1.4   Carbon is in Group IV. With a half-filled (or half-empty) valence shell,
        it is neither strongly electropositive nor strongly electronegative.

1.5   dichloromethane
        (methylene chloride)

$$\begin{array}{ccc} & H & & & H \\ & \overset{..}{H:}\overset{..}{C}:\overset{..}{Cl}: & \text{or} & & H-C-Cl \\ & \overset{..}{:Cl:} & & & Cl \\ & \overset{..}{} & & & \end{array}$$

        trichloromethane
        (chloroform)

$$\begin{array}{ccc} \overset{..}{:Cl:} & & Cl \\ \overset{..}{H:}\overset{..}{C}:\overset{..}{Cl}: & \text{or} & H-C-Cl \\ \overset{..}{:Cl:} & & Cl \\ \overset{..}{} & & \end{array}$$

1.6   If the C—C bond length is 1.54 Å and the Cl—Cl bond length is 1.98
        Å, we expect the C—Cl bond length to be about 1.76 Å: (1.54 + 1.98)/2.
        The C—Cl bond is longer than the C—C bond (1.54 Å).

1.7   propane

$$\begin{array}{ccccc} H & & H & & H \\ | & & | & & | \\ H-C & - & C & - & C-H \\ | & & | & & | \\ H & & H & & H \end{array}$$

1.8 $\overset{\delta+}{N}-\overset{\delta-}{Cl}$ $\overset{\delta+}{S}-\overset{\delta-}{O}$ The first of these is a little risky to predict, because
        the elements (N, Cl) are not in the same horizontal row. But since
        these elements are two groups apart and only one row apart, it is not

too surprising that chlorine is more electronegative than nitrogen.
The polarity of the S — O bond is easy to predict because both elements
are in the same column, and the more electronegative atom appears
nearest the top.

1.9

$$Cl^{\delta-} - \overset{\displaystyle F^{\delta-}}{\underset{\displaystyle F^{\delta-}}{\overset{|}{\underset{|}{C^{\delta+}}}}} - Cl^{\delta-}$$

Both Cl and F are more electronegative than C.

1.10

H—C–Ö̈
(with H's)

Both the C — O and H — O bonds are polar, and the
oxygen is more electronegative than either carbon
or hydrogen.

1.11 (H:)C:::N: H(:C:::)N: H:C(::: :N:)

1.12 a. The carbon shown has 12 electrons around it, 4 more than is
          allowed.
      b. There are 20 valence electrons shown, whereas there should only be
          16 (6 from each oxygen and 4 from the carbon).
      c. There is nothing wrong with this formula, but it does place a formal
          charge of -1 on the "left" oxygen and +1 on the "right" oxygen (see
          Sec. 1.11). This formula is one possible contributor to the reson-
          ance hybrid structure for carbon dioxide (see Sec. 1.12); it is less
          important than the structure with two carbon-oxygen double bonds,
          because it takes energy to separate the + and - charges.

1.13 Formaldehyde, H_2CO. There are 12 valence electrons altogether (C = 4,
      H = 1, and O = 6). A double bond between C and O is necessary to put 8
      electrons around each of these atoms.

                H   H
          .. .. | ..
        H : C :: O :   or   H — C ═ O :

1.14 There are 10 valence electrons, 4 from C and 6 from O. An arrangement
      that puts 8 electrons around each atom is

          : C ::: O :   or   : C ═══ O :

      This structure puts a formal charge of -1 on C and +1 on O (see Sec.
      1.11).

1.15   If the carbon chain is linear, there are two possibilities:

$$
\begin{array}{cccc}
\text{H} & \text{H} & \text{H} & \text{H} \\
| & | & | & | \\
\text{H}-\text{C} =\!\!= \text{C}-\text{C}-\text{C}-\text{H} \\
& & | & | \\
& & \text{H} & \text{H}
\end{array}
\qquad \text{and} \qquad
\begin{array}{cccc}
\text{H} & \text{H} & \text{H} & \text{H} \\
| & | & | & | \\
\text{H}-\text{C}-\text{C} =\!\!= \text{C}-\text{C}-\text{H} \\
| & & & | \\
\text{H} & & & \text{H}
\end{array}
$$

But the carbon chain can be branched, giving a third possibility:

$$
\begin{array}{c}
\text{H}\quad\text{H} \\
\diagdown\,\text{C}\,\diagup \\
\text{H}\diagup\quad\diagdown\text{H} \\
\text{H}\diagdown \quad\diagup \\
\quad\;\text{C}=\text{C} \\
\text{H}\diagup \quad\diagdown \\
\quad\;\text{H} \\
\diagup\,\text{C}\,\diagdown \\
\text{H}\quad\text{H}
\end{array}
$$

1.16   a.
$$
\begin{array}{c}
\text{H} \qquad\quad \text{H} \\
| \qquad\quad \diagup \\
\text{H}-\text{C}-\text{N} \\
| \qquad\quad \diagdown \\
\text{H} \qquad\quad \text{H}
\end{array}
$$
   b.
$$
\begin{array}{c}
\text{H} \\
| \\
\text{H}-\text{C}-\text{O}-\text{H} \\
| \\
\text{H}
\end{array}
$$

1.17   No, it does not. We cannot draw any structure for C_2H_5 that has four bonds to each carbon and one bond to each hydrogen.

1.18   First write the alcohols (compounds with an O — H group)

$$
\begin{array}{ccc}
\text{H} & \text{H} & \text{H} \\
| & | & | \\
\text{H}-\text{C}-\text{C}-\text{C}-\text{OH} \\
| & | & | \\
\text{H} & \text{H} & \text{H}
\end{array}
\qquad \text{and} \qquad
\begin{array}{ccc}
\text{H} & \text{H} & \text{H} \\
| & | & | \\
\text{H}-\text{C}-\text{C}-\text{C}-\text{H} \\
| & | & | \\
\text{H} & | & \text{H} \\
& \text{O}-\text{H}
\end{array}
$$

Then write the structures with a C — O — C bond (ethers)

$$
\begin{array}{cccc}
\text{H} & \text{H} & & \text{H} \\
| & | & & | \\
\text{H}-\text{C}-\text{C}-\text{O}-\text{C}-\text{H} \\
| & | & & | \\
\text{H} & \text{H} & & \text{H}
\end{array}
$$

There are no other possibilities. For example,

$$
\begin{array}{ccc}
& \text{H} & \text{H} & \text{H} \\
& | & | & | \\
\text{H}-\text{O}-\text{C}-\text{C}-\text{C}-\text{H} \\
& | & | & | \\
& \text{H} & \text{H} & \text{H}
\end{array}
\qquad \text{and} \qquad
\begin{array}{ccc}
\text{H} & \text{H} & \text{H} \\
| & | & | \\
\text{H}-\text{C}-\text{C}-\text{C}-\text{H} \\
| & | & | \\
\text{H} & \text{H} & \text{O}-\text{H}
\end{array}
$$

are the same as

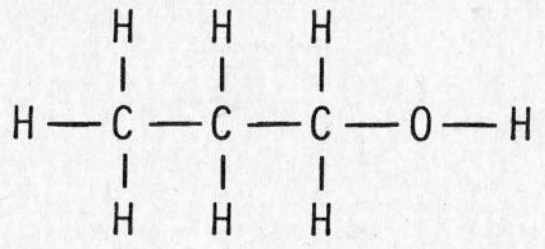

$$
\begin{array}{ccc}
H & H & H \\
| & | & | \\
H-C-C-C-O-H \\
| & | & | \\
H & H & H
\end{array}
$$

They all have the same bond connectivities and represent a single structure. Similarly,

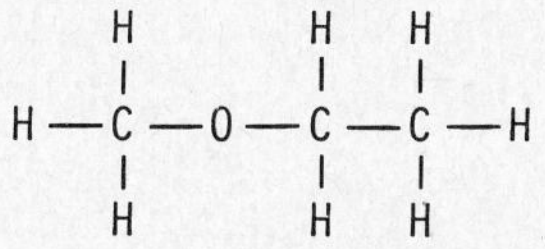

$$
\begin{array}{ccc}
H & H & H \\
| & | & | \\
H-C-O-C-C-H \\
| & | & | \\
H & H & H
\end{array}
\quad \text{is the same as} \quad
\begin{array}{ccc}
H & H & H \\
| & | & | \\
H-C-C-O-C-H \\
| & | & | \\
H & H & H
\end{array}
$$

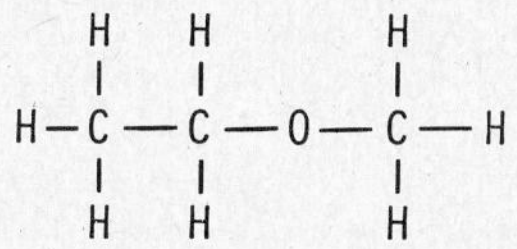

1.19 From left to right: *n*-pentane, isopentane, and isopentane

1.20 a. 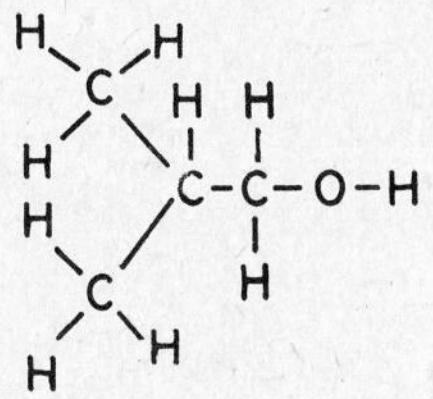 b.

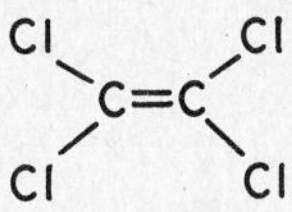

1.21 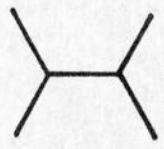   stands for the carbon skeleton

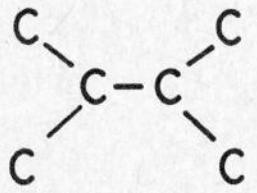

The addition of the appropriate number of hydrogens on each carbon completes the valence of 4.

1.22 ammonia, $H-\ddot{N}-H$ with H below, formal charge = 5 - (2 + 3) = 0

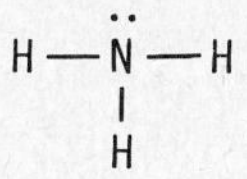

ammonium ion, $\left[H-\overset{H}{\underset{H}{N}}-H \right]^{+}$ formal charge = 5 - (0 + 4) = +1

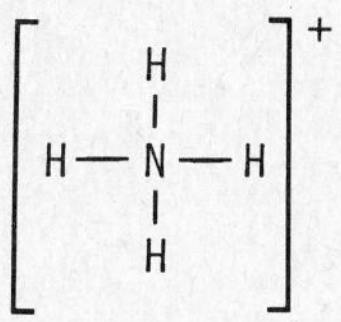

amide ion, $\left[H-\ddot{\underset{\cdot\cdot}{N}}-H \right]^{-}$ formal charge = 5 - (4 + 2) = -1

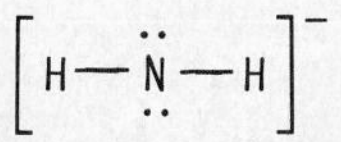

7

1.23

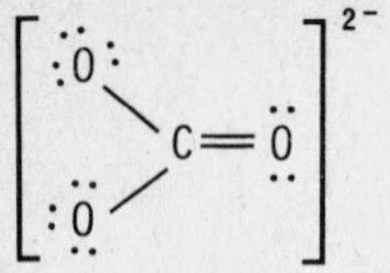

$$\left[\begin{array}{c} :\ddot{O}: \\ :\ddot{O}: \end{array}\hspace{-4pt} \begin{array}{c} \\ C = \ddot{O} \\ \\ \end{array}\right]^{2-}$$

For the singly bonded oxygens, formal charge
= 6 - (6 + 1) = -1.
For the doubly bonded oxygen, formal charge
= 6 - (4 + 2) = 0
For the carbon, formal charge = 4 - (0 + 4) = 0.

1.24 There are 24 valence electrons to use in bonding (6 from each oxygen, 5 from the nitrogen, and one more because of the negative charge). To arrange the atoms with eight valence electrons around each atom, we must have one nitrogen-oxygen double bond:

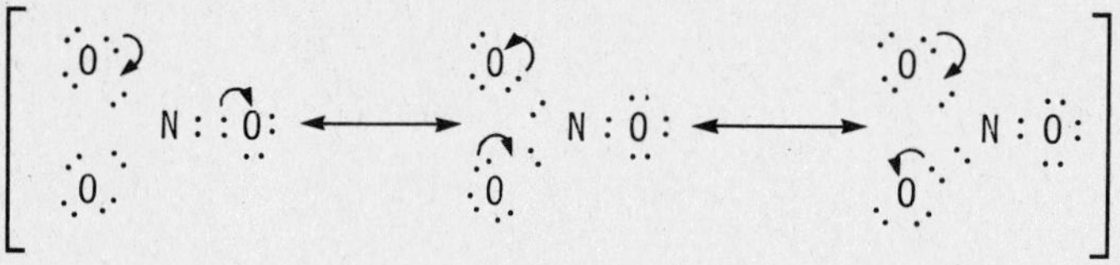

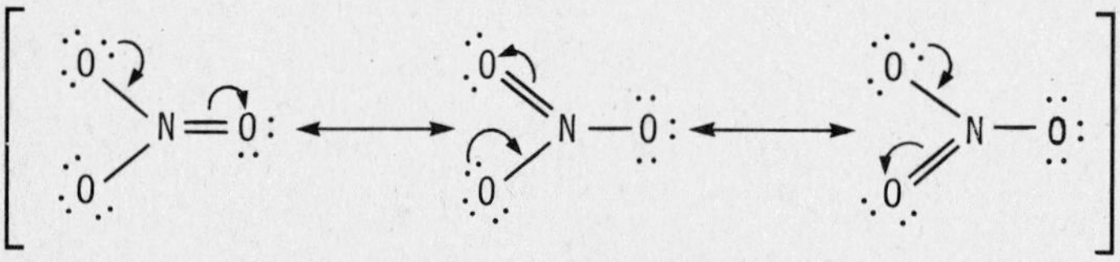

The formal charge on nitrogen is 5 - (0 + 4) = +1.
The formal charge on singly bonded oxygens is 6 - (6 + 1) = -1.
The formal charge on doubly bonded oxygen is 6 - (4 + 2) = 0.

In the resonance hybrid, the formal charge on the nitrogen is +1; on the oxygens, the charge is -2/3 at each oxygen, because each oxygen has a -1 charge in two of the three structures and 0 charge in the third structure.

1.25 In "planar" methane, the H—C—H bond angle would be 90°, whereas in real, tetrahedral methane this angle is 109.5°. Thus the bonding electrons are farther apart and therefore more stable in the tetra-hedral arrangement.

1.26 a. $CH_3CH_2CH_2O^-Na^+ + H_2$

b. $CH_3CHCH_2CH_3 + H_2$
         |
         O^-Na^+

1.27 a. $C = C$, alkene; $O—H$, alcohol
b. $C = O$, carbonyl group, ketone
c. $C = C$, alkene
d. $C = C$, alkene; $C = O$, ketone; $O—H$, alcohol

Additional Problems

1.28 The number of valence electrons is the same as the number in the group of the periodic table (Table 1.3) to which the element belongs.

a. ·Ċ·   b. :F̈·   c. ·Si·   d. ·B̈   e. ·S̈·   f. ·P̈·

1.29 In sodium chloride, Cl is present as chloride ion (Cl^-); Cl^- reacts with Ag^+ to give AgCl, a white precipitate. The C—Cl bonds in CCl_4 are covalent; no Cl^- is present to react with Ag^+.

1.30 a. ionic   b. covalent   c. ionic   d. covalent
     e. covalent   f. ionic   g. covalent   h. covalent

1.31

		Valence electrons	Common valence
a.	O	6	2
b.	H	1	1
c.	Cl	7	1
d.	N	5	3
e.	S	6	2
f.	C	4	4

Note that the *sum* of the number of valence electrons and the common valence is 8 in each case (except for H, where it is 2, the number of electrons in the completed first shell).

1.32

a.
```
     H
     |    ..
H — C — Cl:
     |    ..
     H
```

b.
```
     H   H   H
     |   |   |
H — C — C — C — H
     |   |   |
     H   H   H
```

c.
```
     H   H
     |   |    ..
H — C — C — F:
     |   |    ..
     H   H
```

d.
```
     H       H
     |      /
H — C — N
     |      \
     H       H
```

e.
```
     H   H
     |   |   ..
H — C — C — O — H
     |   |   ..
     H   H
```

f.
```
H       ..
 \
  C = O:
 /
H
```

1.33 a. :Cl̈ — Cl̈:   Since the bond is between identical atoms, it is
                        pure covalent (nonpolar).

     b.   H   Fluorine is more electronegative than carbon.
              |
         H — C$^{\delta+}$—F$^{\delta-}$
              |
              H

     c. $\delta-$ $\delta+$ $\delta-$ The C=O bond is polar, and the oxygen is more
         :O = C = O: electronegative than carbon.

9

d. $\overset{\delta+}{H}\!-\!\overset{\delta-}{\overset{..}{\underset{..}{Br}}}:$ The halogens are more electronegative than hydrogen.

e. $\overset{\delta-}{F}\;\overset{}{F}\delta-$
$\overset{\delta-}{F}\!-\!\overset{\delta+}{\underset{\underset{F}{|}}{S}}\!-\!F\delta-$
$\overset{\delta-}{F}\;\;F\delta-$
Fluorine is more electronegative than sulfur; indeed, it is the *most* electronegative element. Note that the S has 12 electrons around it. Elements below the first full row sometimes have more than 8 valence electrons around them.

f. $\underset{\underset{H}{|}}{\overset{\overset{H}{|}}{H\!-\!C\!-\!H}}$ Carbon and hydrogen have nearly identical electro-negativities, and the bonds are essentially nonpolar.

g. $\overset{\delta-}{:\!\underset{..}{O}}\!=\!\overset{\delta+}{S}\!=\!\overset{\delta-}{\overset{..}{O}}:$ Oxygen is more electronegative than sulfur.

h. $\underset{\underset{H}{|}}{\overset{\overset{H}{|}}{H\!-\!C}}\overset{\delta+}{}\!-\!\overset{\delta-}{O}\!-\!\underset{\underset{H}{|}}{\overset{\overset{H}{|}}{C}}\overset{\delta+}{}\!-\!H$ Oxygen is more electronegative than carbon, so the C — O bonds are polar covalent.

1.34   The O — H bond is polar, with the hydrogen $\delta+$. The reaction is analogous to the reaction of sodium with alcohols.

$$2\ CH_3\overset{\overset{O}{\parallel}}{C}\!-\!OH\ +\ 2\ Na\ \longrightarrow\ 2\ CH_3\overset{\overset{O}{\parallel}}{C}\!-\!O^-Na^+\ +\ H_2$$

1.35   a. C_3H_8 The only possible structure is $CH_3CH_2CH_3$.

b. C_3H_7Cl The chlorine can replace a hydrogen on an end carbon or on the middle carbon in the answer to a.

$CH_3CH_2CH_2Cl$ or $CH_3\underset{\underset{Cl}{|}}{CH}CH_3$

c. $C_2H_4Cl_2$ The chlorines can either be attached to the same carbon or to different carbons.

CH_3CHCl_2 or $ClCH_2\!-\!CH_2Cl$

d. $C_3H_6Br_2$ Be systematic. With one bromine on an end carbon there are three possibilities for the second bromine:

```
     Br  H   H              H   Br  H              H   H   Br
     |   |   |              |   |   |              |   |   |
 H — C — C — C — H     H — C — C — C — H     H — C — C — C — H
     |   |   |              |   |   |              |   |   |
     Br  H   H              Br  H   H              Br  H   H
```

If one bromine is on the middle carbon, the only *new* structure arises with the second bromine also on the middle carbon:

```
     H   Br  H
     |   |   |
 H — C — C — C — H
     |   |   |
     H   Br  H
```

e. C_4H_9F The carbon chain may either be linear or branched. In each case there are two possible positions for the fluorine.

$CH_3CH_2CH_2CH_2F$ $CH_3CHCH_2CH_3$
                                      |
                                      F

                               F
                               |
CH_3CHCH_2F CH_3CCH_3
     | |
    CH_3 CH_3

f. $C_2H_2Cl_2$ The sum of the hydrogens and chlorines is 4, not 6. Therefore there must be a carbon-carbon double bond:

$CCl_2 = CH_2$ or $CHCl = CHCl$

No carbocyclic structure is possible because there are only two carbon atoms.

g. C_3H_6 There must be a double bond, or with three carbons, a ring:

$CH_3CH=CH_2$ or
```
    CH2 ——— CH2
      \     /
       CH2
```

h. $C_4H_{10}O$ With an O—H bond, there are four possibilities:

$CH_3CH_2CH_2CH_2OH$ $CH_3CHCH_2CH_3$
                                                                $|$
                                                                OH

$\qquad\qquad\qquad\qquad\qquad\qquad\qquad CH_3$
$\qquad\qquad\qquad\qquad\qquad\qquad\qquad |$
CH_3CHCH_2OH $CH_3 — C — OH$
$\qquad\quad |$ $\qquad\qquad\qquad\qquad\qquad |$
$\qquad\quad CH_3$ $\qquad\qquad\qquad\qquad\qquad CH_3$

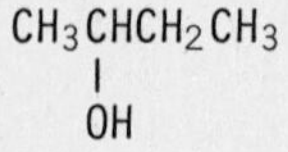

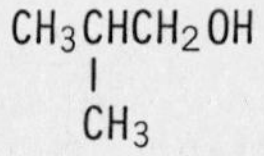

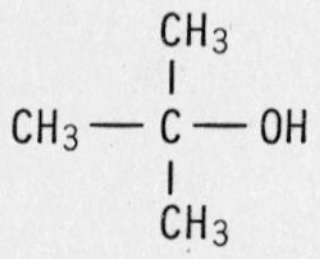

There are also three possibilities in a C—O—C arrangement:

$\qquad\qquad\qquad\qquad\quad\diagup CH_3$
$CH_3OCH_2CH_2CH_3$ CH_3OCH $CH_3CH_2OCH_2CH_3$
$\qquad\qquad\qquad\qquad\quad\diagdown CH_3$

1.36   The problem should be approached systematically. Consider first a chain of six carbons, then a chain of five carbons with a one-carbon branch, and so on.

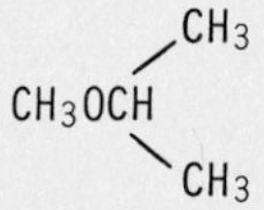

$CH_3CH_2CH_2CH_2CH_2CH_3$

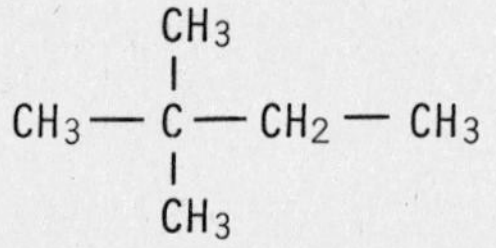

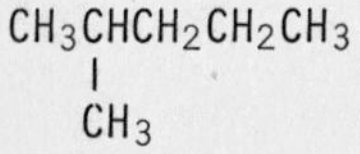

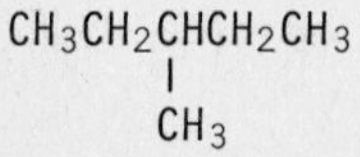

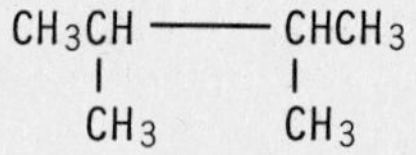

$\qquad\qquad\qquad\qquad\qquad\qquad\qquad CH_3$
$\qquad\qquad\qquad\qquad\qquad\qquad\qquad |$
$CH_3CHCH_2CH_2CH_3$ $CH_3 — C — CH_2 — CH_3$
$\qquad |$ $\qquad\qquad\qquad\qquad\qquad\qquad\qquad |$
$\qquad CH_3$ $\qquad\qquad\qquad\qquad\qquad\qquad CH_3$

$CH_3CH_2CHCH_2CH_3$ $CH_3CH \underline{\qquad} CHCH_3$
$\qquad\qquad |$ $\qquad\qquad\qquad\qquad\quad |\qquad\quad |$
$\qquad\qquad CH_3$ $\qquad\qquad\qquad\qquad CH_3\quad CH_3$

1.37   a.

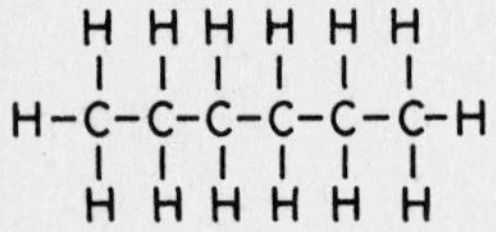

```
    H H H H H H
    | | | | | |
  H-C-C-C-C-C-C-H
    | | | | | |
    H H H H H H
```

b.

```
           H
           |
         H-C-H
      H  |   H H
      |  |   | |
    H-C--C---C-C-H
      |  |   | |
      H  |   H H
         H-C-H
           |
           H
```

c.

```
    H H H
    | | |
  H-C-C-C-H
    | | |
    H | H
      O-H
```

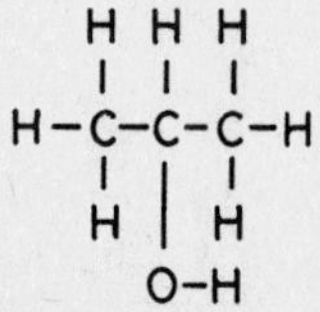

d.

```
    H H   H H
    | |   | |
  H-C-C-S-C-C-H
    | |   | |
    H H   H H
```

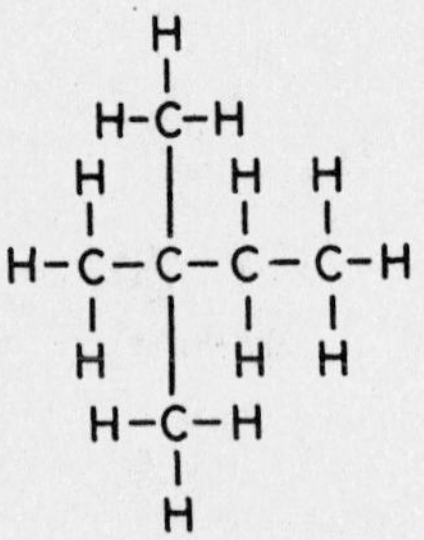

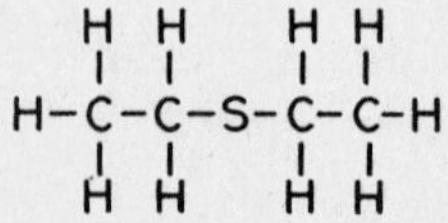

e.

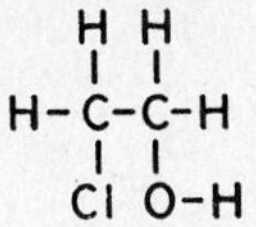

$$H-\overset{\overset{\displaystyle H}{|}}{\underset{\underset{\displaystyle Cl}{|}}{C}}-\overset{\overset{\displaystyle H}{|}}{\underset{\underset{\displaystyle O-H}{|}}{C}}-H$$

f.

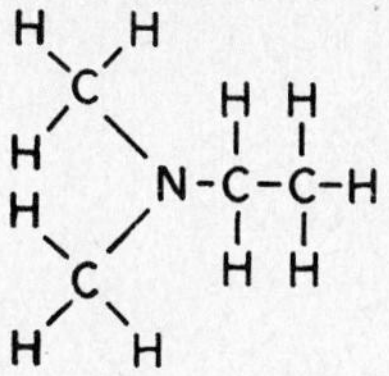

1.38   a. $CH_2{=}CHCH_2CH_3$

b.

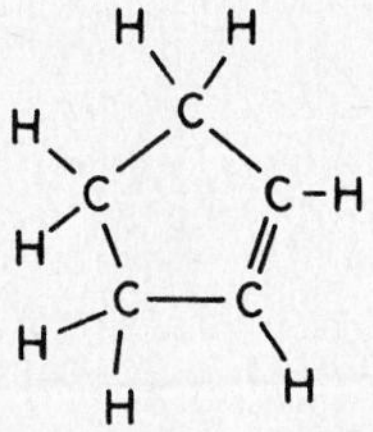

c. $(CH_3)_2CHCH(CH_3)_2$

d. 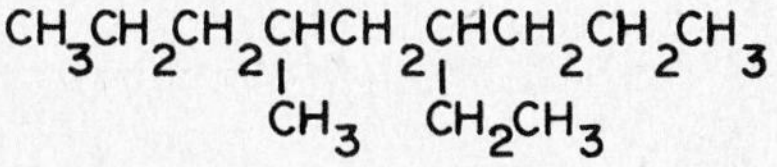 $CH_3CH_2CH_2\overset{\overset{\displaystyle }{|}}{\underset{\underset{\displaystyle CH_3}{|}}{C}}HCH_2\overset{\overset{\displaystyle }{|}}{\underset{\underset{\displaystyle CH_2CH_3}{|}}{C}}HCH_2CH_2CH_3$

e. $CH_3CH_2OCH_2CH_3$

f. $(CH_3)_3C{-}C(CH_3)_3$

g.

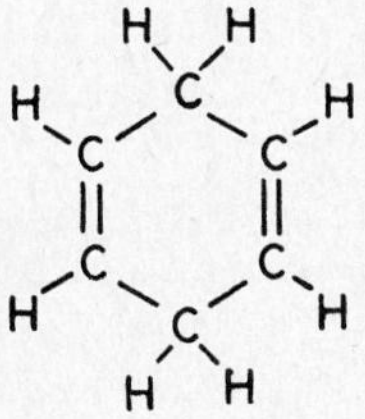

h. 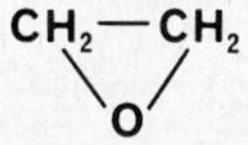 $\underset{\displaystyle O}{CH_2{-}CH_2}$

1.39   a. 10   b. $C_{10}H_{18}O$ c. $(CH_3)_2C{=}CHCH_2CH_2\overset{\overset{\displaystyle }{|}}{\underset{\underset{\displaystyle CH_3}{|}}{C}}{=}CHCH_2OH$

1.40   First count the carbons, then the hydrogens, and finally the remaining atoms.
a. $C_{16}H_{30}O$ b. C_6H_6 c. $C_{19}H_{28}O_2$
d. $C_{10}H_{14}N_2$ e. $C_{10}H_{16}$ f. $C_5H_5N_5$

1.41 a. HONO   First determine the total number of valence electrons; H = 1, O = 2 x 6 = 12, N = 5, for a total of 18. These must be arranged in pairs so that the hydrogen has 2 and the other atoms have 8 electrons around them.

$$H : \overset{..}{\underset{..}{O}} : \overset{..}{N} : : \overset{..}{O} : \quad \text{or} \quad H - \overset{..}{\underset{..}{O}} - \overset{..}{N} = \overset{..}{O} :$$

Using the formula for formal charge given in Sec. 1.11, it can be determined that none of the atoms has a formal charge.

b. $HONO_2$ $H : \overset{..}{\underset{..}{O}} : N$ (24 valence electrons)

  $\overset{..}{O} \leftarrow (-)$

  $(+)$ $\overset{..}{\underset{..}{O}}$

The nitrogen has a +1 formal charge [5 - (0 + 4) = +1], and the singly bonded oxygen has a -1 formal charge [6 - (6 + 1) = -1]. The whole molecule is neutral.

c. H_2CO

  H
  $H : \overset{..}{C} : : \overset{..}{O} :$ There are no formal charges.

d. NH_4^+

  H
  $H : \overset{..}{N} : H$ The nitrogen has a +1 formal charge; see
  H $\searrow (+)$ the answer to Problem 1.22.

e. CN^- There are 10 valence electrons (C = 4, N = 5, plus 1 more because of the negative charge).

  $^{\ominus} : C : : : N :$

The carbon has a -1 formal charge [4 - (2 + 3) = -1].

f. CO   There are 10 valence electrons.

  $: C : : : O :$

The carbon has a -1 formal charge, and the oxygen has a +1 formal charge. Carbon monoxide is isoelectronic with cyanide ion but has no net charge (-1 + 1 = 0).

g. SO_4^{2-} There are 32 valence electrons (6 each from the sulfur and oxygens, plus 2 more because of the double negative charge).

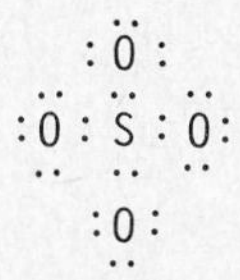

$$\ddot{:}\overset{\displaystyle\cdot\cdot}{\underset{\displaystyle\cdot\cdot}{O}}\ddot{:}$$

$$:\ddot{O}:\overset{\displaystyle\cdot\cdot}{\underset{\displaystyle\cdot\cdot}{S}}:\ddot{O}:$$

$$:\overset{\displaystyle\cdot\cdot}{\underset{\displaystyle\cdot\cdot}{O}}:$$

Each oxygen has a formal charge of -1 $[6 - (6 + 1) = -1]$. The sulfur has a formal charge of $+2$ $[6 - (0 + 4) = +2]$.

h. BF_3 There are 24 valence electrons (B = 3, F = 7). The structure is usually written with only 6 electrons around the boron.

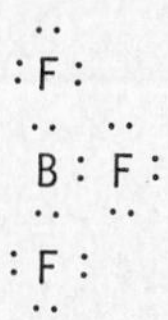

$$:\ddot{F}:$$

$$B:\ddot{F}:$$

$$:\ddot{F}:$$

In this case, there are no formal charges. This structure shows that BF_3 is a Lewis acid, which readily accepts an electron pair to complete the octet around the boron.

i. H_2O_2 There are 14 valence electrons and an $O—O$ bond.

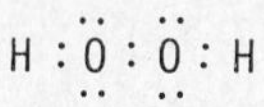

$H:\ddot{O}:\ddot{O}:H$ There are no formal charges.

j. HCO_3^- There are 24 valence electrons involved. The hydrogen is attached to an oxygen, not to carbon.

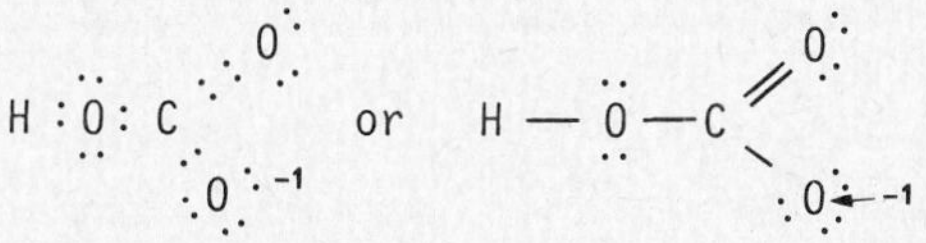

$$H:\ddot{O}:C\overset{\displaystyle O}{\underset{\displaystyle \ddot{O}\,^{-1}}{}} \quad \text{or} \quad H — \ddot{O} — C\overset{\displaystyle O}{\underset{\displaystyle \ddot{O}\,^{-1}}{}}$$

The indicated oxygen has a formal charge: $6 - (6 + 1) = -1$. All other atoms are formally neutral.

1.42 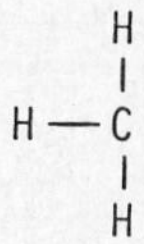
$$H—\overset{\displaystyle H}{\underset{\displaystyle H}{C}}$$
  formal charge $= 4 - (0 + 3) = +1$

This is a methyl carbocation, CH_3^+

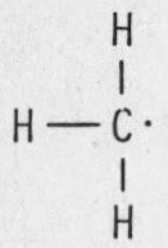

```
    H
    |
H — C·        formal charge = 4 - (1 + 3) = 0
    |
    H
```

This is a methyl free radical, $CH_3\cdot$.

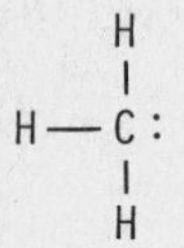

```
    H
    |
H — C :       formal charge = 4 - (2 + 3) = -1
    |
    H
```

This is a methyl carbanion, $CH_3:^-$

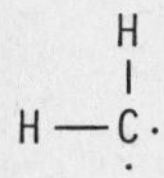

```
    H
    |
H — C·        formal charge = 4 - (2 + 2) = 0
    ·
```

This is a methylene or carbene, CH_2:

All of these fragments are extremely reactive. They may act as intermediates in organic reactions.

1.43   There are 18 valence electrons (6 from each of the oxygens, 5 from the nitrogen, and 1 from the negative charge).

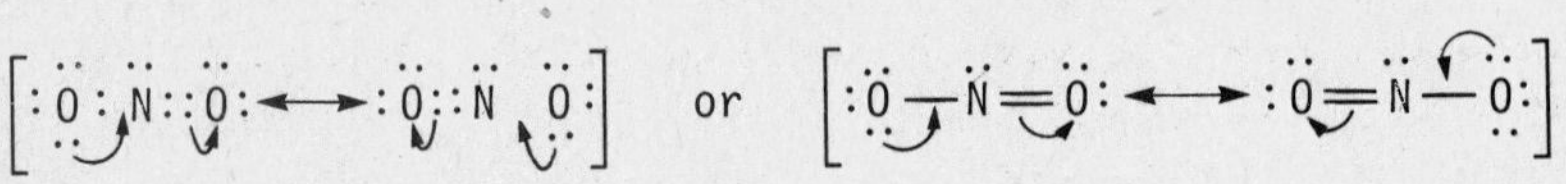

$$\left[:\ddot{O}:\ddot{N}::\ddot{O}:\longleftrightarrow:\ddot{O}::\ddot{N}\ \ddot{O}:\right] \quad \text{or} \quad \left[:\ddot{O}-\ddot{N}=\ddot{O}:\longleftrightarrow:\ddot{O}=\ddot{N}-\ddot{O}:\right]$$

The negative charge in each contributor is on the singly bonded oxygen [6 - (6 + 1) = -1]. The other oxygen and the nitrogen have no formal charge. In the resonance hybrid, the negative charge is spread equally over the two oxygens; each is -1/2.

1.44   a.   There are 16 valence electrons (3 x 5 + 1).

$$\left[:\ddot{N}=N=\ddot{N}:\longleftrightarrow:N\equiv N-\ddot{N}:\right]$$

```
    -1  +1  -1          0  +1  -2
```

The formal charges on each nitrogen are shown below the structures.

b.

$$\left[\begin{array}{ccc} & \overset{H}{\underset{|}{\text{H}-\text{C}-\text{C}}} & \overset{:\overset{..}{\text{O}}:}{\diagup} \\ & \underset{|}{\overset{|}{\text{H}}} & \diagdown\overset{..}{\text{O}}:^{-1} \end{array} \longleftrightarrow \begin{array}{ccc} & \overset{H}{\underset{|}{\text{H}-\text{C}-\text{C}}} & \overset{:\overset{..}{\text{O}}:^{-1}}{\diagup} \\ & \underset{|}{\overset{|}{\text{H}}} & \diagdown\overset{..}{\text{O}}: \end{array} \right]$$

The singly bonded oxygen carries the negative charge in each contributor. In the resonance hybrid, this charge is spread equally
over both oxygens.

1.45

$$\left[\underset{\text{CH}_3-\text{C}-\text{NH}_2}{\overset{:\overset{..}{\text{O}}:}{\overset{\|}{}}} \longleftrightarrow \underset{\text{CH}_3-\text{C}=\overset{+}{\text{NH}}_2}{\overset{:\overset{..}{\text{O}}:^{-}}{\overset{|}{}}} \right]$$

Each atom in both structures has a complete valence shell of electrons.
There are no formal charges in the initial structure, but in the second
structure the oxygen is formally negative and the nitrogen formally
positive.

1.46

In the first structure, there are no formal charges. In the second
structure, the oxygen is formally +1, and the ring carbon bearing the
unshared electron pair is formally -1. (Don't forget to count the
hydrogen that is attached to each ring carbon except the one that is
doubly bonded to oxygen.)

1.47

$$\text{CH}_3-\overset{..}{\text{N}}\text{H}_2 + \text{CH}_3-\overset{\overset{:\overset{..}{\text{O}}}{\|}}{\text{C}}-\text{OCH}_3 \longrightarrow \underset{\underset{\text{H}_2\overset{+1}{\text{N}}-\text{CH}_3}{|}}{\overset{:\overset{..}{\text{O}}:^{-1}}{\underset{|}{\text{CH}_3-\text{C}-\text{OCH}_3}}}$$

1.48 a. CH₃ONa

$$\underset{\overset{..}{\underset{\text{H}}{|}}}{\overset{\text{H}}{\underset{..}{\text{H}:\text{C}:\overset{..}{\underset{..}{\text{O}}}:^{-}}}} \qquad \text{Na}^{+}$$

b. NH₄Cl

$$\left[\underset{\overset{..}{\underset{\text{H}}{}}}{\overset{\text{H}}{\text{H}:\overset{..}{\text{N}}:\text{H}}} \right]^{+} \qquad :\overset{..}{\underset{..}{\text{Cl}}}:^{-}$$

1.49   a. $CH_3CH_2\overset{..}{\underset{..}{O}}H$ b. $CH_3\overset{\displaystyle :\!O\!:}{\overset{\|}{C}}-\overset{..}{\underset{..}{O}}H$

c. $CH_3-\overset{..}{\underset{\underset{\displaystyle H}{|}}{N}}-CH_3$ ·   d. $CH_3\overset{..}{\underset{..}{O}}CH_2CH_2\overset{..}{\underset{..}{O}}H$

1.50   If the s and p orbitals were hybridized to sp^3, two electrons would go into one of these orbitals and one electron would go into each of the remaining three orbitals.

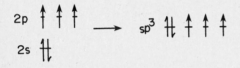

The predicted geometry of ammonia would then be tetrahedral, with one hydrogen at each of three corners, and the unshared pair at the fourth corner.   In fact, ammonia has a pyramidal shape--a somewhat flattened tetrahedron.   The bond angle is 107° rather than 109°28'.

1.51   The ammonium ion is in fact isoelectronic with (has the same arrangement of electrons as) methane, and consequently has the same geometry. Four sp^3 orbitals of nitrogen each contain one electron.   These orbitals then overlap with the $1s$ hydrogen orbitals, as in Figure 1.9.

1.52   The bonding is exactly as in methane.   The geometry of silane is tetrahedral.

1.53

The geometry is tetrahedral at carbon.   It does not matter whether we draw the wedge bonds to the right or to the left of the carbon, or indeed "up" or "down":

1.54 Many correct answers are possible; a few are given here.

a.  $CH_3CH_2\underset{\underset{O}{\|}}{C}CH_2CH_3$ $CH_2{=}CH{-}\underset{\underset{OH}{|}}{CH}{-}CH_2CH_3$

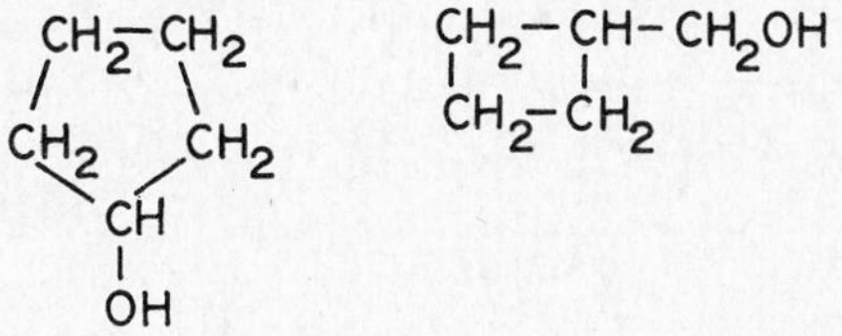

b.

$\begin{matrix} CH_2{-}CH{-}CH_2OH \\ |\quad\quad\quad| \\ CH_2{-}CH_2 \end{matrix}$

c.
$CH_2{-}CHCH_2CH_2CH_3$

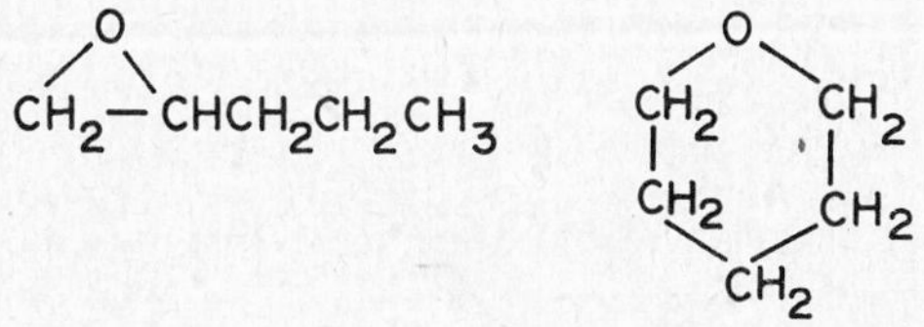

1.55 Compounds a, c, e, and g all have hydroxyl groups and belong to the class of compounds called alcohols. Compounds b, h, and i are ethers; they all have a $C{-}O{-}C$ bond. Compounds d and f are both hydrocarbons.

1.56

1.57 The more common functional groups are listed in Table 1.5. Often more than one answer is possible.

a. $CH_3CH_2CH_2CH_2OH$, $CH_3CH(OH)CH_2CH_3$, $(CH_3)_2CHCH_2OH$, and $(CH_3)_2COH$

b. $CH_3OCH_2CH_3$

c. $CH_3CH_2CH{=}O$

d. $CH_3\underset{\underset{O}{\|}}{C}CH_2CH_3$

e. $CH_3CH_2CO_2H$

f. $\overset{O}{\overset{\|}{HCOCH_2CH_2CH_2CH_3}}$, $\overset{O}{\overset{\|}{HCOCHCH_2CH_3}}$, $\overset{O}{\overset{\|}{HCOCH_2CH(CH_3)_2}}$,
|
CH_3

$\overset{O}{\overset{\|}{HCOC(CH_3)_3}}$, $\overset{O}{\overset{\|}{CH_3COCH_2CH_2CH_3}}$, $\overset{O}{\overset{\|}{CH_3COCH(CH_3)_2}}$

$\overset{O}{\overset{\|}{CH_3CH_2COCH_2CH_3}}$, $\overset{O}{\overset{\|}{CH_3CH_2CH_2COCH_3}}$, and $\overset{O}{\overset{\|}{(CH_3)_2CHCOCH_3}}$

g. See part f.

h. $(CH_3)_3N$, $CH_3NHCH_2CH_3$, $CH_3CH_2CH_2NH_2$, and $(CH_3)_2CHNH_2$

CHAPTER SUMMARY

Hydrocarbons contain only carbon and hydrogen atoms. **Alkanes** are
saturated acyclic hydrocarbons (contain only single bonds); **cycloalkanes**
are similar but have carbon rings.

Alkanes have the general molecular formula C_nH_{2n+2}. The first four
members of this **homologous series** are **methane, ethane, propane,** and **butane;**
each differs from the next by a $- CH_2 -$, or **methylene** group. The **IUPAC**
(International Union of Pure and Applied Chemistry) nomenclature system is
used worldwide to name organic compounds. The IUPAC rules for naming alkanes
are described in Secs. 2.4-2.6. **Alkyl groups,** alkanes minus one hydrogen
atom, are named similarly except that the -*ane* ending is changed to -*yl*. The
letter **R** stands for any alkyl group.

The two main natural sources of alkanes are **natural gas** and **petroleum.**
Alkanes are insoluble in and less dense than water. Their boiling points
increase with molecular weight and, for isomers, decrease with chain
branching.

Conformations are different structures that are interconvertible by
rotation about single bonds. For ethane (and in general) the **staggered**
conformation is more stable than the **eclipsed** conformation (Figure 2.3).

The prefix *cyclo-* is used to name cycloalkanes. **Cyclopropane** is planar,
but larger carbon rings are puckered. **Cyclohexane** exists mainly in a **chair
conformation** with all bonds on adjacent carbons staggered. One bond on each
carbon is **axial** (perpendicular to the mean carbon plane), the other is
equatorial (roughly in that plane). The conformations can be intercon-
verted by "flipping" the ring, which only requires bond rotation and occurs
rapidly at room temperature for cyclohexane. Ring substituents usually
prefer the less crowded, equatorial position.

Stereoisomers have the same order of atom attachments but different
arrangements of the atoms in space. *Cis-trans* isomerism is one kind of
stereoisomerism. For example, two substituents on a cycloalkane can be
either on the same (*cis*) or opposite (*trans*) sides of the mean ring plane.
Stereoisomers can be divided into two groups, **conformational isomers** (interconvertible by bond rotation) and **configurational isomers** (not interconvertible by bond rotation). *Cis-trans* isomers belong to the latter class.

Alkanes are fuels; they burn in air if ignited. Complete combustion gives carbon dioxide and water; less complete combustion gives carbon monoxide or other less oxidized forms of carbon.

Alkanes react with **halogens** (chlorine or bromine) in a reaction initiated by heat or light. One or more hydrogens can be replaced by halogens. This **substitution reaction** occurs by a **free-radical chain mechanism.**

REACTION SUMMARY

Combustion

$$C_nH_{2n+2} + \left(\frac{3n + 1}{2}\right) O_2 \longrightarrow n\ CO_2 + (n + 1)\ H_2O$$

Halogenation (substitution)

$$R-H + X_2 \xrightarrow[\text{light}]{\text{heat or}} R-X + H-X \quad (X = Cl, Br)$$

MECHANISM SUMMARY

Halogenation

initiation $:\ddot{X}-\ddot{X}: \longrightarrow 2\ :\ddot{X}\cdot$

propagation $R-H + :\ddot{X}\cdot \longrightarrow R\cdot + H-\ddot{X}:$

                $R\cdot + :\ddot{X}-\ddot{X}: \longrightarrow R-\ddot{X}: + :\ddot{X}\cdot$

termination $2\ :\ddot{X}\cdot \longrightarrow :\ddot{X}-\ddot{X}:$

                $2\ R\cdot \longrightarrow R-R$

                $R\cdot + :\ddot{X}\cdot \longrightarrow R-\ddot{X}:$

LEARNING OBJECTIVES

1. Know the meaning of: saturated hydrocarbon, alkane, cycloalkane, homologous series, methylene group.

2. Know the meaning of: conformation, staggered, eclipsed, Newman projection, "sawhorse" projection, rotational isomers, rotamers.

3. Know the meaning of: chair conformation of cyclohexane, equatorial, axial, geometric or *cis-trans* isomerism, conformational and configurational isomerism.

4. Know the meaning of: substitution reaction, halogenation, chlorination, bromination, free-radical chain reaction, chain initiation, propagation, termination, combustion.

5. Given the IUPAC name of an alkane or cycloalkane, or a halogen-substituted alkane or cycloalkane, draw its structural formula.

6. Given the structural formula of an alkane or cycloalkane or a halogenated derivative, write the correct IUPAC name.

7. Know the common names of the alkyl groups, cycloalkyl groups, methylene halides, and haloforms.

8. Tell whether two hydrogens in a particular structure are identical or different from one another, by determining whether they give the same or different products by monosubstitution with some group X.

9. Know the relationship between boiling points of alkanes and (a) their molecular weights and (b) the extent of chain branching.

10. Write all steps in the free-radical chain reaction between a halogen and an alkane, and identify the initiation, propagation, and termination steps.

11. Write a balanced equation for the complete combustion of an alkane or cycloalkane.

12. Draw, using sawhorse or Newman projection formulas, the important conformations of ethane, propane, butane, and various halogenated derivatives of these alkanes.

13. Recognize, draw, and name *cis-trans* isomers of substituted cycloalkanes.

14. Draw the chair conformation of cyclohexane, and show clearly the distinction between axial and equatorial bonds.

15. Identify the more stable conformation of a monosubstituted cyclohexane; also, identify substituents as axial or equatorial when the structure is "flipped" from one chair conformation to another.

16. Classify a pair of isomers as structural (constitutional) isomers or stereoisomers, and if the latter, as conformational or configurational (see Figure 2.6).

ANSWERS TO PROBLEMS

Problems Within the Chapter

2.1 $C_{20}H_{42}$; use the formula C_nH_{2n+2}, where n = 20.

2.2   The formulas in parts b and d fit the general formula C_nH_{2n+2} and are
       alkanes. C_8H_{16} (part a) has two fewer hydrogens than called for by
       the alkane formula, and must be either an alkene or a cycloalkane.
       C_7H_{18} (part c) is an impossible molecular formula; it has too many
       hydrogens for the number of carbons.

2.3   a. 2-methylbutane (number the longest chain from left to right)
       b. 2-methylbutane (number the longest chain from right to left)
       The structures in parts a and b are identical, as the names show.
       c. 2,2-dimethylpropane

2.4 $CH_3CH_2CH_2CH_2CH_2-$ 1-pentyl group
       $CH_3CHCH_2CH_2CH_3$ 2-pentyl group
          |

2.5   a. $CH_3CH_2CH_2I$ b. CH_3CHCH_3 or $(CH_3)_2CHCl$
                                                  |
                                                  Cl

       c. CH_3CHCH_3 d. $(CH_3)_3C-I$
              |
              Cl

       e. $(CH_3)_2CHCH_2Br$ f. RF; the letter R stands for any
                                             alkyl group

2.6   a. 2-fluoropropane   b. 4-chloro-2,2-dimethylpentane (*not*
                                           2-chloro-4,4-dimethylpentane, which
                                           has higher numbers for the
                                           substituents)

2.7 CH_3
                   |
       $CH_3CH_2CCH_2CH_3$
                   |
                 CH_3

2.8 $CH_2CH_2CHCH_3$ $\overset{4}{C}H_2\overset{3}{C}H_2\overset{2}{C}H\overset{1}{C}H_3$
                  | | $\overset{5}{|}$ |
                  Cl   Cl CH_3 CH_3

          1,3-dichlorobutane   correct name: 2-methylpentane

24

2.9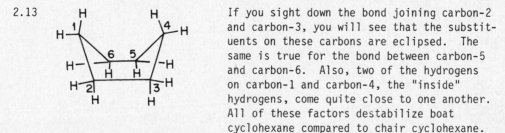

Carbon-2 and carbon-3 in the chain each have two hydrogens and a methyl group attached. The staggered conformation at the left is the more stable. In the staggered conformation at the right, the two methyl substituents (large groups) come close to one another, a less stable arrangement than in the structure to the left. These conformations are referred to as *anti* and *gauche*, respectively.

2.10 a.

CH$_3$
|
CH
CH$_2$ CH$_2$ or
CH—CH$_2$
CH$_3$

b.

ClCH———CHCl
CH   or   Cl⟍△⟍Cl
|
Cl   Cl

2.11 a. ethylcyclobutane   b. 1,1-dichlorocyclopropane

2.12 Since each ring carbon is tetrahedral, the H—C—H plane and the C—C—C plane at any ring carbon are mutually perpendicular.

2.13

If you sight down the bond joining carbon-2 and carbon-3, you will see that the substituents on these carbons are eclipsed. The same is true for the bond between carbon-5 and carbon-6. Also, two of the hydrogens on carbon-1 and carbon-4, the "inside" hydrogens, come quite close to one another. All of these factors destabilize boat cyclohexane compared to chair cyclohexane.

25

2.14   The *tert*-butyl group is much larger than a methyl group. Therefore
the conformational preference is essentially 100% for an equatorial
tert-butyl group:

H
—C(CH₃)₃

2.15   a.

Cl   Cl
H   H

cis-1,2-dichlorocyclopropane

Cl   H
H   Cl

trans-1,2-dichlorocyclopropane

b.

Cl
H
Br
H

cis

H
Cl
Br
H

trans

2.16   III, because both methyl substituents are equatorial

2.17   a.

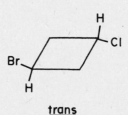

CH₃
4   H
CH₃ 1
H
I(ea)

CH₃
1
H
4 CH₃
H
I(ae)

⇌

The two conformers have *equal* energies, because they each have one
axial and one equatorial methyl substituent.

b.

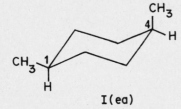

CH₃
4   H
H 1
CH₃
II(aa)

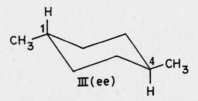

H
1
CH₃
4 CH₃
H
III(ee)

The equilibrium favors the (*e*,*e*) conformer

2.18 a. structural (constitutional) isomers
      b. conformational isomers (same bond pattern; stereoisomers inter-
          convertible by bond rotations)
      c. configurational isomers (same bond pattern, but not interconver-
          tible by bond rotations)
      d. conformational isomers

2.19 Follow eq. 2.15, but replace chloro with bromo:

CH_3Br bromomethane (methyl bromide)

CH_2Br_2 dibromomethane (methylene bromide)

$CHBr_3$ tribromomethane (bromoform)

CBr_4 tetrabromomethane (carbon tetrabromide)

2.20 $CH_2CH_2CH_2CH_2CH_3$ 1-bromopentane
      |
      Br

      $CH_3CHCH_2CH_2CH_3$ 2-bromopentane
            |
            Br

      $CH_3CH_2CHCH_2CH_3$ 3-bromopentane
              |
              Br

2.21 Four monochloro products can be obtained from octane, but only one
      monochloro product is obtained from cyclooctane.

      C-C-C-C-C-C-C-C   C-C-C-C-C-C-C-C
      | |
      Cl   Cl

      C-C-C-C-C-C-C-C   C-C-C-C-C-C-C-C
          | |
          Cl   Cl

2.22 *initiation* Cl_2 $\xrightarrow[\text{or heat}]{\text{light}}$ 2 $:\ddot{C}l\cdot$

      propagation $:\ddot{C}l\cdot$ + CH_4 $\longrightarrow$ $CH_3\cdot$ + HCl

                          $CH_3\cdot$ + Cl_2 $\longrightarrow$ CH_3Cl + $:\ddot{C}l\cdot$

termination
$$2 \;\; :\!\overset{\cdot\cdot}{\underset{\cdot\cdot}{Cl}}\!\cdot \longrightarrow Cl_2$$

$$:\!\overset{\cdot\cdot}{\underset{\cdot\cdot}{Cl}}\!\cdot \; + \; CH_3\cdot \longrightarrow CH_3Cl$$

$$2 \; CH_3\cdot \longrightarrow CH_3 - CH_3$$

2.23   The last termination step in the answer to Problem 2.22 accounts for the formation of ethane in the chlorination of methane. Ethane can also be chlorinated, which explains the formation of small amounts of chloroethane in this reaction.

Additional Problems

2.24   a. 3-methylpentane: First, note the root of the name (in this case, *pent*), and write down and number the carbon chain.

$$\overset{1}{C} - \overset{2}{C} - \overset{3}{C} - \overset{4}{C} - \overset{5}{C}$$

Next, locate the substituents (3-methyl).

$$\overset{1}{C} - \overset{2}{C} - \underset{\underset{CH_3}{|}}{\overset{3}{C}} - \overset{4}{C} - \overset{5}{C}$$

Finally, fill in the remaining hydrogens.

$$CH_3 - CH_2 - \underset{\underset{CH_3}{|}}{CH} - CH_2 - CH_3$$

b. $$CH_3 - \underset{\underset{CH_3}{|}}{CH} - \underset{\underset{CH_3}{|}}{CH} - CH_3$$

c. $$CH_3 - CH_2 - \underset{\underset{CH_3\;\;CH_2CH_3}{|\quad\;\;|}}{\overset{\overset{CH_3}{|}}{C}} - CH - CH_2 - CH_3$$

d. $$CH_3 - \underset{\underset{Cl}{|}}{CH} - \underset{\underset{CH_3}{|}}{CH} - CH_2 - CH_3$$

e. $$CH_3 - \underset{\underset{CH_3\;\;CH_3}{|\quad\;\;|}}{\overset{\overset{CH_3}{|}}{C}} - CH - CH_3$$

f. $CH_3 - CH - CH_3$
        |
        Br

g. 1,1-dichlorocyclopropane: The root *prop* indicates three carbons; the prefix *cyclo* designates that they form a ring.

Next, the substituents are placed.

Finally, the hydrogens are filled in.

h. Cl Cl
        | |
    $Cl - CH - CH_2 - CH - Cl$ or $CHCl_2CH_2CHCl_2$

i.

j.   (Two isomers, *cis-* and *trans-*, are possible.)

2.25 a. $CH_3CH_2CH_2CH_2CH_3$ b. $\overset{1}{C}H_3\overset{2}{C}H\overset{3}{C}H_2\overset{4}{C}H_3$
                                              |
         pentane CH_3

                                          2-methylbutane

29

c.

$$CH_3$$
$$|$$
$$CH_3CH_2CCH_2CH_3$$
$$|$$
$$CH_3$$

3,3-dimethylpentane

d.

$$CH_3$$
$$|$$
$$CH_3CH_2CH_2CCH_3$$
$$|$$
$$CH_3$$

2,2-dimethylpentane

e. $CH_3CH_2CHCH_3$
        $|$
        Br

2-bromobutane

f.

$$Cl \quad Br$$
$$| \quad |$$
$$CH_3C - C - Br$$
$$| \quad |$$
$$Cl \quad Br$$

1,1,1-tribromo-2,2-dichloropropane

(The placement of commas and hyphens is important; this answer shows clearly how these punctuation marks are to be used.)

g.

$$CH_2CH_3$$
$$|$$
$$CH_3CH_2CCH_2CH_3$$
$$|$$
$$CH_2CH_3$$

3,3-diethylpentane

h.

$$\overset{2}{CH_2} - \overset{1}{CH_2}$$
$$| \qquad |$$
$$Cl \qquad Br$$

1-bromo-2-chloroethane

(The priority of the halogens follows alphabetical order.)

i.

$$\overset{1}{CH_2} - \overset{2}{CH} - \overset{3}{CH} - \overset{4}{CH_3}$$
$$| \qquad | \qquad |$$
$$Br \quad CH_3 \quad CH_3$$

1-bromo-2,3-dimethyl-butane

j.

$$CH_2$$
$$CH_2 \quad CH_2$$
$$CH_2 - CH_2$$

cyclopentane

k. CH_3I

iodomethane

l. CH_3CHCH_3
        $|$
        Br

2-bromopropane

2.26

Common	IUPAC
a. methyl iodide	iodomethane
b. ethyl chloride	chloroethane
c. methylene chloride	dichloromethane
(CH_2 = methylene)	

d.	bromoform	tribromomethane
e.	*n*-propyl chloride	1-chloropropane
f.	isopropyl bromide	2-bromopropane
g.	chloroform	trichloromethane
h.	cyclobutyl chloride	chlorocyclobutane
i.	*sec*-butyl iodide	2-iodobutane
j.	*tert*-butyl chloride	2-chloro-2-methylpropane

2.27 a. $CH_2CH_2CH_2CH_2CH_3$
          |
          CH_3

The longest consecutive chain has six carbon atoms; the correct
name is hexane.

b.
$$\overset{1}{CH_3} - \overset{2}{CH} - \overset{3}{CH_2} - \overset{4}{CH_3}$$
          |
          CH_2
          |
          CH_3

The longest chain was not selected when the compound was numbered.
The correct numbering is

$$CH_3 - \overset{3}{CH} - \overset{4}{CH_2} - \overset{5}{CH_3} \quad \text{3-methylpentane}$$
          |
          $^2CH_2$
          |
          $^1CH_3$

c. The numbering started at the wrong end. The name should be
    1,2-dichloropropane.

$$\overset{1}{C} - \overset{2}{C} - \overset{3}{C}$$
  | |
  Cl   Cl

d. The ring was numbered the wrong way to give the lowest substituent
    numbers. The correct name is 1,2-dimethylcyclobutane.

C   C
 \ $_1$ $_2$ /
   C — C
  $_4$| |$_3$
   C — C

e. The longest chain was not selected. The correct name is
2-methylpentane.

```
1
C
|2  3   4
C — C — C
|       |5
C       C
```

f. The correct name for the compound is 1-bromo-2-methylpropane
(use the lower number).

```
        C
3   |2  1
C — C — C
        |
        Br
```

2.28 The root of the name, *heptadec*, indicates a 17-carbon chain. The
correct formula is

$$CH_3CHCH_2CH_2CH_2CH_2CH_2CH_2CH_2CH_2CH_2CH_2CH_2CH_2CH_2CH_3$$
$$|$$
$$CH_3$$

or $(CH_3)_2CH(CH_2)_{14}CH_3$

2.29 The molecular formulas are derived using C_nH_{2n+2}. They are $C_{102}H_{206}$,
$C_{150}H_{352}$, $C_{198}H_{398}$, $C_{246}H_{494}$, and $C_{390}H_{782}$. Using approximate atomic
weights of C = 12 and H = 1, the molecular weight of the C_{390} compound
is 5462. However, using more precise atomic weights of C = 12.01 and
H = 1.008, the molecular weight is 5472.2. (The small fractions in
each atomic weight become significant when so many atoms are present
in a single molecule.)

2.30 Approach each problem systematically. Start with the longest possible
carbon chain and shorten it one carbon at a time until no further
isomers are possible. To conserve space, the formulas below are
written in condensed form, but you should write them out as expanded
formulas.

a. $CH_3(CH_2)_2CH_3$ butane
     $(CH_3)_3CH$ 2-methylpropane

b. $CH_3CH_2CH_2CH_2Br$ 1-bromobutane
     $CH_3CHBrCH_2CH_3$ 2-bromobutane
     $(CH_3)_2CHCH_2Br$ 1-bromo-2-methylpropane
     $(CH_3)_3CBr$ 2-bromo-2-methylpropane

c. $CH_3(CH_2)_4CH_3$ hexane
    $CH_3CH(CH_3)CH_2CH_2CH_3$ 2-methylpentane
    $CH_3CH_2CH(CH_3)CH_2CH_3$ 3-methylpentane
    $CH_3CH(CH_3)CH(CH_3)CH_3$ 2,3-dimethylbutane
    $CH_3C(CH_3)_2CH_2CH_3$ 2,2-dimethylbutane

d. $CH_3CH_2CHBr_2$ 1,1-dibromopropane
    $CH_3CHBrCH_2Br$ 1,2-dibromopropane
    $CH_2BrCH_2CH_2Br$ 1,3-dibromopropane
    $CH_3CBr_2CH_3$ 2,2-dibromopropane

e. The two hydrogens can be either on the same or on different
    carbon atoms:
    CH_2BrCCl_3 1-bromo-2,2,2-trichloroethane
    $CH_2ClCBrCl_2$ 1-bromo-1,1,2-trichloroethane
    $CHBrClCHCl_2$ 1-bromo-1,2,2-trichloroethane

f. $CHBrClCH_2CH_3$ 1-bromo-1-chloropropane
    $CH_2BrCHClCH_3$ 1-bromo-2-chloropropane
    $CH_2BrCH_2CH_2Cl$ 1-bromo-3-chloropropane
    $CH_2ClCHBrCH_3$ 2-bromo-1-chloropropane
    $CH_3CBrClCH_3$ 2-bromo-2-chloropropane

2.31 A cycloalkane has two fewer hydrogens than the corresponding alkane.
      Thus the general formula for a cycloalkane is C_nH_{2n}.

2.32 a. Start with a five-membered ring, then proceed to the four- and
          three-membered rings:

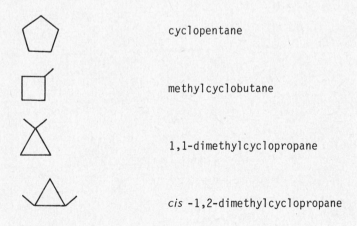

          cyclopentane

          methylcyclobutane

          1,1-dimethylcyclopropane

          cis -1,2-dimethylcyclopropane

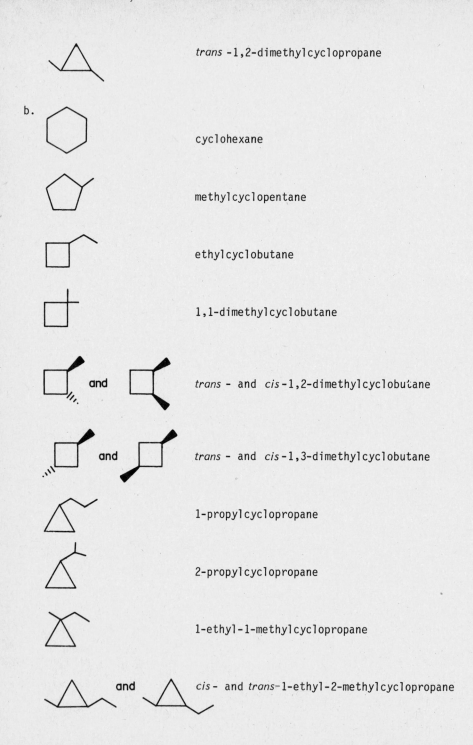

trans-1,2-dimethylcyclopropane

b.

cyclohexane

methylcyclopentane

ethylcyclobutane

1,1-dimethylcyclobutane

and   *trans*- and *cis*-1,2-dimethylcyclobutane

and   *trans*- and *cis*-1,3-dimethylcyclobutane

1-propylcyclopropane

2-propylcyclopropane

1-ethyl-1-methylcyclopropane

and   *cis*- and *trans*-1-ethyl-2-methylcyclopropane

1,1,2-trimethylcyclopropane

cis,cis-1,2,3-trimethylcyclopropane

cis,trans-1,2,3-trimethylcyclopropane

2.33 The higher the molecular weight and the less branching of the chain, the higher the boiling point. On these grounds, the expected order from the lowest to highest boiling point should be e, d, c, a, b. The actual boiling points are as follows: e, 2-methylpentane (60°C); d, *n*-hexane (69°C); c, 3,3-dimethylpentane (86°C); a, 2-methylhexane (90°C); b, *n*-heptane (98.4°C).

2.34 The four conformations are as follows:

A: most stable staggered conformation

B: less stable staggered conformation (large groups are closer together)

C: less stable than the staggered, but more stable than the eclipsed conformation with two methyls eclipsed

D: least stable of all four conformations

Staggered conformations are more stable than eclipsed conformations. Therefore A and B are more stable than C or D. Within each pair, CH_3-CH_3 interactions (for methyls on adjacent carbons) are avoided because of the large size of these groups.

35

2.35

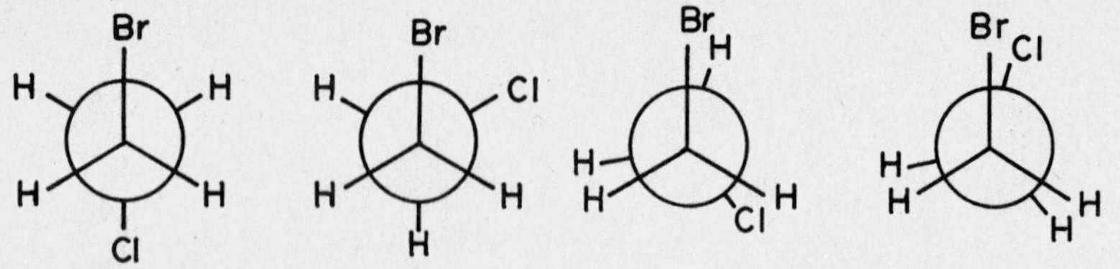

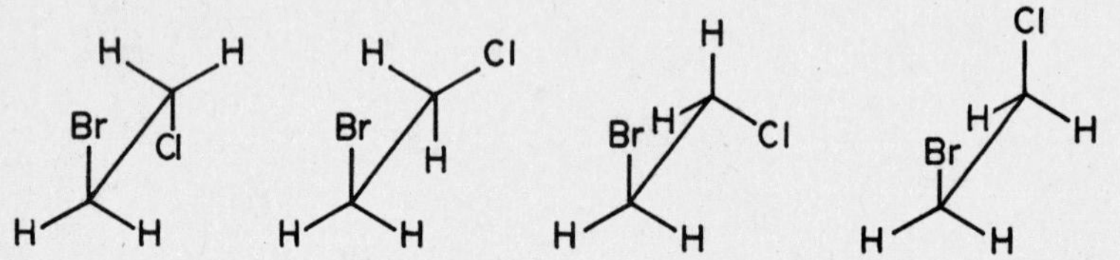

The stability decreases from left to right in the structures shown above.

2.36   a.

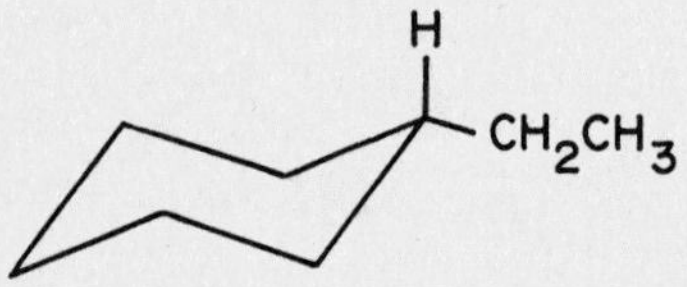

The ethyl group is equatorial. If we were to draw out the ethyl group, we would arrange its substituents in a staggered conformation.

b.

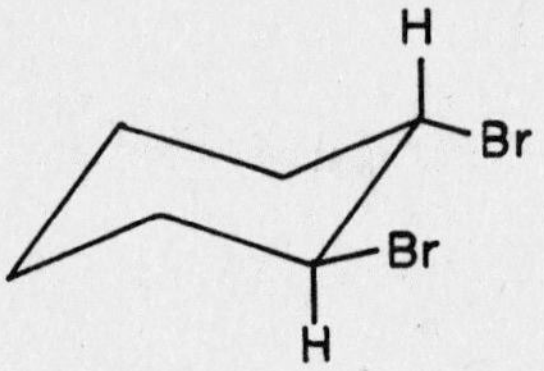

The bromines are *trans*, and both are equatorial.

c.

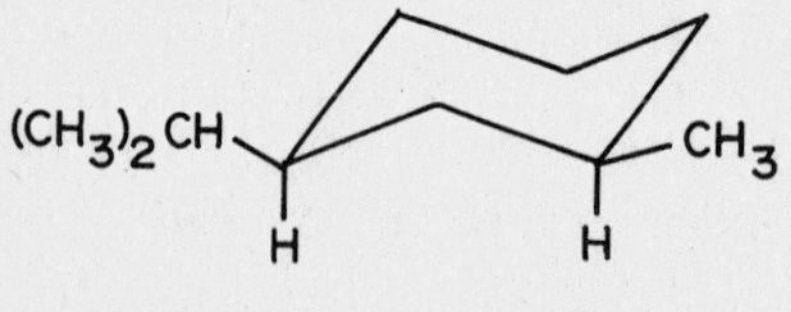

Both alkyl groups are equatorial and *cis*.

d.

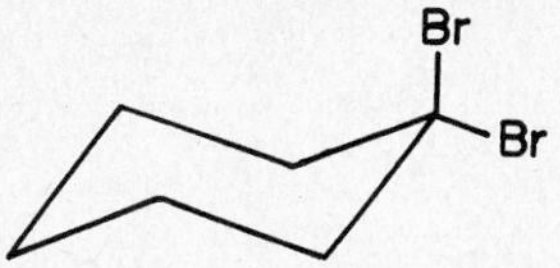

One bromine is axial; one is equatorial.

2.37   In each case, the formula at the left is named; the right-hand structure
shows the other isomer.
a.   *trans* -1-chloro-3-methylcyclobutane
b.   *cis* -1-chloro-3-methylcyclohexane
c.   *trans* -1,2-dichlorocyclohexane
d.   *cis* -1,3-dibromo-1,3-dichlorocyclopentane

2.38   *cis* -1,3-Dimethylcyclohexane can exist in a conformation in which both
methyl substituents are equatorial:

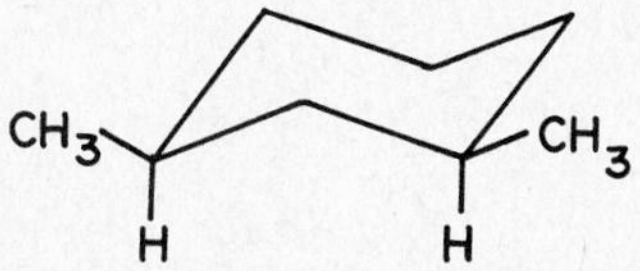

In the *trans* isomer, one methyl group must be axial. On the other hand,
in 1,2- or 1,4-dimethylcyclohexane, only the *trans* isomer can have both
methyls equatorial.

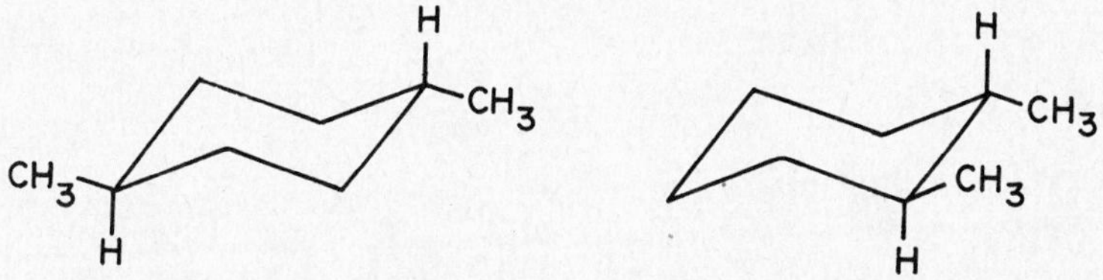

2.39   The *trans* isomer is very much more stable than the *cis* isomer, because
both *t*-butyl groups, which are very large, can be equatorial.

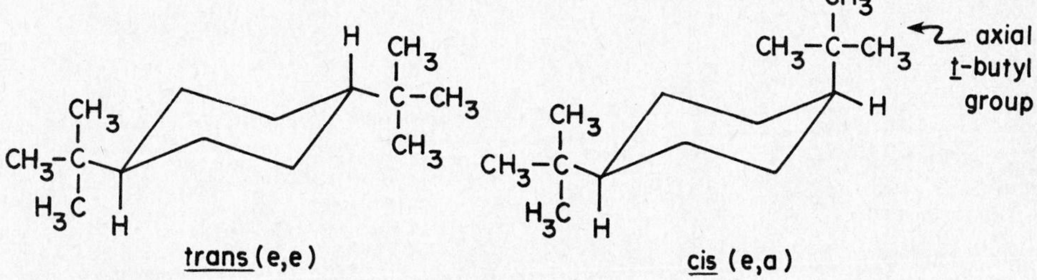

trans (e,e)   cis (e,a)

2.40   a.   All pairs have the same bond patterns and are stereoisomers.
Since they are *not* interconvertible by bond rotations, they
are configurational isomers.

37

b. conformational isomers

c. conformational isomers

d. The bond patterns are *not* the same; the first Newman projection is for 1,1-dichloropropane, and the second is for 1,2-dichloropropane. Thus, these are structural (or constitutional) isomers.

e. The structures are *identical* (both represent 2-methylbutane) and are not isomers at all.

2.41 We can obtain 1,1- or 1,2- or 1,3-dichlorocyclopentane. Of these, the last two can exist as *cis-trans* isomers.

1,1-dichlorocyclopentane

cis -1,2-dichlorocyclopentane

trans -1,2-dichlorocyclopentane

cis -1,3-dichlorocyclopentane

trans -1,3-dichlorocyclopentane

2.42

The isomer in which each chlorine is *trans* to both adjacent chlorines will have all six chlorines equatorial. This isomer is predicted to be the most stable. All other isomers must have at least one axial chlorine atom.

2.43   a. $CH_3CH_2CH_2CH_2CH_3 + 8O_2 \longrightarrow 5\ CO_2 + 6\ H_2O$

b. $2 \pentagon + 15\ O_2 \longrightarrow 10\ CO_2 + 10\ H_2O$

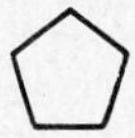

c. $CH_3CH_2CH_3 + Br_2 \longrightarrow CH_3CH_2CH_2Br$ or $CH_3\underset{\underset{Br}{|}}{C}HCH_3 + HBr$

d.

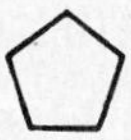

$+ Cl_2 \longrightarrow$ (H, Cl substituted cyclopentane) $+ HCl$

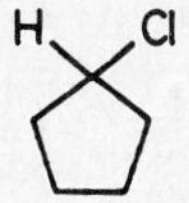

e. $CH_3CH_3 + 6Cl_2 \xrightarrow{h\nu} CCl_3CCl_3 + 6HCl$

2.44   The four possible structures are

$CH_3CH_2CHCl_2$ $CH_3\underset{\underset{Cl}{|}}{C}H-\underset{\underset{Cl}{|}}{C}H_2$

1,1-dichloropropane   1,2-dichloropropane

$\underset{\underset{Cl}{|}}{C}H_2\ CH_2\ \underset{\underset{Cl}{|}}{C}H_2$ $CH_3-\overset{\overset{Cl}{|}}{\underset{\underset{Cl}{|}}{C}}-CH_3$

1,3-dichloropropane

                                            2,2-dichloropropane

Only the last of these has all hydrogens equivalent and can give only
one trichloro compound. This must therefore be C:

$CH_3CCl_2CH_3 \xrightarrow{\ Cl_2\ } CH_3CCl_2CH_2Cl$

1,3-Dichloropropane has only two different "kinds" of hydrogens.
It must be D:

$\underset{\underset{Cl}{|}}{C}H_2\ CH_2\ \underset{\underset{Cl}{|}}{C}H_2 \xrightarrow{\ Cl_2\ } \underset{\underset{Cl}{|}}{C}H_2-\underset{\underset{Cl}{|}}{C}H-\underset{\underset{Cl}{|}}{C}H_2$ or $\underset{\underset{Cl}{|}}{C}H_2CH_2CHCl_2$

Next, A must be capable of giving 1,2,2-trichloropropane (the product from C). This is not possible for the 1,1 isomer, since it already has two chlorines on carbon-1. Therefore, A must be

$$\begin{array}{l} CH_3\underset{\displaystyle Cl}{\underset{|}{C}}H\underset{\displaystyle Cl}{\underset{|}{C}}H_2 \end{array}$$

since it can give the 1,2,2-trichloro product (as well as 1,1,2- and 1,2,3-). By elimination, B is $CH_3CH_2CHCl_2$.

2.45 The equations follow the same pattern as in eqs. 2.18 through 2.23.

$$Cl_2 \xrightarrow[\text{sunlight}]{\text{heat or}} 2\,:\ddot{\underset{..}{Cl}}\cdot \qquad \textit{initiation}$$

$$\left.\begin{array}{l} CH_3CH_3 + Cl\cdot \longrightarrow CH_3CH_2\cdot + HCl \\ CH_3CH_2\cdot + Cl_2 \longrightarrow CH_3CH_2Cl + Cl\cdot \end{array}\right\} \qquad \textit{propagation}$$

$$\left.\begin{array}{l} CH_3CH_2\cdot + :\ddot{\underset{..}{Cl}}\cdot \longrightarrow CH_3CH_2Cl \\ 2\,:\ddot{\underset{..}{Cl}}\cdot \longrightarrow Cl_2 \\ 2\ CH_3CH_2\cdot \longrightarrow CH_3CH_2CH_2CH_3 \end{array}\right\} \qquad \textit{termination}$$

The trace products resulting from the chain-termination steps would be butane and any chlorination products derived from it.

CHAPTER SUMMARY

Alkenes have a carbon-carbon double bond; alkynes have a carbon-carbon triple bond. Nomenclature rules are given in Sec. 3.2.

Each carbon of a double bond is trigonal, or connected to only *three* other atoms. These atoms lie in a plane with bond angles of 120°. Ordinarily, rotation around double bonds is restricted. All six atoms of ethylene lie in a single plane. The $C = C$ bond length is 1.34 Å, shorter than a $C - C$ bond (1.54 Å). These facts can be explained by an orbital model with three sp^2 hybrid orbitals (one electron in each) and one p orbital perpendicular to these (containing the fourth electron). The double bond is formed by end-on overlap of sp^2 orbitals to form a σ bond and lateral overlap of aligned p orbitals to form a π bond (Figures 3.4 and 3.5). Since rotation around the double bond is restricted, *cis-trans* isomerism is possible if each carbon atom of the double bond has two different groups attached to it.

Alkenes react mainly by addition. Typical reagents that add to the double bond are halogens, hydrogen (metal catalyst required), water (acid catalyst required), and various acids. If either the alkene or the reagent is symmetric (Table 3.2), only one product is possible. If *both* the alkene and reagent are unsymmetric, however, two products are possible, in principle; in this case, Markovnikov's rule (Secs. 3.12 through 3.14) allows us to predict the product obtained.

Electrophilic additions occur by a two-step mechanism. In the first step, the electrophile adds in such a way as to form the most stable carbocation (the stability order is tertiary > secondary > primary). Then the carbocation combines with a nucleophile to give the product. Bromine addition occurs by a similar path, but the intermediate is a cyclic bromonium ion instead of a carbocation.

Conjugated dienes have alternating single and double bonds. They may undergo 1,2- or 1,4-addition. They also undergo cycloaddition reactions with alkenes (Diels-Alder reaction), a useful synthesis for six-membered rings.

Addition to double bonds may also occur by a free-radical mechanism. Vinyl polymers such as polyethylene (Table 3.3) can be made in this way from vinyl monomers and are important commercial materials. Natural rubber is a polymer of isoprene. The most common synthetic rubber is a copolymer of styrene and butadiene.

Alkenes undergo a number of other reactions, such as **hydroboration,** permanganate oxidation, and **ozonolysis.**

Triple bonds are **linear;** the carbons are *sp*-hybridized (Figure 3.10). Alkynes, like alkenes, undergo addition reactions. A hydrogen connected to a triply-bonded carbon is weakly acidic and can be removed by a very strong base such as **sodamide, NaNH$_2$,** to give **acetylides.**

REACTION SUMMARY

Additions to the Double Bond

$$\underset{/}{\overset{\backslash}{C}} = \underset{\backslash}{\overset{/}{C}} \quad + \quad X_2 \quad \longrightarrow \quad -\underset{|}{\overset{|}{\underset{X}{C}}}-\underset{|}{\overset{|}{\underset{X}{C}}}- \quad (X = Cl\ or\ Br)$$

$$+ \quad H_2 \quad \xrightarrow{\ \text{Pt or}\ \ \text{Ni}\ } \quad -\underset{|}{\overset{|}{\underset{H}{C}}}-\underset{|}{\overset{|}{\underset{H}{C}}}-$$

$$+ \quad H-OH \quad \xrightarrow{\ H^+\ } \quad -\underset{|}{\overset{|}{\underset{H}{C}}}-\underset{|}{\overset{|}{\underset{OH}{C}}}- \quad (an\ alcohol)$$

$$C = C \quad + \quad H-X \quad \longrightarrow \quad \overset{H}{\underset{|}{C}}-\overset{}{\underset{X}{C}}$$

$$(X = -F, \quad -Cl, \quad -Br, \quad -I,$$

$$\overset{O}{\overset{\|}{-SO_3H, \quad -OCR})}$$

Conjugated Dienes

$$C = C - C = C \quad \xrightarrow{\ X_2\ } \quad \underset{|}{\overset{|}{\underset{X}{C}}}-\underset{|}{\overset{|}{\underset{X}{C}}}-C = C \quad + \quad \underset{|}{\overset{|}{\underset{X}{C}}}-C = C -\underset{|}{\overset{|}{\underset{X}{C}}}$$

$$\xrightarrow{\ H-X\ } \quad \underset{|}{\overset{|}{\underset{H}{C}}}-\underset{|}{\overset{|}{\underset{X}{C}}}-C = C \quad + \quad \underset{|}{\overset{|}{\underset{H}{C}}}-C = C -\underset{|}{\overset{|}{\underset{X}{C}}}$$

(X = Cl or Br)   1,2-addition   1,4-addition

Cycloaddition (Diels-Alder)

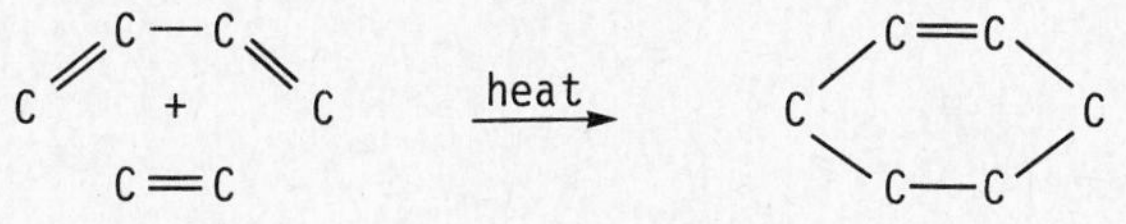

$$C \overset{C-C}{\underset{C=C}{+}} C \quad \xrightarrow{\text{heat}} \quad C \overset{C=C}{\underset{C-C}{\diagup\diagdown}} C$$

Polymerization (see Table 3.3)

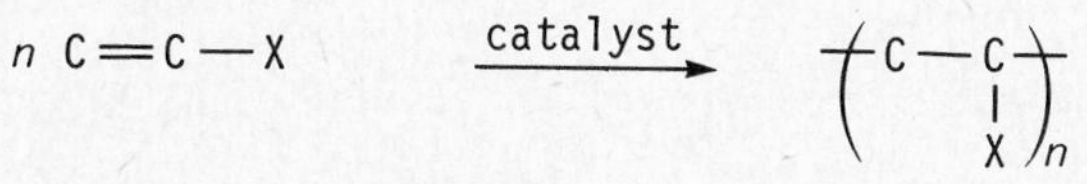

$$n\ C=C-X \quad \xrightarrow{\text{catalyst}} \quad \left(\begin{array}{c} C-C \\ | \\ X \end{array} \right)_n$$

Hydroboration-Oxidation

$$R-CH=CH_2 \quad \xrightarrow{BH_3} \quad (RCH_2CH_2)_3B \quad \xrightarrow[OH^-]{H_2O_2} \quad RCH_2CH_2OH$$

Permanganate Oxidation

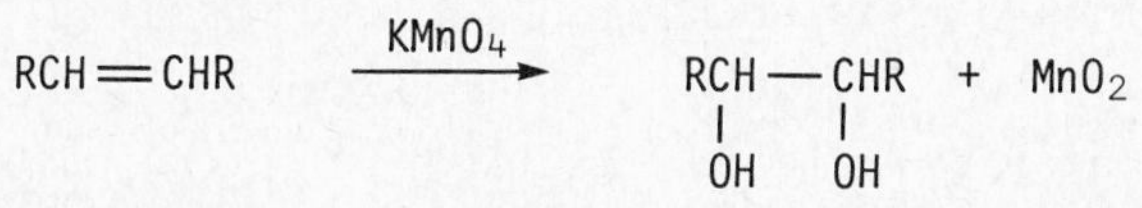

$$RCH=CHR \quad \xrightarrow{KMnO_4} \quad \underset{\underset{OH}{|}}{RCH}-\underset{\underset{OH}{|}}{CHR} \ +\ MnO_2$$

Ozonolysis

$$\diagup\diagdown{C}=C\diagup\diagdown \ +\ O_3 \quad \longrightarrow \quad \diagup\diagdown{C}=O \ +\ O=C\diagup\diagdown$$

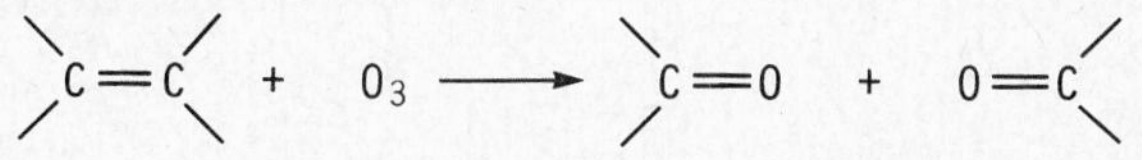

Additions to the Triple Bond

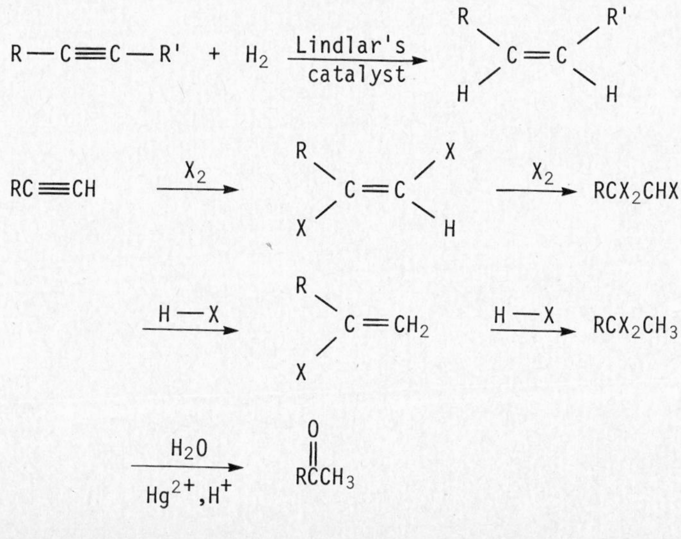

$$R-C\equiv C-R' \ +\ H_2 \quad \xrightarrow[\text{catalyst}]{\text{Lindlar's}} \quad \underset{H}{\overset{R}{\diagup}}C=C\underset{H}{\overset{R'}{\diagdown}} \quad (\textit{cis addition})$$

$$RC\equiv CH \quad \xrightarrow{X_2} \quad \underset{X}{\overset{R}{\diagup}}C=C\underset{H}{\overset{X}{\diagdown}} \quad \xrightarrow{X_2} \quad RCX_2CHX_2$$

$$\xrightarrow{H-X} \quad \underset{X}{\overset{R}{\diagup}}C=CH_2 \quad \xrightarrow{H-X} \quad RCX_2CH_3$$

$$\xrightarrow[Hg^{2+},H^+]{H_2O} \quad \overset{O}{\overset{\|}{RCCH_3}}$$

Alkyne Acidity

$$RC \equiv CH \quad + \quad NaNH_2 \quad \xrightarrow{NH_3} \quad RC \equiv C^-Na^+ \quad + \quad NH_3$$

MECHANISM SUMMARY

Electrophilic Addition (E^+ = electrophile; Nu: = nucleophile)

carbocation

1,4-Addition

Free-Radical Polymerization

etc.

LEARNING OBJECTIVES

1. Know the meaning of: alkene, alkyne, and diene; conjugated, cumu-
    lated and isolated double bonds; vinyl, allyl, and propargyl groups.

2. Know the meaning of: trigonal carbon, sp^2-hybridization, restricted
    rotation, σ and π bond, *cis* and *trans* double bond isomers.

3. Know the meaning of: addition reaction, Markovnikov's rule, electro-
    phile, nucleophile, symmetric and unsymmetric double bonds and reagents,
    carbocation, bromonium ion.

4. Know the meaning of: 1,2-addition, 1,4-addition, cycloaddition, diene,
    dienophile, polymer, monomer, polymerization, vinyl polymer, polyethylene,
    polypropylene, polyvinyl chloride, polystyrene, Orlon, Teflon, natural and
    synthetic rubber, isoprene, isoprene unit.

5. Know the meaning of: hydroboration, glycol, ozone, ozonolysis.

6. Know the meaning of: petroleum, cracking, octane number, tetraethyllead,
    isomerization, alkylation, platforming.

7. Given the structure of an acyclic or cyclic alkene, alkyne, diene, etc.,
    state the IUPAC name.

8. Given the IUPAC name of an alkene, alkyne, diene, etc., write the struc-
    tural formula.

9. Given the molecular formula of a hydrocarbon and the number of double
    bonds, triple bonds, or rings, draw the possible structures.

10. Given the name or abbreviated structure of an unsaturated compound, tell
    whether it can exist in *cis* and *trans* isomeric forms, and if so, how many.
    Draw them.

11. Given an alkene, alkyne, or diene, and one of the following reagents,
    draw the structure of the product (reagents: acids such as HCl, HBr,
    HI, and H_2SO_4; water in the presence of an acid catalyst; halogens such
    as Br_2 and Cl_2; hydrogen).

12. Given the structure or name of a compound that can be prepared by an
    addition reaction, deduce what unsaturated compound and what reagent
    react to form it.

13. Write the steps in the mechanism of an electrophilic addition reaction.

14. Given an unsymmetrical alkene and an unsymmetrical electrophilic reagent,
    give the structure of the predominant product (i.e., apply Markovnikov's
    rule).

15. Given a conjugated diene and a reagent that adds to it, write the structures of the 1,2- and 1,4-addition products.

16. Given a diene and dienophile, write the structure of the resulting cycloaddition (Diels-Alder) adduct.

17. Given the structure of a cyclic compound that can be synthesized by the Diels-Alder reaction, deduce the structures of the required diene and dienophile.

18. Given an alkyne, write the structures of products obtained by adding one or two moles of a particular reagent to it.

19. Given a vinyl compound, draw a structure for the corresponding polyvinyl compound. Also, given the structure of a polymer segment, deduce the structure of the monomer from which it can be prepared.

20. Write the steps in the mechanism of vinyl polymerization catalyzed by a free radical.

21. Write the structure of the alcohol product from the hydroboration-oxidation sequence when applied to a particular alkene.

22. Given an alkene or cycloalkene (or diene, and so on), write the structures of the expected ozonolysis products.

23. Given the structures of ozonolysis products, deduce the structure of the unsaturated hydrocarbon that produced them.

24. Draw orbital pictures for a double bond and a triple bond.

25. Draw conventional structures for the contributors to the resonance hybrid of an allyl cation.

26. Describe simple chemical tests that can distinguish an alkane from an alkene or alkyne.

27. Know the meaning of cracking, alkylation, isomerization, platforming, and octane number, as applied to petroleum refining.

ANSWERS TO PROBLEMS

Problems Within the Chapter

3.1 The formula C_4H_6 corresponds to C_nH_{2n-2}. The possibilities are one triple bond, two double bonds, one double bond and one ring, or two rings.

$HC{\equiv}CCH_2CH_3$ $CH_2{=}CH{-}CH{=}CH_2$

$CH_3C{\equiv}CCH_3$ $CH_2{=}C{=}CHCH_3$ } **Acyclic**

Cyclic

3.2 Compounds b and c have alternating single and double bonds, and are conjugated. In a and d the double bonds are isolated.

3.3 a. $\overset{1}{C}lCH{=}\overset{2}{C}H\overset{3}{C}H_3$ 1-chloro-1-propene (or simply 1-chloropropene)

b. $\overset{1}{C}H{-}\overset{2}{C}{=}\overset{3}{C}{-}\overset{4}{C}H_3$ with CH_3 CH_3 substituents   2,3-dimethyl-2-butene

c. $\overset{1}{C}H_2{=}\overset{2}{C}{-}\overset{3}{C}H{=}\overset{4}{C}H_2$ with CH_3 substituent   2-methyl-1,3-butadiene (also called isoprene)

d.

1-methylcyclopentene

e. $\overset{1}{C}H_2{=}\overset{2}{C}{-}\overset{3}{C}H_3$ with Cl substituent   2-chloro-1-propene, or 2-chloropropene

f. $\overset{1}{H}\overset{23}{C}\equiv\overset{4}{CC}H_2\overset{5}{C}H_2\overset{6}{C}H_2CH_3$ 1-hexyne

3.4 a. First write out the five-carbon chain, with a double bond between carbon-2 and carbon-3:

$\overset{1}{C}-\overset{2}{C}=\overset{3}{C}-\overset{4}{C}-\overset{5}{C}$

Add the substituents:

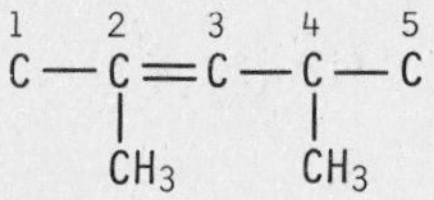

$\overset{1}{C}-\overset{2}{C}=\overset{3}{C}-\overset{4}{C}-\overset{5}{C}$
       | |
      CH_3 CH_3

Fill in the hydrogens:

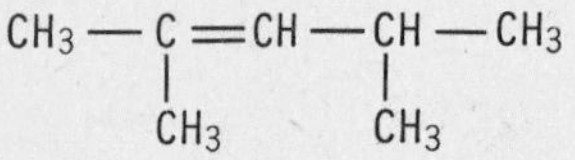

$CH_3-C=CH-CH-CH_3$ or $(CH_3)_2C=CHCH(CH_3)_2$
       | |
      CH_3 CH_3

b. $CH_3C\equiv CCH_2CH_2CH_3$

c.

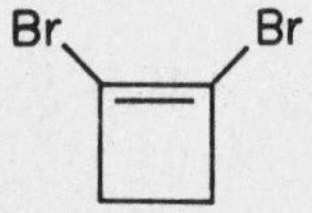

d. 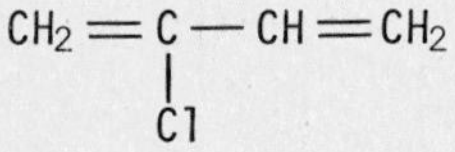 $CH_2=C-CH=CH_2$
        |
        Cl

3.5 a. 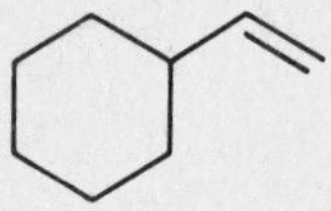   or

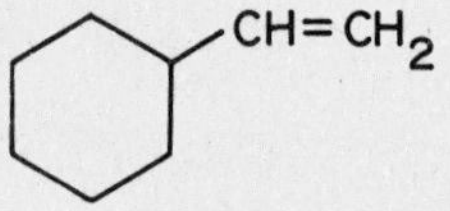

b. 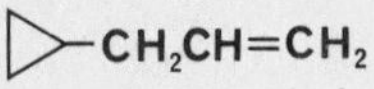 $-CH_2CH=CH_2$

c. $HC\equiv CCH_2I$

3.6 Compounds a and d have only one possible structure, because in each case one of the carbons of the double bond has two identical substituents:

a. $CH_2=CHCH_3$

b

$$CH_3CH_2 \quad CH_2CH_3$$
$$C=C$$
$$H \quad\quad H$$

and

$$CH_3CH_2 \quad\quad H$$
$$C=C$$
$$H \quad\quad CH_2CH_3$$

<u>cis</u>-3-hexene

<u>trans</u>-3-hexene

c.

$$CH_3 \quad CH_2CH_2CH_3$$
$$C=C$$
$$H \quad\quad H$$

and

$$CH_3 \quad H$$
$$C=C$$
$$H \quad\quad CH_2CH_2CH_3$$

<u>cis</u>-2-hexene

<u>trans</u>-2-hexene

d. $(CH_3)_2C=CHCH_3$

3.7 The electron pair in a σ bond lies directly between the nuclei it joins. In a π bond, the electron pair is further from the two nuclei that it joins. Therefore more energy is required to break a σ bond than a π bond.

3.8 a. One bromine atom adds to each doubly bonded carbon, and the double bond in the starting material becomes a single bond in the product:

$$CH_2=CHCH_2CH_3 \;+\; Br_2 \longrightarrow \underset{\underset{Br}{|}}{CH_2}-\underset{\underset{Br}{|}}{CH}-CH_2CH_3$$

b.

$$CH_3-\underset{\underset{CH_3}{|}}{C}=CHCH_3 \;+\; Br_2 \longrightarrow CH_3-\overset{\overset{Br}{|}}{\underset{\underset{CH_3}{|}}{C}}-\overset{\overset{Br}{|}}{C}HCH_3$$

3.9 a.

$$CH_2=\underset{\underset{CH_3}{|}}{C}-CH_3 \;+\; H_2 \xrightarrow{\text{catalyst}} CH_3-\overset{\overset{H}{|}}{\underset{\underset{CH_3}{|}}{C}}-CH_3$$

b.

+ H₂ →(catalyst)→

3.10 a. $CH_3CH=CHCH_3$ + H$-$OH $\xrightarrow{H^+}$ $CH_3CH_2\underset{\underset{OH}{|}}{C}HCH_3$

b.

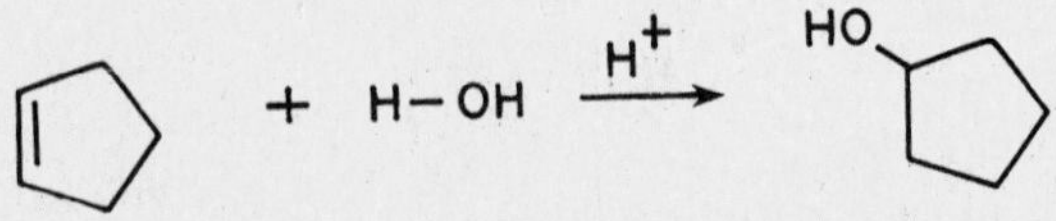

$+$ H$-$OH $\xrightarrow{H^+}$

3.11 a. $CH_3CH=CHCH_3$ + HI $\longrightarrow$ $CH_3\underset{\underset{I}{|}}{C}HCH_2CH_3$

b.

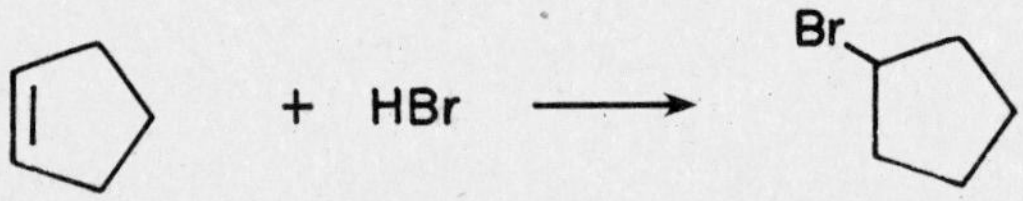

$+$ HBr $\longrightarrow$

3.12 a. $CH_2=CHCH_2CH_3$
$\uparrow$

This doubly bonded carbon has the most hydrogens.

H$-$Cl
$\uparrow$

This is the more electropositive part of the reagent.

$CH_2=CHCH_2CH_3$ $+$ $\overset{\delta+\quad\delta-}{H-Cl}$ $\longrightarrow$ $CH_3-\underset{\underset{Cl}{|}}{C}HCH_2CH_3$

b. $CH_3-\underset{\underset{CH_3}{|}}{C}=CH-CH_3$ $+$ $\overset{\delta+\quad\delta-}{H-OH}$ $\longrightarrow$ $CH_3-\underset{\underset{CH_3}{|}}{\overset{\overset{OH}{|}}{C}}-CH_2CH_3$

3.13 $CH_3CH=CHCH_2CH_3$ $+$ HCl $\longrightarrow$ $CH_3CH_2\underset{\underset{Cl}{|}}{C}HCH_2CH_3$

$+$ $CH_3\underset{\underset{Cl}{|}}{C}HCH_2CH_2CH_3$

Since both carbons of the double bond have the same number of attached hydrogens (one), application of Markovnikov's rule is ambiguous. Both products are formed, and the reaction is *not* regiospecific.

3.14   eq. 3.19:

$$CH_3-\underset{\underset{CH_3}{|}}{C}=CH_2 \ + \ H^+ \ \longrightarrow \ CH_3-\overset{+}{\underset{\underset{CH_3}{|}}{C}}-CH_3$$

$$CH_3-\overset{+}{\underset{\underset{CH_3}{|}}{C}}-CH_3 \ + \ H\ddot{O}H \ \longrightarrow \ CH_3-\underset{\underset{CH_3}{|}}{\overset{\overset{\overset{H}{\underset{}{\oplus\ddot{O}}}{}}{|}}{C}}-CH_3$$

$$CH_3-\underset{\underset{CH_3}{|}}{\overset{\overset{\overset{H\quad H}{\underset{}{\oplus\ddot{O}}}}{|}}{C}}-CH_3 \ \longrightarrow \ CH_3-\underset{\underset{CH_3}{|}}{\overset{\overset{:\ddot{O}H}{|}}{C}}-CH_3 \ + \ H^+$$

The intermediate carbocation is tertiary. If, in the first step, the proton had added to the other carbon of the double bond, the carbocation produced would have been primary:

$$CH_3-\underset{\underset{CH_3}{|}}{C}=CH_2 \ + \ H^+ \ \xrightarrow{\ \ //\ \ } \ CH_3-\underset{\underset{CH_3}{|}}{\overset{\overset{H}{|}}{C}}-CH_2{}^+$$

eq. 3.20:

The intermediate carbocation is tertiary. Had the proton added to the methyl-bearing carbon, the intermediate carbocation would have been secondary:

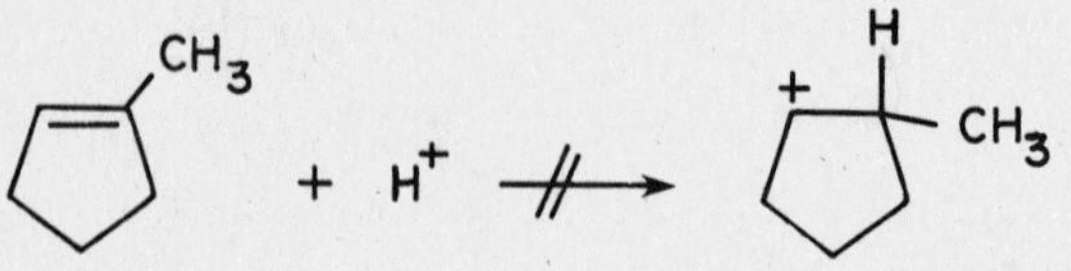

3.15

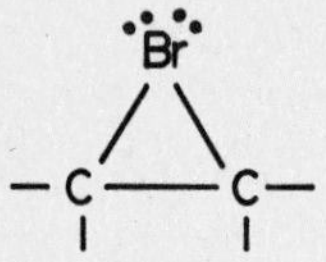

Using the formula in Sec. 1.11, the formal charge on the bromine is $7 - (4 + 2) = +1$. At each carbon atom, the formal charge is $4 - (0 + 4) = 0$.

3.16 The bromonium ion is formed on one face of the cyclopentene ring:

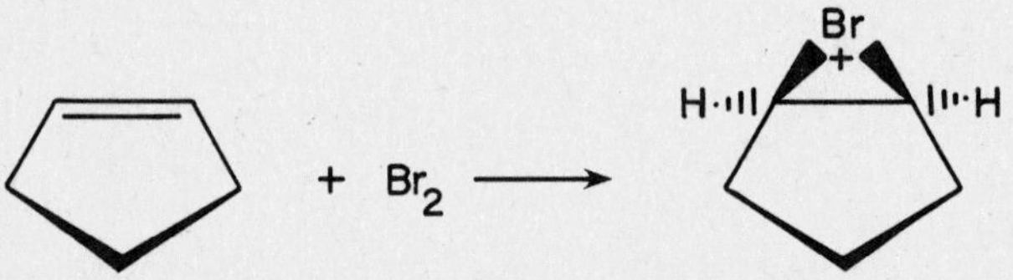

The second bromine then adds from the opposite face to give the *trans* product:

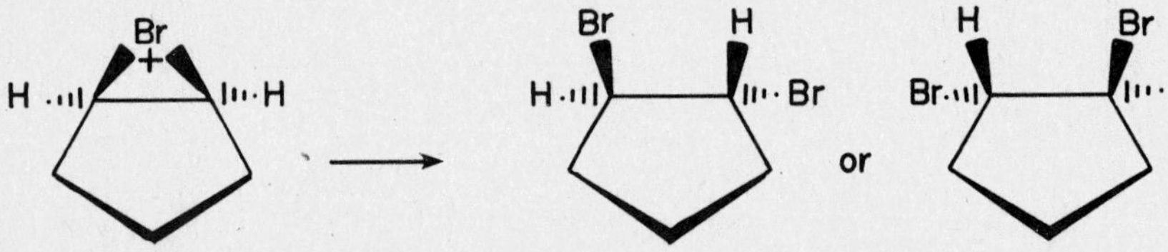

3.17 $CH_2 = CH — CH = CH_2$ + H^+ $\xrightarrow[\text{C-1}]{\text{adds to}}$ $CH_3 — \overset{+}{C}H — CH = CH_2$

allylic carbocation, stabilized by resonance

$CH_2 = CH — CH = CH_2$ + H^+ $\xrightarrow[\text{C-2}]{\text{adds to}}$ $\overset{+}{C}H_2 — CH_2 — CH = CH_2$

primary carbocation, much less stable than the allylic carbocation

3.18 $CH_2=CH-CH=CH_2$ + Br_2 ⟶

$CH_2-CH-CH=CH_2$ + $CH_2-CH=CH-CH_2$

  | | | |
  Br   Br   Br   Br

1,2-addition   1,4-addition

3.19 To find the structures of the diene and dienophile, break the cyclohexene ring just beyond the ring carbons that are connected to the double bond (the allylic carbons):

The equation is:

3.20 a.

b.

3.21 a.

b.

3.22 $CH_3-C=CHCH_3$ $\xrightarrow[\text{2. } H_2O_2,\ OH^-]{\text{1. } BH_3}$ $CH_3-\overset{\displaystyle OH}{\underset{\displaystyle CH_3}{CH}-CHCH_3}$

      |
      CH_3

The boron adds to the *less* substituted carbon of the double bond, and in the oxidation the boron is replaced by an OH group. Note that acid-catalyzed hydration of the same alkene would give the alcohol $(CH_3)_2CCH_2CH_3$ instead, according to Markovnikov's rule.

        |
        OH

3.23

$-CH=CH_2$

3.24 $(CH_3)_2C = C(CH_3)_2$

3.25 a. $CH_3C \equiv CH$ + Cl_2 $\longrightarrow$ $CH_3CCl = CHCl$

b. $CH_3C \equiv CH$ + $2 Cl_2$ $\longrightarrow$ $CH_3CCl_2 - CHCl_2$

c. $HC \equiv CCH_2CH_3$ + HBr $\longrightarrow$ $CH_2 = CBrCH_2CH_3$

$HC \equiv CCH_2CH_3$ + $2 HBr$ $\longrightarrow$ $CH_3CBr_2CH_2CH_3$

d. $HC \equiv CCH_2CH_3$ + $H-OH$ $\xrightarrow[Hg^{2+}]{H^+}$ $CH_3\overset{\overset{\displaystyle O}{\|}}{C}CH_2CH_3$

3.26 Follow eq. 3.53 as a guideline:

$CH_3CH_2C \equiv CH$ + $Na^+NH_2^-$ $\xrightarrow{NH_3}$ $CH_3CH_2C \equiv C^-Na^+$ + NH_3

3.27 2-Butyne has no hydrogens on the triple bond:

$CH_3C \equiv CCH_3$

Therefore, it does not react with sodium amide.

Additional Problems

3.28 In each case, start with the longest possible carbon chain and
determine all possible positions for the double bond. Then shorten
the chain by one carbon and repeat, and so on.

a. $CH_2 = CHCH_2CH_3$ 1-butene
$CH_3CH = CHCH_3$ 2-butene (*cis* and *trans*)
$CH_2 = C - CH_3$ 2-methylpropene
          $|$
         CH_3

b. $CH_2 = CHCH_2CH_2CH_3$ 1-pentene
$CH_3CH = CHCH_2CH_3$ 2-pentene (*cis* and *trans*)
$CH_2 = C - CH_2CH_3$ 2-methyl-1-butene
          $|$
         CH_3

$CH_2 = CHCHCH_3$ 3-methyl-1-butene
        |
        CH_3

$CH_3C = CHCH_3$ 2-methyl-2-butene
   |
   CH_3

c. $CH_2 = C = CH - CH_2CH_3$ 1,2-pentadiene
    $CH_2 = CH - CH = CH - CH_3$ 1,3-pentadiene
    $CH_2 = CH - CH_2 - CH = CH_2$ 1,4-pentadiene
    $CH_3 - CH = C = CH - CH_3$ 2,3-pentadiene
    $CH_2 = C = C - CH_3$ 3-methyl-1,2-butadiene
          |
          CH_3

    $CH_2 = C - CH = CH_2$ 2-methyl-1,3-butadiene
        |
        CH_3

d. $HC \equiv C - CH_2CH_2CH_3$ 1-pentyne
    $CH_3C \equiv CCH_2CH_3$ 2-pentyne
    $HC \equiv CCHCH_3$ 3-methyl-1-butyne
          |
          CH_3

3.29 a. 2-pentene   b. 2-methyl-2-butene (number the chain from the end with the methyl substituent)

     c. 1,2-dimethylcyclopentene (number the ring "through" the double bond)   d. 2-pentyne

     e. 2-chloro-1,3-butadiene (number from the end with the chlorine substituent)   f. *trans*-2-hexene

                                        g. *cis*-2-hexene

3.30 In general, write the longest carbon chain or ring and number it, then locate the double bond, place the substituents, and finally, write the correct number of hydrogens on each carbon atom.

               3
     a. $CH_3CH_2CH = CHCH_2CH_3$ b. $CH = CH$
                                      | |
                                   $CH_2 - CH_2$

       1   3   1   3
     c. $CH_2 - CH = C - CH_3$ d. $HC \equiv C - CH - CH_2CH_3$
         | | |
        Br   Br CH_3

55

e. $\overset{1}{C}H_2=CHCH_2\overset{4}{C}H=CHCH_3$ f. $CH_2=CHBr$ ($CH_2=CH-$ is vinyl.)

g. $CH_2=CHCH_2Cl$ ($CH_2=CHCH_2-$ is allyl.)

h.

i.

Number "through" the double bond.

j.

Number "through" both double bonds.

3.31 a. $\overset{1}{C}H_3\overset{2}{C}H=\overset{3}{C}H\overset{4}{C}H_3$ 2-butene; use the lower of the two numbers for the double bond.

b. $\overset{1}{C}H_3\overset{2}{C}\equiv\overset{3}{C}\overset{4}{C}H_2\overset{5}{C}H_3$ 2-pentyne; number the chain from the other end.

c. $\overset{1}{C}H_2=\overset{2}{C}-CH_3$ 2-methyl-1-butene; number the
$\quad\quad\quad\overset{3|}{}\;\overset{4}{}$ longest chain.
$\quad\quad\quad CH_2CH_3$

d.

1-methylcyclopentene, *not*

e. $\overset{1}{C}H_2\!=\!\overset{2}{C}\!-\!\overset{3}{C}H\!=\!CH_2$
       $\underset{CH_3}{|}$

2-methyl-1,3-butadiene; number to give the substituent the lowest possible number.

f. $\overset{4}{C}H_2\!-\!\overset{3}{C}H\!=\!\overset{2}{C}H\!-\!\overset{1}{C}H_3$
     $\underset{CH_3}{\overset{5}{|}}$

2-pentene; the "1-methyl" substituent lengthens the chain.

3.32   a.   The average values are 1.54 Å, 1.34 Å, and 1.21 Å, respectively.

b.   These single bonds are shorter than the usual 1.54 Å because they are between sp^2-sp^2 (1.47 Å), sp^2-sp (1.43 Å), and sp-sp (1.37 Å) hybridized carbons. The more s character the orbitals have, the more closely the electrons are pulled in toward the nuclei and the shorter the bonds.

3.33   Review Sec. 3.5 if you have difficulty with this question.

a.   Only one structure is possible, since one of the doubly bonded carbons has two identical groups (hydrogens):

b.

cis – 2 – pentene   trans- 2 – pentene

c.

cis–1–chloropropene   trans –1– chloropropene

d. Only one structure:

e.

cis-1,3,5-hexatriene   trans-1,3,5-hexatriene

Only the central double bond has two different groups (a vinyl group and a hydrogen) attached to each carbon.

f. The ring is large enough to accommodate a *trans* double bond.

cis-1,2-dibromocyclodecene   trans-1,2-dibromocyclodecene

Normally, rings must be at least eight-membered to accommodate a *trans* double bond.

3.34

$$\overset{13}{H}C\equiv\overset{12}{C}-\overset{11}{C}\equiv\overset{10}{C}-\overset{9}{C}H=\overset{8}{C}=\overset{7}{C}H-\overset{6}{C}H=\overset{5}{C}H-\overset{4}{C}H=\overset{3}{C}H-\overset{2}{C}H_2-\overset{1}{C}\diagup_{OH}^{\diagup O}$$

a. The 3-4, 5-6, and 7-8 double bonds are conjugated. Also, the 8-9, 10-11, and 12-13 multiple bonds are conjugated (alternate single and multiple bonds).

b. The 7-8 and 8-9 double bonds are cumulated.

c. Only the C=O bond is isolated (separated from the nearest multiple bond by two single bonds).

3.35 a. $CH_3CH = CHCH_3$ + Br_2 ⟶ $CH_3CH - CHCH_3$
                                                    | |
                                                    Br   Br

2,3-dibromobutane

b. $CH_2 = CHBr$ + Br_2 ⟶ CH_2CHBr_2
                                         |
                                         Br

1,1,2-tribromoethane

c.

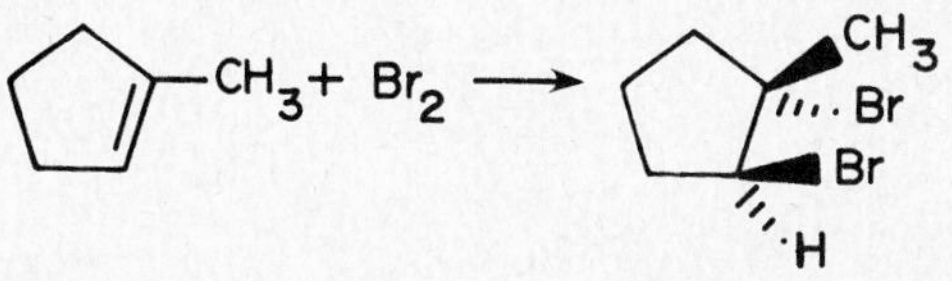

**trans−1,2−dibromo−1−
methylcyclopentane**

The bromine adds in a *trans* manner; compare with eq. 3.26.

d.

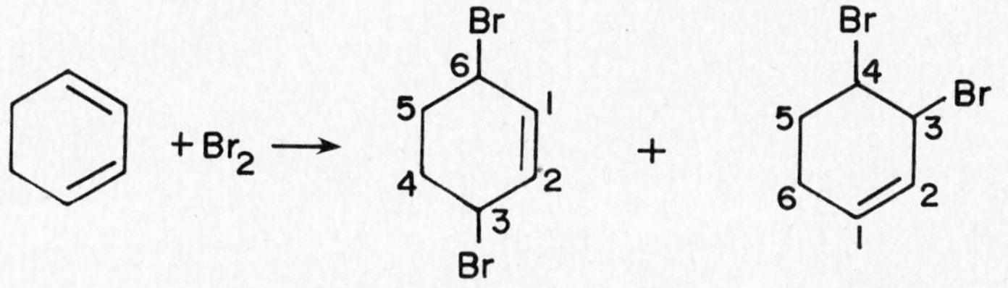

3,6−dibromocyclohexene   3,4−dibromocyclohexene
(1,4−addition)   (1,2−addition)

Compare with eq. 3.27. The 1,4-addition product predominates.
The geometry of the products (bromines *cis* or *trans*) is not shown,
because a mixture is formed.

e.

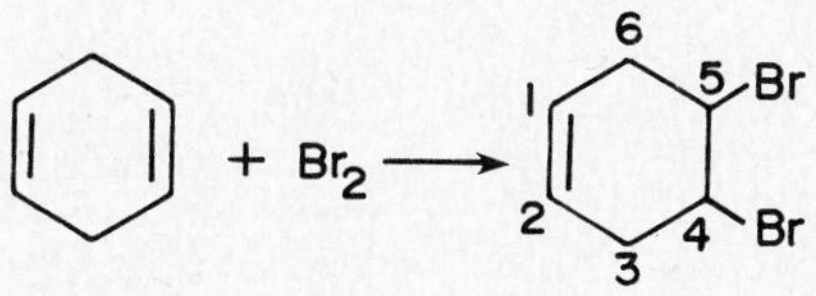

4,5−dibromocyclohexene

The double bonds are not conjugated, so only 1,2-addition is
possible.

f. $CH_3-C\overset{\shortmid}{=}C-CH_3$ + Br_2 $\longrightarrow$ $CH_3-\overset{Br}{\underset{CH_3}{\overset{|}{C}}}-\overset{Br}{\underset{CH_3}{\overset{|}{C}}}-CH_3$
  with CH_3 CH_3 below the left carbons

2,3-dibromo-2,3-dimethylbutane

3.36   a. $CH_2=CHCH_2CH_3$ + Cl $\longrightarrow$ $CH_2-\underset{Cl}{\overset{|}{CH}}-CH_2CH_3$
with Cl under CH_2 and Cl under CH

b.   Follow Markovnikov's rule:

     $CH_2=CHCH_2CH_3$ + $H-Cl$ $\longrightarrow$ $CH_3\underset{Cl}{\overset{|}{CH}}CH_2CH_3$

c. $CH_2=CHCH_2CH_3$ + H_2 $\xrightarrow{Pt}$ $CH_3CH_2CH_2CH_3$

d. $CH_2=CHCH_2CH_3$ $\xrightarrow[\text{2. } Zn, H^+]{\text{1. } O_3}$ $CH_2=O$ + $O=CHCH_2CH_3$

e.   Follow Markovnikov's rule:

     $CH_2=CHCH_2CH_3$ + $H-OH$ $\xrightarrow{H^+}$ $CH_3-\underset{OH}{\overset{|}{CH}}-CH_2CH_3$

f. $CH_2=CHCH_2CH_3$ $\xrightarrow[\text{2. } H_2O_2, OH^-]{\text{1. } B_2H_6}$ $HOCH_2CH_2CH_2CH_3$

     The product is the regioisomer of that obtained by acid-catalyzed addition of water (part e).

3.37   To work this kind of problem, try to locate (on adjacent carbons) the atoms or groups that must have come from the small molecule or reagent. Then remove them from the structure and insert the multiple bond appropriately.

a. $CH_3CH=CHCH_3$ + Br_2

b. $CH_2=\underset{CH_3}{\overset{|}{C}}-CH_3$ + HOH + H^+

c. $CH_2=CHCH_3$ + $H-OSO_3H$

d. 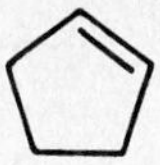 + H—Br

e. $CH_2{=}CH{-}CH{=}CH_2$ + H—Br (1,4-addition)

f. $CH_3C{\equiv}CCH_3$ + $2Cl_2$

g. 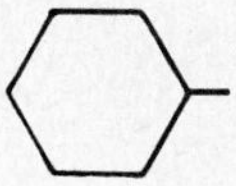 —$CH{=}CH_2$ + HBr

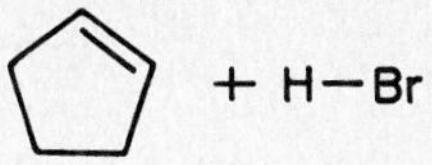

3.38 If the saturated hydrocarbon contained no rings, it would have a molecular formula $C_{40}H_{82}(C_nH_{2n+2})$. Since there are four fewer hydrogens in $C_{40}H_{78}$, it must have two rings. Since β-carotene absorbed 11 moles of H_2 ($C_{40}H_{56}$ + 11 H_2 = $C_{40}H_{78}$), it must have 11 double bonds and two rings. For the correct structure, see p. 208 of the text.

3.39 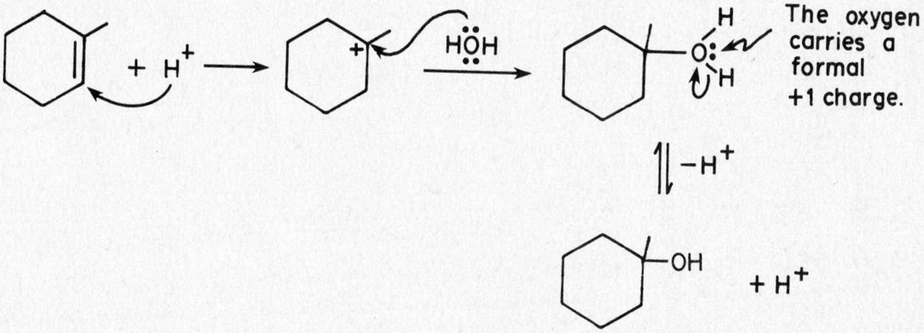
+ H+
HÖH
H
O:
H
The oxygen carries a formal +1 charge.
−H+
OH
+ H+

In the first step, the proton adds in such a way as to give the tertiary carbocation intermediate.

3.40 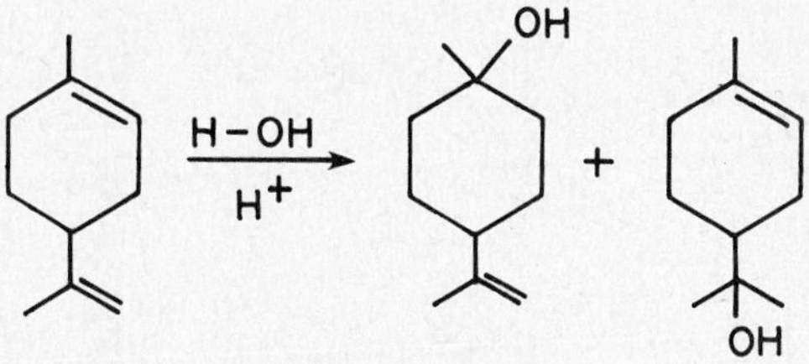
H−OH
H+
OH
+
OH

In each case, water adds according to Markovnikov's rule (via a tertiary carbonium ion).

The diol is

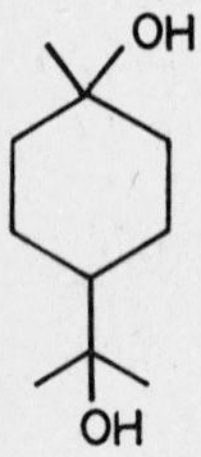

terpin

3.41   The reaction begins with the formation of a bromonium ion:

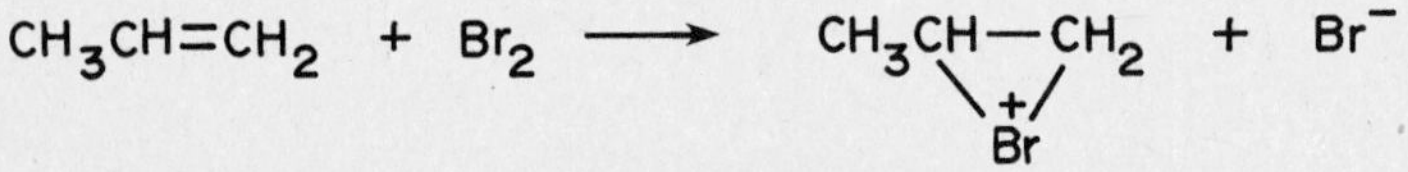

$$CH_3CH{=}CH_2 \ + \ Br_2 \ \longrightarrow \ CH_3\overset{\displaystyle}{CH}\!\!-\!\!CH_2 \ + \ Br^-$$

(with $\overset{+}{Br}$ bridging)

The bromonium ion then reacts either with bromide ion or with methanol:

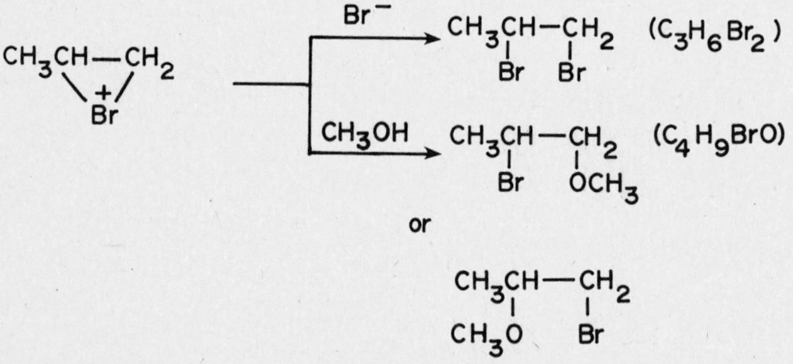

$CH_3CH{-}CH_2$ with $\overset{+}{Br}$ bridge

→ Br⁻ → $CH_3\underset{\displaystyle Br}{CH}\!\!-\!\!\underset{\displaystyle Br}{CH_2}$ $(C_3H_6Br_2)$

→ CH_3OH → $CH_3\underset{\displaystyle Br}{CH}\!\!-\!\!\underset{\displaystyle OCH_3}{CH_2}$ (C_4H_9BrO)

or

$CH_3\underset{\displaystyle CH_3O}{CH}\!\!-\!\!\underset{\displaystyle Br}{CH_2}$

The intermediate bromonium ion can react with either nucleophile, to give the observed products.

3.42 $CH_3CH{=}CH{-}CH{=}CHCH_3 \ + \ HBr \ \longrightarrow \ CH_3CH_2\underset{\displaystyle Br}{CH}\!\!-\!\!CH{=}CHCH_3$

1,2-addition

and $CH_3CH_2CH{=}CH\!-\!\underset{\displaystyle Br}{CH}CH_3$

1,4-addition

Mechanism:

$$CH_3CH=CH-CH=CHCH_3 \xrightarrow{H^+} \left[\begin{array}{c} \overset{+}{C}H_3CH_2CHCH=CHCH_3 \\ \updownarrow \\ CH_3CH_2CH=CH-\overset{+}{C}HCH_3 \end{array} \right]$$ allylic carbocation resonance contributors

$$\underset{Br}{CH_3CH_2\underset{|}{C}HCH=CHCH_3} \quad + \quad CH_3CH_2CH=CH\underset{Br}{\underset{|}{C}HCH_3} \xleftarrow{Br^-}$$

In the final step, the nucleophile Br^- can react with the allylic carbocation at either of the two positive carbons.

3.43   a.   b.

3.44   In each case, work backwards from the cyclohexene double bond:

a.   b.

An alternative, though less satisfactory, answer for part b is

However, it is usually better to start with electron-withdrawing substituents on the dienophile rather than on the diene.

3.45   a.   Orlon $\left(\!\!-CH_2-\underset{\underset{CN}{|}}{CH}-\!\!\right)_n$ b.   polyvinyl chloride $\left(\!\!-CH_2-\underset{\underset{Cl}{|}}{CH}-\!\!\right)_n$

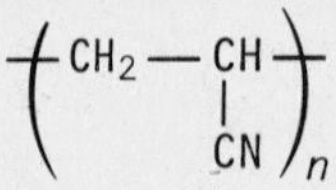

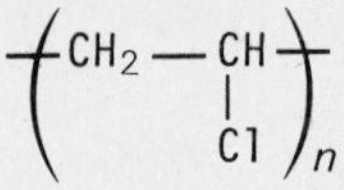

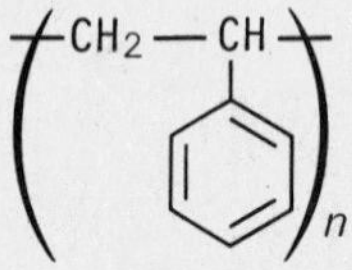

c.   polystyrene $\left(\!\!-CH_2-\underset{\underset{\bigcirc}{|}}{CH}-\!\!\right)_n$ d.   Saran $\left(\!\!-CH_2-\overset{\overset{Cl}{|}}{\underset{\underset{Cl}{|}}{C}}-\!\!\right)_n$

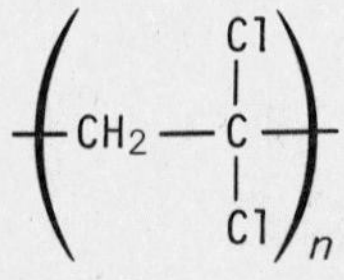

3.46   The reaction may be initiated by addition of a free radical from
the catalyst to either butadiene or styrene:

$$R\cdot \;+\; CH_2\!=\!CHCH\!=\!CH_2 \longrightarrow \left[\begin{array}{c} RCH_2CH\!-\!CH\!=\!CH_2 \\ \updownarrow \\ RCH_2CH\!=\!CH\overset{\cdot}{C}H_2 \end{array}\right]$$

$$R\cdot \;+\; CH_2\!=\!CHC_6H_5 \longrightarrow RCH_2\overset{\cdot}{C}HC_6H_5$$

These radicals may then add either to butadiene or to styrene; the
allylic radical from butadiene may add in either a 1,2- or a
1,4-manner (shown below is only one of several alternative sequences
of addition):

$$RCH_2CH\!=\!CH\overset{\cdot}{C}H_2 \;+\; CH_2\!=\!CHCH\!=\!CH_2 \longrightarrow \left[\begin{array}{c} RCH_2CH\!=\!CHCH_2CH_2\overset{\cdot}{C}HCH\!=\!CH_2 \\ \updownarrow \\ RCH_2CH\!=\!CHCH_2CH_2CH\!=\!CH\overset{\cdot}{C}H_2 \end{array}\right]$$

$$RCH_2CH\!=\!CHCH_2CH_2CH\!=\!CH\overset{\cdot}{C}H_2 \;+\; CH_2\!=\!CHC_6H_5 \longrightarrow$$

$$RCH_2CH\!=\!CHCH_2CH_2CH\!=\!CHCH_2CH_2\overset{\cdot}{C}HC_6H_5 \xrightarrow{\;CH_2=CHCH=CH_2\;}$$

$$RCH_2CH\!=\!CHCH_2CH_2CH\!=\!CHCH_2CH_2\overset{\overset{C_6H_5}{|}}{C}HCH_2CH\!=\!CH\overset{\cdot}{C}H_2, \text{ and so on}$$

3.47 $\left(\!\!-CH_2\underset{\underset{Cl}{|}}{C}\!=\!CHCH_2-\!\!\right)_n$

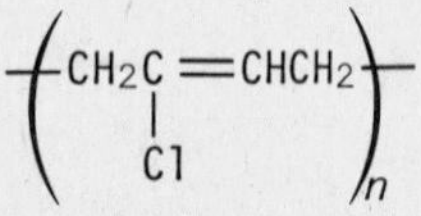

The structure is analogous to that of natural rubber, except that the
methyl group in each isoprene unit is replaced by a chlorine atom.

3.48　a.　$CH_2\!=\!C\!-\!CH_2CH_3$ $\xrightarrow{BH_3}$ $B\!\left(\!CH_2CHCH_2CH_3\atop \right)_3$
　　　　　　　$|$
　　　　　　　CH_3

 with CH_3 branch

$HOCH_2CHCH_2CH_3$ $\xleftarrow[OH^-]{H_2O_2}$
　　　$|$
　　CH_3

b.

In the first step, the boron and hydrogen add to the same face of the double bond, in a *cis* manner. In the second step, the OH takes the identical position occupied by the boron. Thus, in the product the H and OH groups are *cis*.

3.49　a.

$=CH_2$ + H–OH $\xrightarrow{H^+}$ product with CH_3 and OH

Addition follows Markovnikov's rule and the reaction proceeds via a tertiary carbocation intermediate.

b.

$=CH_2$ $\xrightarrow[2.H_2O_2,OH^-]{1.BH_3}$ $-CH_2OH$

The boron adds to the less substituted, or CH_2, carbon. In the second step, the boron is replaced by the hydroxyl group.

3.50 Cyclohexene will rapidly decolorize a dilute solution of bromine in carbon tetrachloride (Sec. 3.8), and will be oxidized by potassium permanganate, resulting in a color change from the purple of $KMnO_4$ to the brown solid MnO_2 (Sec. 3.21). Cyclohexane, being saturated, does not react with either of these reagents.

3.51 The alkene that gave the particular aldehyde or ketone can be deduced by joining the two carbons attached to oxygens by a $C{=}C$ double bond:

a. $CH_3CH_2CH{=}CHCH_2CH_3$ c. $CH_2{=}CHCH(CH_3)_2$

b. $(CH_3)_2C{=}CHCH_3$ d. $\begin{array}{ll} CH_2 - CH \\ | \quad\;\; || \\ CH_2 - CH \end{array}$

In the case of compound a, where *cis* and *trans* isomers are possible, either isomer gives the same ozonolysis products.

3.52 The structure of natural rubber is:

It will be cleaved by ozone at each double bond. The product of ozonolysis is therefore:

or $\begin{array}{c} O \quad\quad O \\ || \quad\quad\; || \\ HCCH_2CH_2CCH_3 \end{array}$
levulinic aldehyde

3.53 a. $CH_3C{\equiv}CCH_2CH_3 \;+\; 2\,Cl_2 \longrightarrow \begin{array}{c} Cl \;\; Cl \\ | \quad\; | \\ CH_3C - CCH_2CH_3 \\ | \quad\; | \\ Cl \;\; Cl \end{array}$

b.

$CH_3CH_2C{\equiv}CCH_2CH_3 \;+\; H_2 \xrightarrow[\text{catalyst}]{\text{Lindlar's}}$

This catalyst limits the addition to 1 mole of H_2, which adds in a *cis* manner.

c. $CH_3C \equiv CH + H_2O \xrightarrow[HgSO_4]{H^+} \left[CH_3\overset{\overset{\textstyle OH}{|}}{C} = CH_2 \right] \longrightarrow CH_3\overset{\overset{\textstyle O}{||}}{C}CH_3$

Compare with eq. 3.52 where R = CH_3.

d. $CH_3C \equiv CH + NaNH_2 \xrightarrow{NH_3} CH_3C \equiv C^-Na^+ + NH_3$

Compare with eq. 3.53 where R = CH_3.

3.54   a. $CH_3CH_2C \equiv CH \xrightarrow{2HBr} CH_3CH_2\overset{\overset{\textstyle Br}{|}}{\underset{\underset{\textstyle Br}{|}}{C}} - CH_3 \xleftarrow{2HBr} CH_3C \equiv CCH_3$

Either 1- or 2-butyne will add HBr in the manner shown.

b. $CH_3C \equiv CCH_3 + 2\ Br_2 \longrightarrow CH_3\overset{\overset{\textstyle Br}{|}}{\underset{\underset{\textstyle Br}{|}}{C}} - \overset{\overset{\textstyle Br}{|}}{\underset{\underset{\textstyle Br}{|}}{C}}CH_3$

The triple bond must be between C-2 and C-3 if the bromines are to be attached to those carbons in the product.

CHAPTER FOUR   AROMATIC COMPOUNDS

CHAPTER SUMMARY

Benzene is the parent of the family of **aromatic hydrocarbons.** Its six carbons lie in a plane at the corners of a regular hexagon, and each carbon has one hydrogen attached. Benzene is a resonance hybrid of two contributing Kekulé structures:

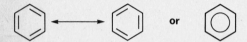

In orbital terms, each carbon is sp^2-hybridized; these orbitals form σ bonds to the hydrogen and the two neighboring carbons, all in the ring plane. A p orbital at each carbon is perpendicular to this plane, and the six electrons, one from each carbon, form an electron cloud of π orbitals that lie above and below the ring plane.

The bond angles in benzene are 120°. All C—C bond distances are equal (1.39 Å). The compound is more stable than either of the contributing Kekulé structures and has a **resonance** or **stabilization energy** of ~36 kcal/mol.

The nomenclature of benzene derivatives is described in Sec. 4.7. Common names and structures to be memorized include those of **toluene, styrene, phenol, aniline,** and **xylene.** Monosubstituted benzenes are named as benzene derivatives (bromobenzene, nitrobenzene, etc.). Disubstituted benzenes are named *ortho-* (1,2-), *meta-* (1,3-), or *para-* (1,4-), depending on the relative positions of the substituents on the ring. Two important groups are **phenyl** (C_6H_5—) and **benzyl** ($C_6H_5CH_2$—).

Aromatic compounds react mainly by **electrophilic aromatic substitution,** in which one or more ring hydrogens are replaced by various electrophiles. Typical reactions are **chlorination, bromination, nitration, sulfonation,** and **alkylation** (the Friedel-Crafts reaction). The mechanism involves two steps: addition of the electrophile to a ring carbon, to produce an intermediate **benzenonium ion,** followed by proton loss to again achieve the (now substituted) aromatic system.

Substituents already present on the ring affect the rate of further substitution and the position taken by the next substituent. Most groups are either *meta-directing and ring-deactivating* or *ortho,para-directing and ring-activating* (Table 4.1). An exception is the halogens, which are *ortho,para-directing* but ring-deactivating. These effects must be taken into account in devising syntheses of aromatic compounds.

Benzene is a major commercial chemical--a source of styrene, phenol, other aromatics, and cyclohexane.

Polycyclic aromatic hydrocarbons, which are built of *fused* benzene rings, include **naphthalene, anthracene,** and **phenanthrene.** Some are, such as benzo[a]pyrene, carcinogens.

The **Hückel rule** extends the concept of aromaticity (that is, the special stability of certain cyclic conjugated systems). It states that planar monocyclic conjugated π systems with $(4n + 2)$ π electrons will be aromatic (n is zero or an integer). Similar systems with $4n$ π electrons are **antiaromatic.**

The carbons in an aromatic ring may be replaced by one or more **heteroatoms** (oxygen, nitrogen, sulfur, etc.) to give **heterocyclic aromatic compounds.** The heteroatom may contribute one electron to the aromatic π system (as in **pyridine**), or it may contribute two such electrons (as in **pyrrole**).

REACTION SUMMARY

Electrophilic Aromatic Substitution

halogenation

$$\text{C}_6\text{H}_6 + X_2 \xrightarrow{\text{FeX}_3} \text{C}_6\text{H}_5X + HX$$

(X = Cl, Br)

nitration

$$\text{C}_6\text{H}_6 + \text{HONO}_2 \xrightarrow{\text{H}_2\text{SO}_4} \text{C}_6\text{H}_5\text{NO}_2 + \text{H}_2\text{O}$$

nitric acid   nitrobenzene

sulfonation

$$\text{C}_6\text{H}_6 + \text{HOSO}_3\text{H} \xrightarrow{\text{heat}} \text{C}_6\text{H}_5\text{OSO}_3\text{H} + \text{H}_2\text{O}$$

sulfuric acid   benzenesulfonic acid

alkylation

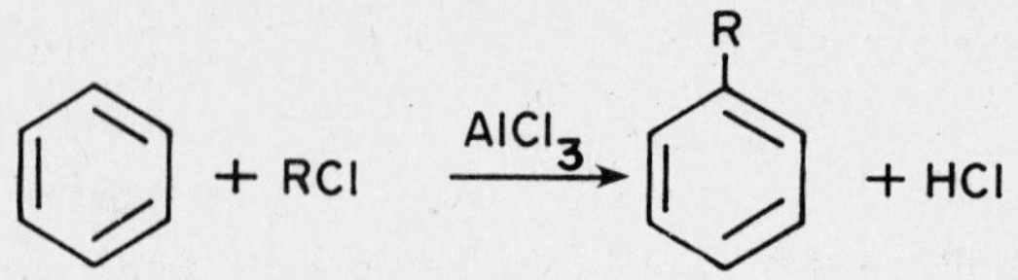

Friedel—Crafts reaction; R = an alkyl group

alkylation

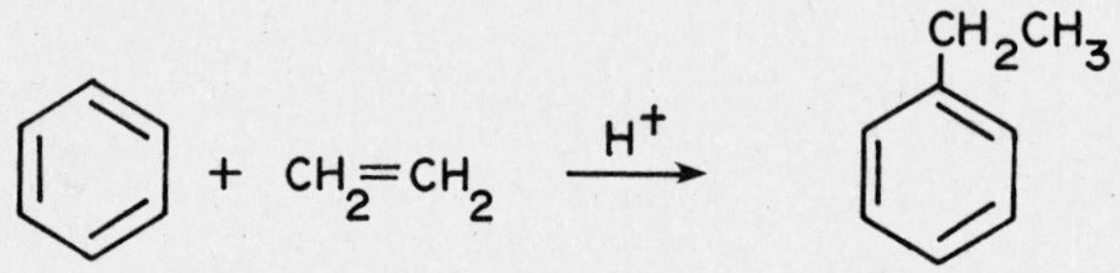

Catalytic Hydrogenation

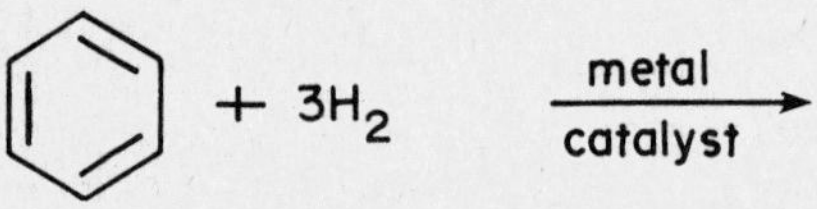

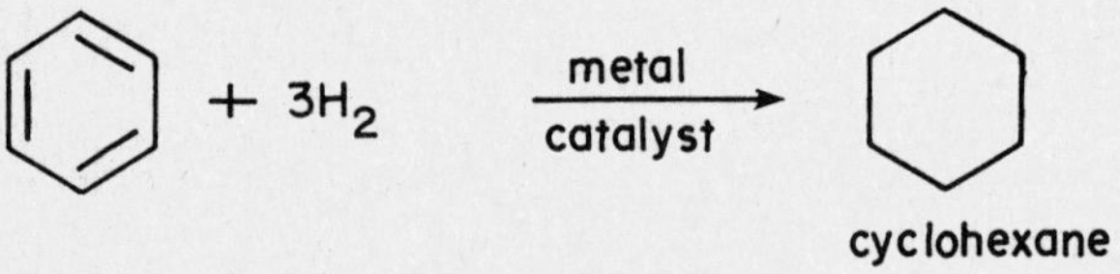

cyclohexane

MECHANISM SUMMARY

Electrophilic Aromatic Substitution

benzenonium ion

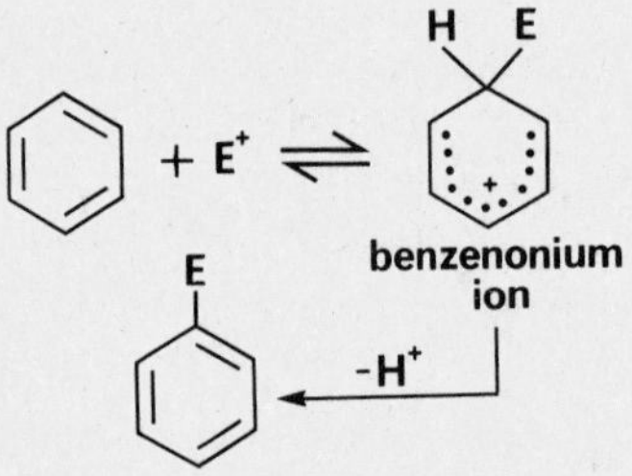

LEARNING OBJECTIVES

1. Know the meaning of: Kekulé structure, benzene resonance hybrid, resonance or stabilization energy.

2. Know the meaning of: *ortho, meta, para,* phenyl group, benzyl group, Ar——, benzene, toluene, styrene, phenol, aniline, xylene, arene.

3. Know the meaning of: electrophilic aromatic substitution, halogenation, nitration, sulfonation, alkylation, Friedel-Crafts reaction.

4. Know the meaning of: benzenonium ion, *ortho,para*-directing group, *meta*-directing group, ring-activating substituent, ring-deactivating substituent.

5. Know the meaning of: polycyclic aromatic hydrocarbon, naphthalene, anthracene, phenanthrene, carcinogenic, Hückel rule, heterocyclic aromatic compound, pyridine, pyrrole.

6. Name and write the structures for aromatic compounds, especially mono-substituted and disubstituted benzenes and toluenes.

7. Given the reactants, write the structures of the main organic products of the common electrophilic aromatic substitution reactions (halogenation, nitration, sulfonation, and alkylation).

8. Write the steps in the mechanism for an electrophilic aromatic substitution reaction.

9. Draw the structures of the main contributors to the benzenonium ion resonance hybrid.

10. Draw the structures of the main contributors to substituted benzenonium ions and tell whether the substituent stabilizes or destabilizes the ion.

11. Know which groups are *ortho,para*-directing, which are *meta*-directing, and explain why each group directs the way it does.

12. Given two successive electrophilic aromatic substitution reactions, write the structure of the product, with substituents in the correct locations on the ring.

13. Given a disubstituted or trisubstituted benzene, deduce the correct sequence in which to carry out electrophilic substitutions to give the product with the desired orientation.

14. Apply the Hückel rule to tell whether or not a particular cyclic unsaturated compound is aromatic.

15. Be able to explain why compounds of the pyridine type are basic, whereas those of the pyrrole type are much less so.

16. Predict whether a particular nitrogen in a given heterocyclic ring will be strongly or weakly basic.

ANSWERS TO PROBLEMS

Problems Within the Chapter

4.1 The formula C_6H_6 (or C_nH_{2n-6}) corresponds to any of the following possibilities: two triple bonds; one triple bond and two double bonds; one triple bond, one double bond, and one ring; one triple bond and two rings; four double bonds; three double bonds and one ring; two double bonds and two rings; one double bond and three rings; four rings. Obviously there are very many possibilities. One example from each of the above categories is shown below.

$$HC \equiv C - C \equiv C - CH_2CH_3$$

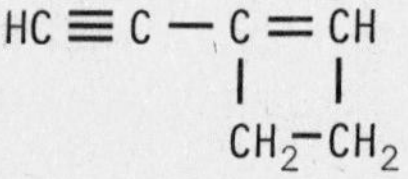

$$HC \equiv C - \underset{\underset{CH_2-CH_2}{|}}{C} = \underset{|}{CH}$$

$$HC \equiv C - CH = CH - CH = CH_2$$

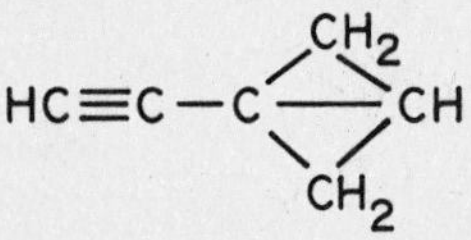

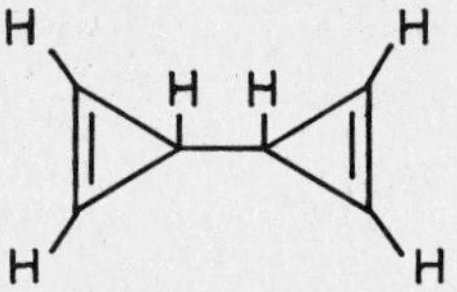

$$CH_2 = C = CH - CH = C = CH_2$$

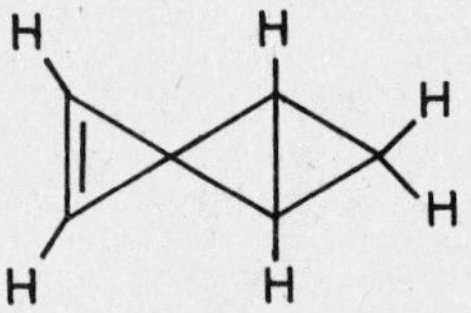

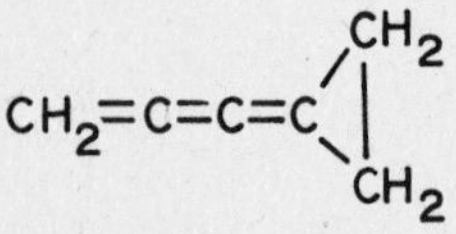

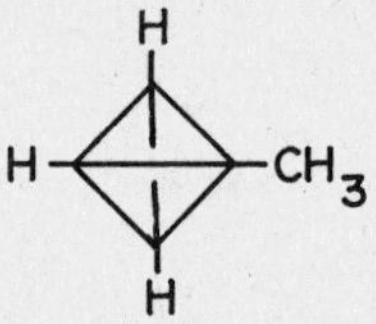

4.2   eq. 4.2:

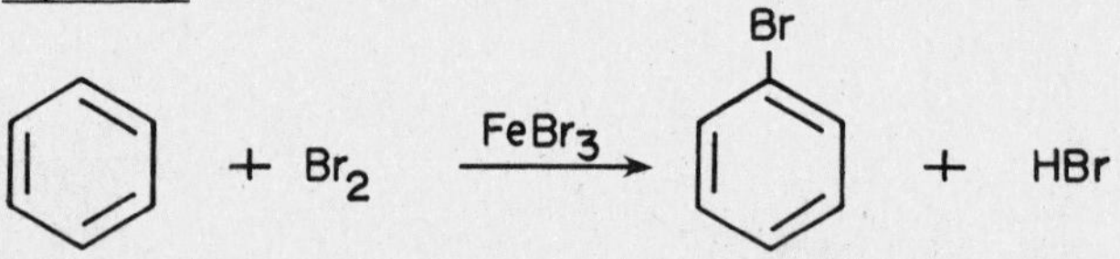

only one possible structure.

eq. 4.4:

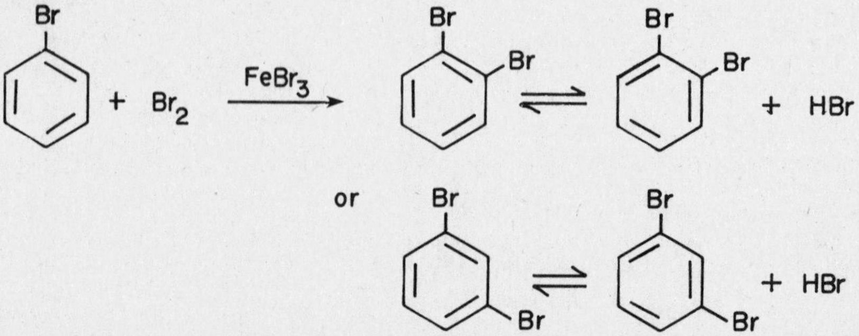

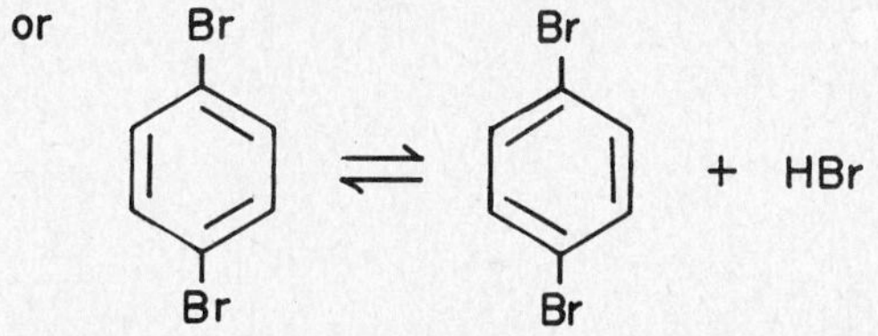

or

$$\text{Br} \quad \rightleftharpoons \quad \text{Br} \quad + \text{ HBr}$$

Yes, Kekulé's explanation accounts for one monobromobenzene and three dibromobenzenes, if the equilibrium between the two Kekulé structures for each dibromo isomer is taken into account. As you will see in Sec. 4.4, however, this picture is not quite correct.

4.3   Kekulé would have said that the structures are in such rapid equilibrium with one another that they cannot be separated. We know now, however, that there is only one such structure, which is not accurately represented by either Kekulé formula.

4.4

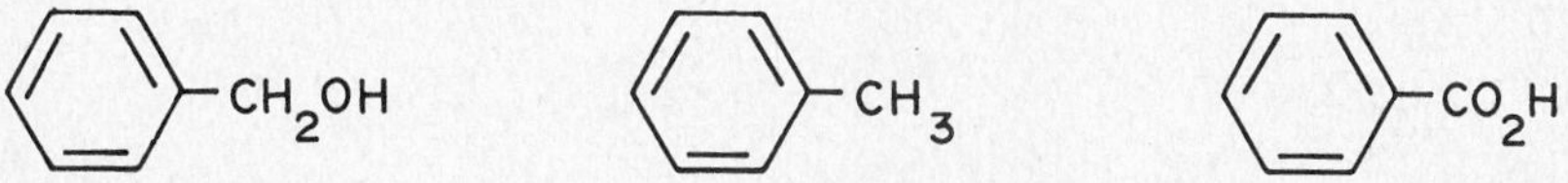

benzyl alcohol   toluene   benzoic acid

4.5

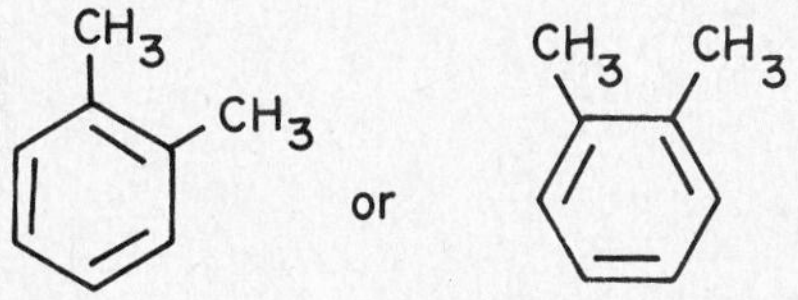

or

ortho –xylene

It does not matter whether we write the benzene ring standing on one corner or lying on a side.

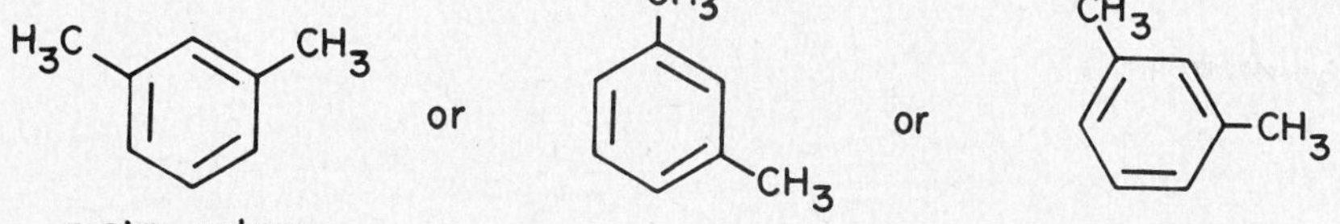

meta – xylene   or   or

The important feature of the structure is that the two methyl substituents are in a 1,3-relationship.

4.6 a.

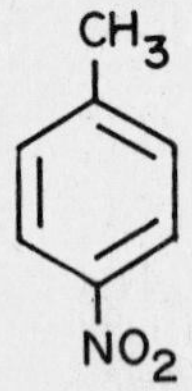

CH₃ ... NO₂ (4-nitrotoluene structure)

b.

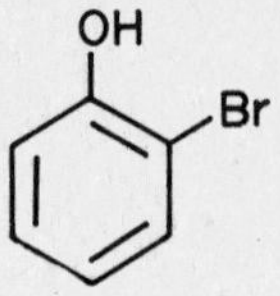

OH ... Br (2-bromophenol structure)

c.

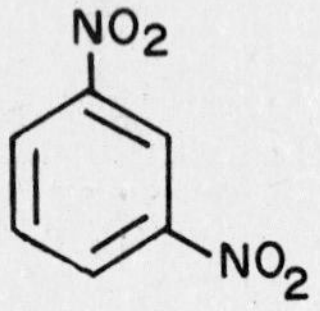

NO₂ ... NO₂ (m-dinitrobenzene structure)

d.

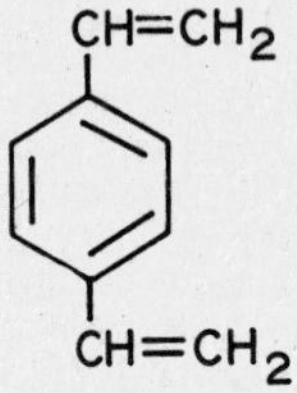

CH=CH₂ ... CH=CH₂ (p-divinylbenzene structure)

4.7 a.

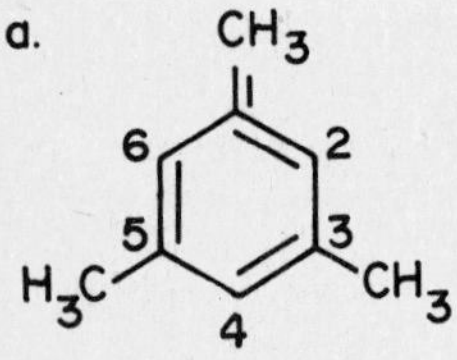

CH₃
6   2
5   3
H₃C   CH₃
4

b.

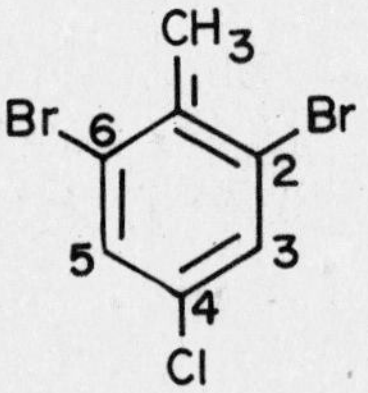

CH₃
Br 6   2 Br
5   3
4
Cl

4.8 a. 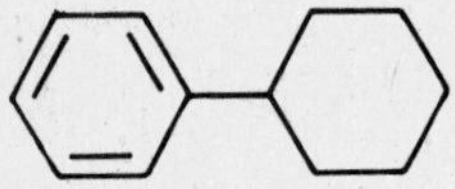 (phenylcyclohexane structure)

b. 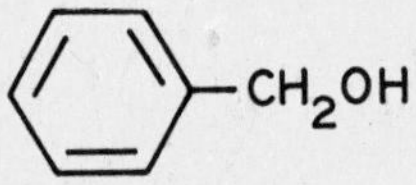 ⟨benzene⟩—CH₂OH

c. 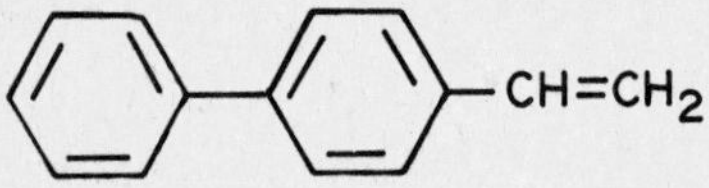 (biphenyl)—CH=CH₂

d. 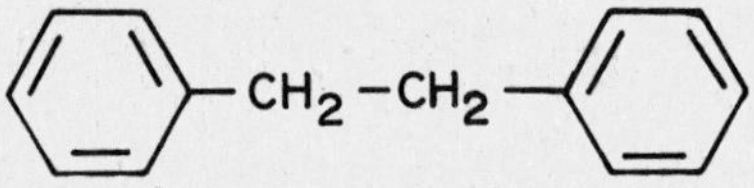 ⟨benzene⟩—CH₂—CH₂—⟨benzene⟩

4.9 a. phenylcyclopentane   b. *o*-benzylaniline

4.10. The electrophile is formed according to the following equilibria, beginning with the protonation of one sulfuric acid molecule by another:

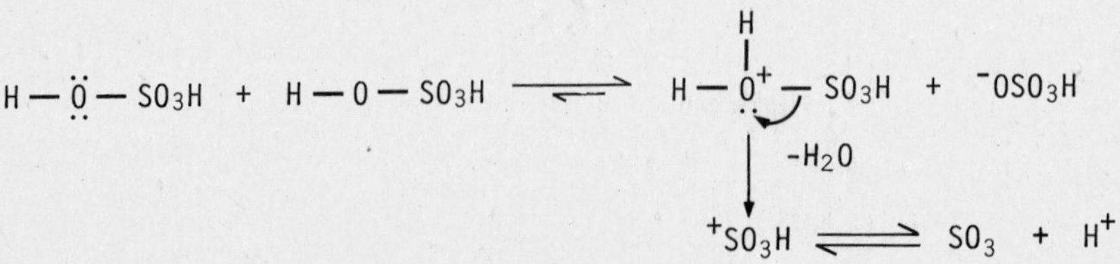

$$H-\underset{\cdot\cdot}{\overset{\cdot\cdot}{O}}-SO_3H \;+\; H-O-SO_3H \;\rightleftharpoons\; H-\underset{\cdot\cdot}{\overset{H}{O^+}}-SO_3H \;+\; {}^-OSO_3H$$

$$\downarrow -H_2O$$

$$^+SO_3H \;\rightleftharpoons\; SO_3 \;+\; H^+$$

Using $^+SO_3H$ as the electrophile, we can write the sulfonation mechanism as follows:

4.11   The product would be isopropylbenzene, because the proton of the acid catalyst would add to propylene according to Markovnikov's rule to give the isopropyl cation:

$$CH_3CH=CH_2 \ + \ H^+ \longrightarrow CH_3\overset{+}{C}HCH_3$$

4.12   **ortho**

meta

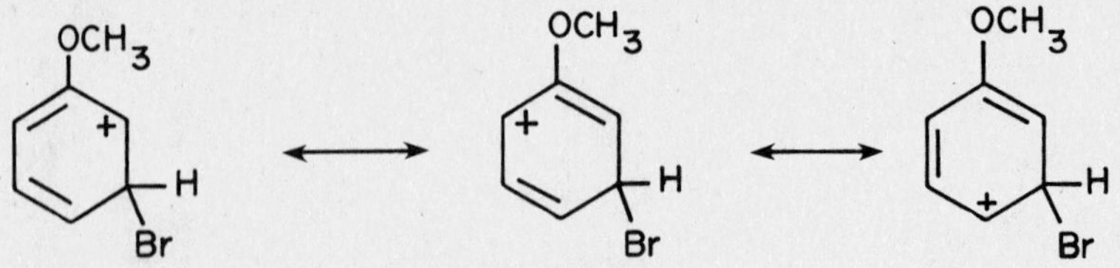

para

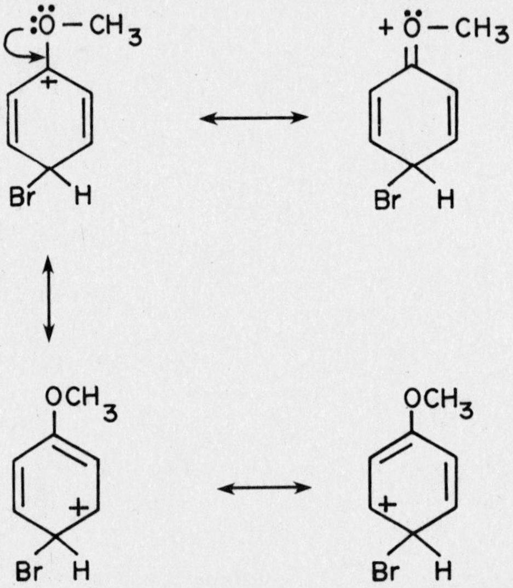

Note that in *ortho* or *para* substitution the carbocation can be stabilized by delocalization of the positive charge to the oxygen atom. This is not possible for *meta* substitution. Therefore, *ortho, para* substitution is preferred.

4.13   **ortho**

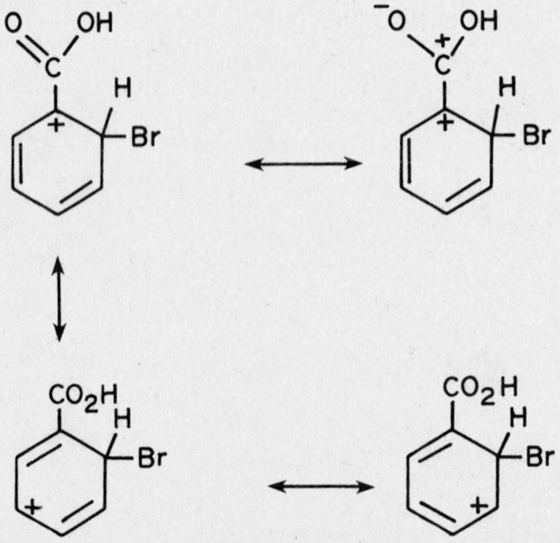

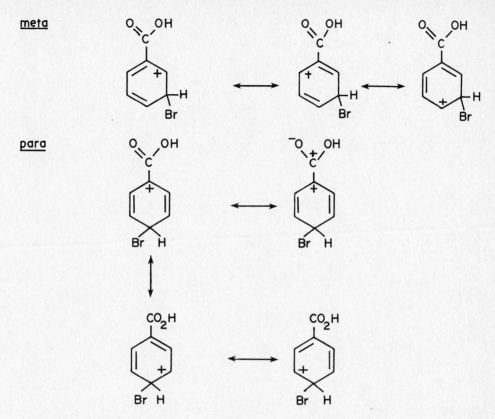

meta

para

Note that in *ortho* or *para* substitution the positive charge of the benzenonium ion is adjacent to the partially positive carboxyl carbon. It is energetically unfavorable to have two adjacent like charges. In *meta* substitution, this unfavorable possibility does not exist. Thus, *meta* substitution predominates.

4.14   a.   Since the substituents are *meta* in the product, we must introduce the *meta*-directing substituent first:

b.

$$\text{benzene} + CH_3Cl \xrightarrow{AlCl_3} \text{toluene} + HCl$$

$$\xrightarrow[\text{H}_2\text{SO}_4]{\text{HONO}_2} \text{p-nitrotoluene} \quad (+ \text{ some ortho isomer})$$

The methyl substituent is *ortho,para*-directing; the two isomers obtained in the final step would have to be separated. Usually the *para* isomer, in which the two substituents are furthest apart, predominates.

4.15 We could not make *m*-bromochlorobenzene in good yield this way, because both Br and Cl are *ortho,para*-directing. Similarly, we could not prepare *p*-nitrobenzenesulfonic acid directly, because both the nitro and sulfonic acid substituents are *meta*-directing.

4.16 Only two such contributors are possible:

$$\left[\text{naphthalene-NO}_2 \text{ contributor} \leftrightarrow \text{naphthalene-NO}_2 \text{ contributor} \right]$$

Any additional resonance contributors disrupt the benzenoid structure in the "left" ring. Since the intermediate carbocation for nitration of naphthalene at C-1 is more stable, substitution at that position is preferred.

4.17 The carbon-to-hydrogen ratios are: benzene = 1; naphthalene = 1.25; anthracene = 1.40; pyrene = 1.60. The percentage of carbon in a structure increases with the number of fused aromatic rings.

4.18

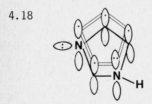

The hydrogen-bearing nitrogen contributes two electrons to the aromatic system of six π electrons; the other nitrogen contributes only one such electron and has an unshared electron pair in an orbital that lies in the ring plane.

4.19 The formula for imidazole is:

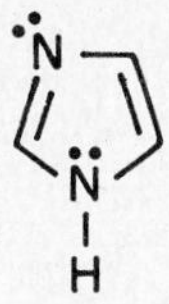

The two electrons on the N—H nitrogen are involved in the six π electron aromatic system, just as in pyrrole. Thus, it is the other nitrogen, with its unshared electron pair, that is basic:

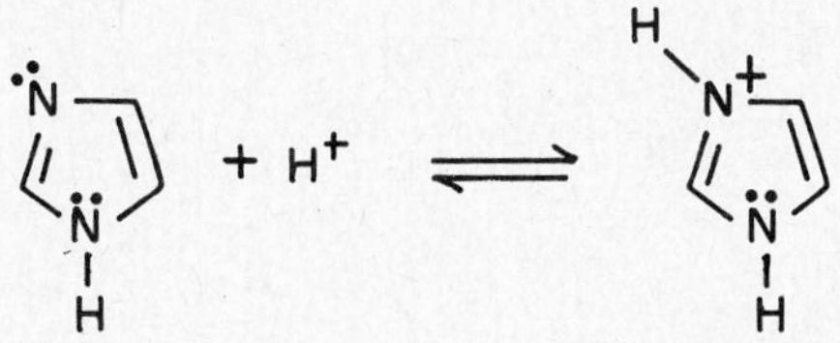

the imidazolium ion

Additional Problems

4.20 a.

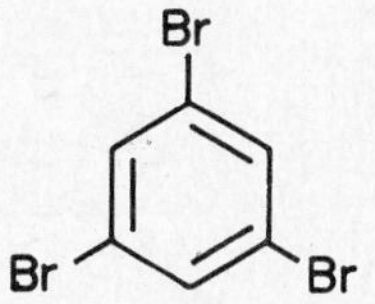

b.

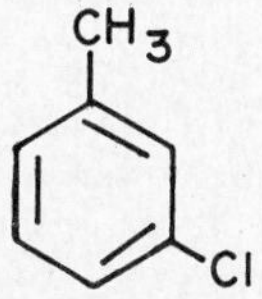

c.

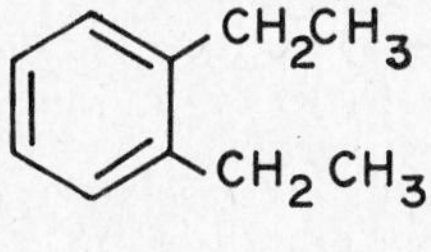

d.

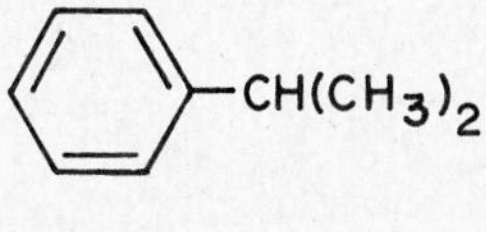

e.

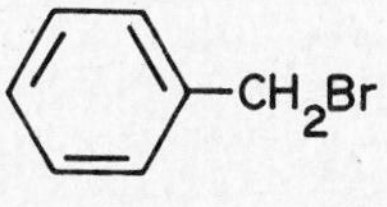

f.

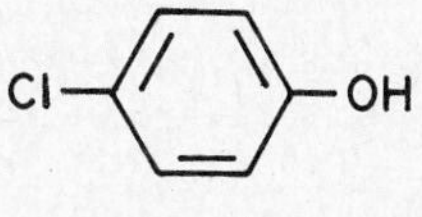

g.

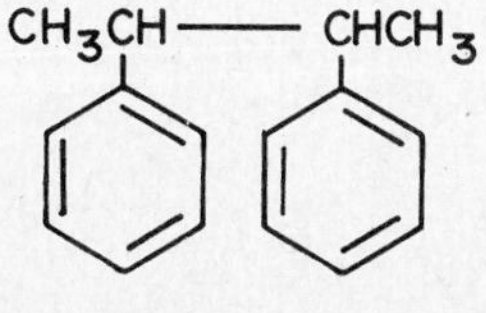

h.

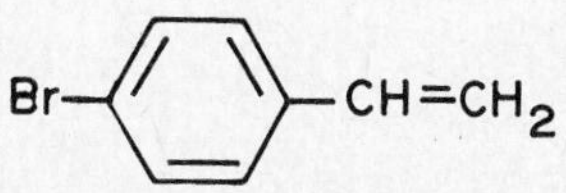

i.

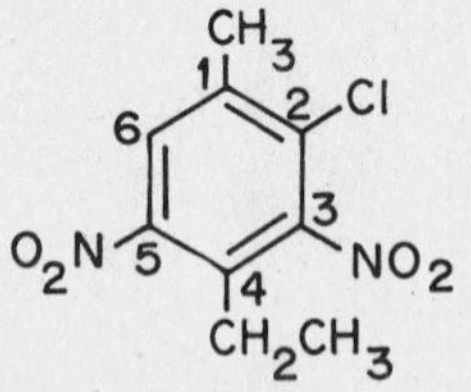

j.

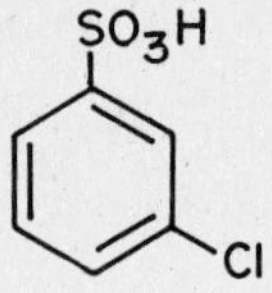

k.

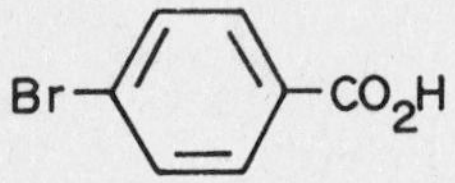

l.

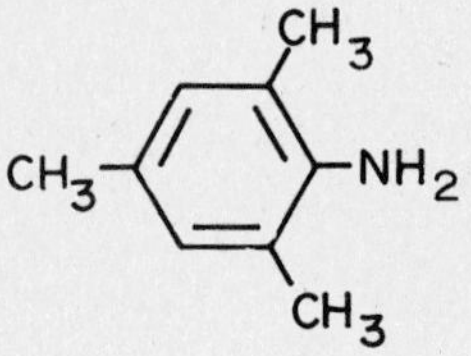

4.21   a. *n*-propylbenzene (or 1-phenylpropane)
       b. *m*-bromochlorobenzene (alphabetical order)
       c. 1,8-dibromonaphthalene
       d. 2,5-dichlorotoluene; number as shown in the following structure:

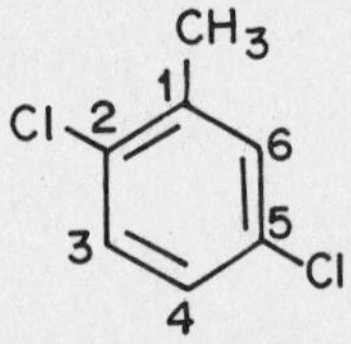

       e. *p-t*-butylphenol
       f. *o*-chlorotoluene
       g. 3,5-dibromostyrene
       h. hexamethylbenzene (No numbers are necessary, since all possible positions on the benzene ring are substituted.)
       i. 1-methyl-1-phenylcyclopropane (substituents in alphabetical order)
       j. triphenylmethane

4.22   a.

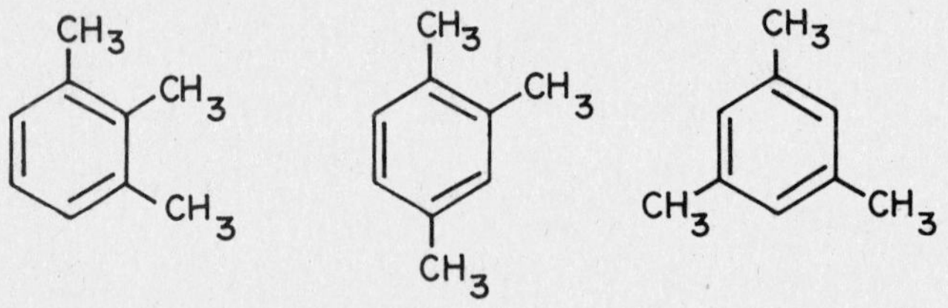

1,2,3-trimethyl benzene   1,2,4   1,3,5

b.

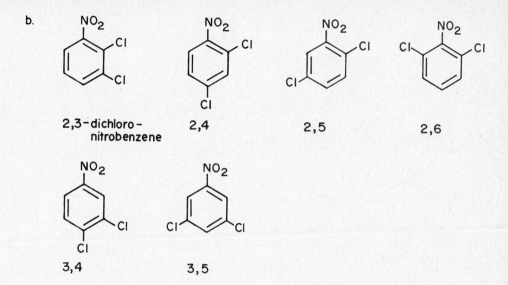

2,3-dichloro-nitrobenzene   2,4   2,5   2,6

3,4   3,5

4.23   The three possible structures are:

Only in the *para* isomer are all four remaining hydrogens equivalent. Therefore, it can give only *one* mononitro derivative, and must be A:

$$HONO_2 \xrightarrow{H^+}$$

A, mp 87°C

The *ortho* isomer can give only *two* mononitration products, and must be B:

$$HONO_2 \xrightarrow{H^+}$$ NO$_2$ and   NO$_2$

The *meta* isomer is therefore C:

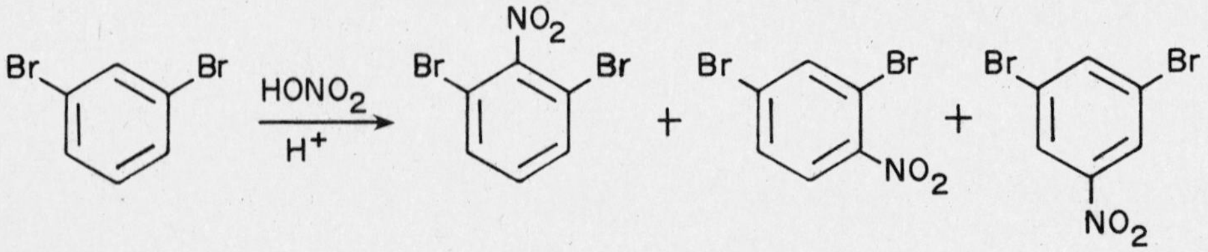

4.24 In each case six carbons are required for the benzene ring; the remaining carbons must be present as alkyl substituents:

a.

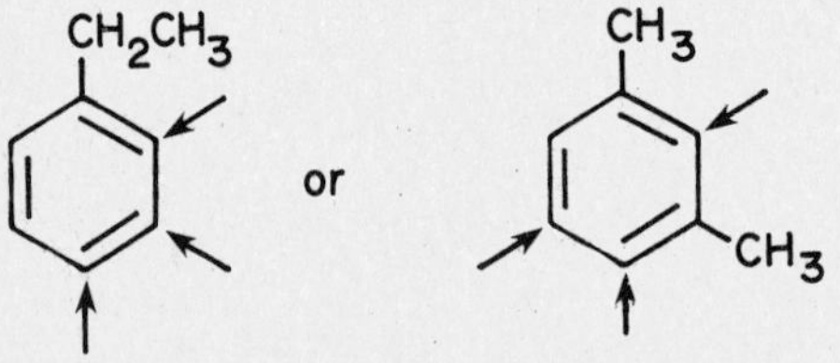

or

Each compound gives three monobromo derivatives, as shown by the arrows.

b.

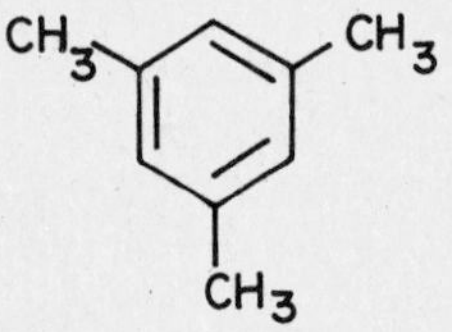

c.

The structure is symmetric, and all three positions for aromatic substitution are equivalent.

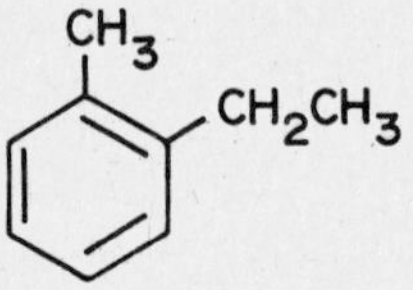

Substitution at each unoccupied ring position gives a different product.

4.25 The energy released on hydrogenating a carbon-carbon double bond is 26-30 kcal/mol (eq. 4.5). With four double bonds, we can calculate that 104-120 kcal/mol should be liberated when cyclooctatetraene is hydrogenated. The observed value (110 kcal/mol) falls within this range and suggests that cyclooctatetraene has no appreciable resonance energy. One reason is that cyclooctatetraene is not planar, and its tublike shape (see p. 129) prevents overlap of the *p* orbitals around the ring.

4.26   The nitro group has two main contributing structures:

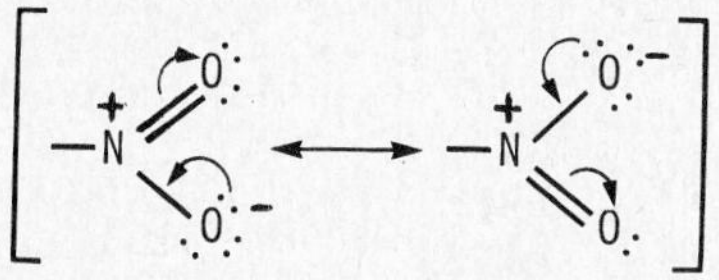

Since they are identical and contribute equally, there is only one type of nitrogen-oxygen bond, intermediate between double and single in length.

4.27 NO_2^+: There are 16 valence electrons available (N = 5, and 2 x O = 2 x 6 = 12 for a total of 17, but we must subtract 1 electron since the ion is positive).

The structure with the positive charge on the nitrogen is preferred because each atom has an octet of electrons. In the structure with the positive charge on the oxygen, the oxygen atom has only six electrons around it. Note that in aromatic nitrations, it is the nitrogen atom of NO_2^+ that attacks and becomes attached to the aromatic ring.

4.28   a. The electrophile, $^+NO_2$, is formed as in eq. 4.20. Then the mechanism follows the same steps as the solution to Example 4.2.

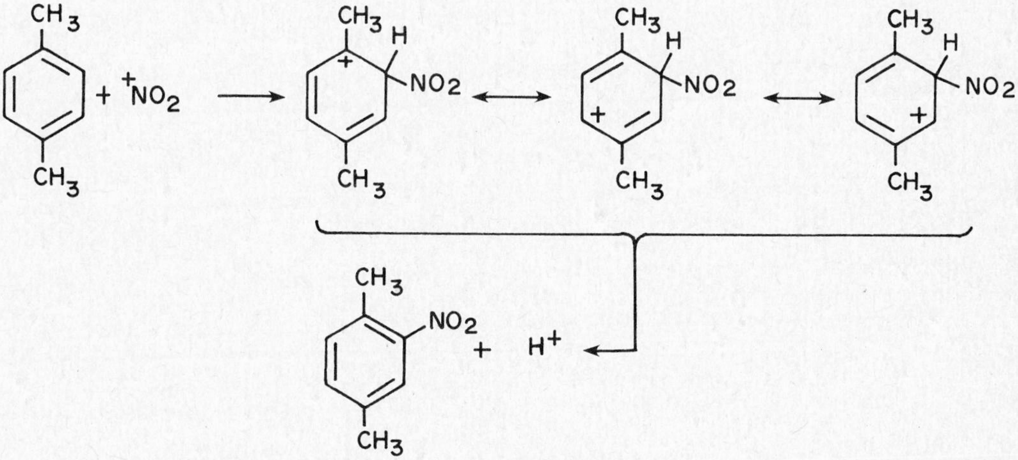

b. The electrophile is formed as in eq. 4.21:

$$(CH_3)_3CCl + AlCl_3 \rightleftharpoons (CH_3)_3C^+ + AlCl_4^-$$

Then

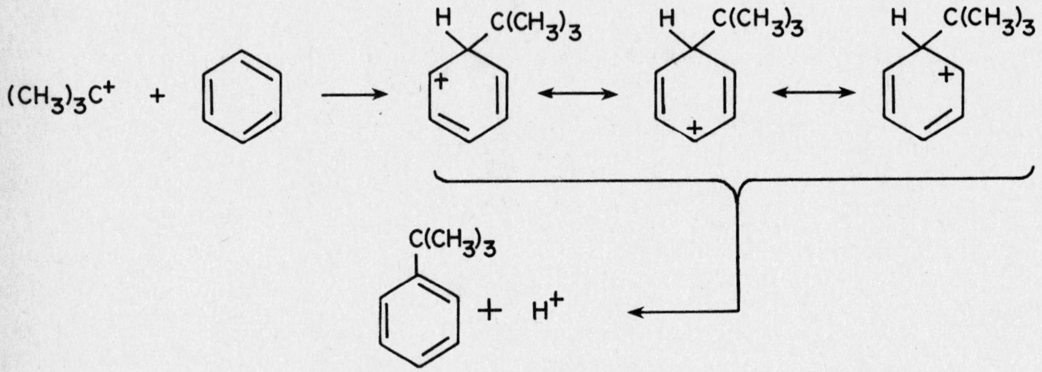

4.29   **ortho**

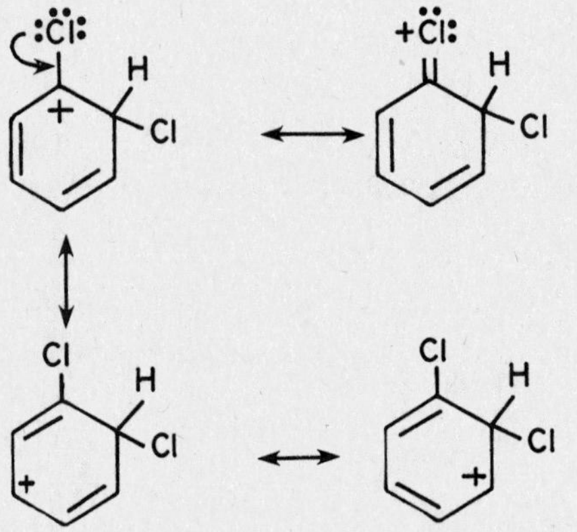

meta

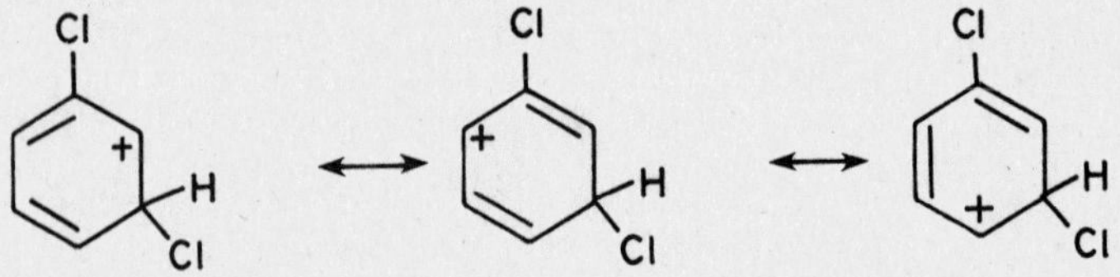

para

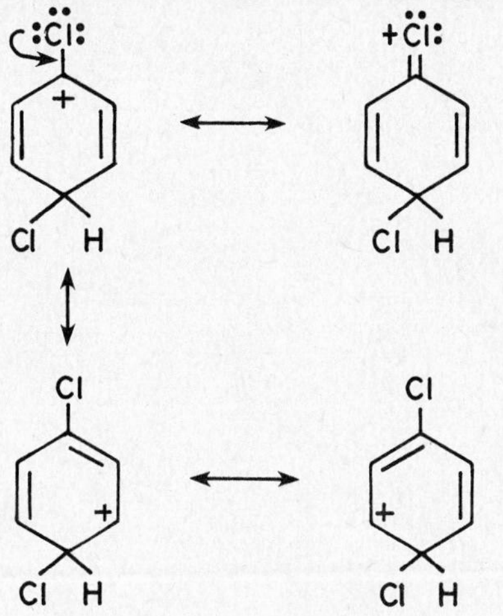

In the intermediate for *ortho* or *para* substitution, the positive charge can be delocalized to the initial chloro substituent. This delocalization is not possible with *meta* substitution. Therefore, *ortho,para*-substitution predominates.

4.30 **ortho**

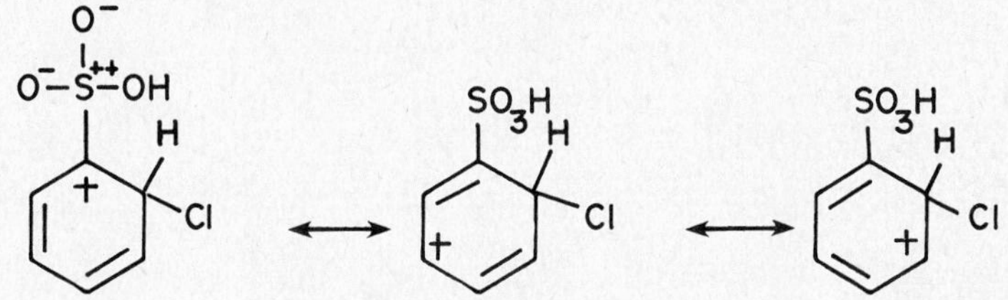

meta

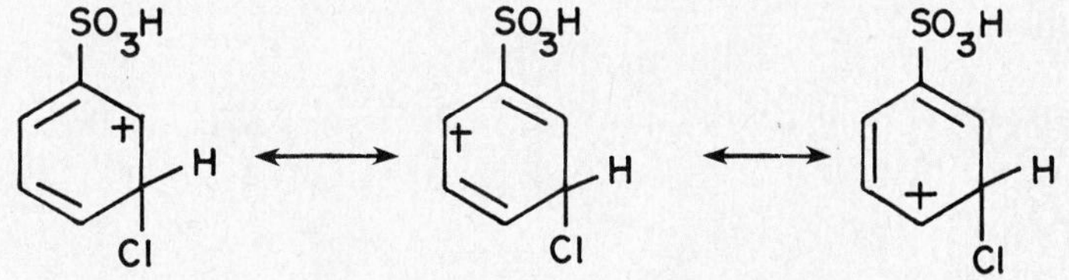

para

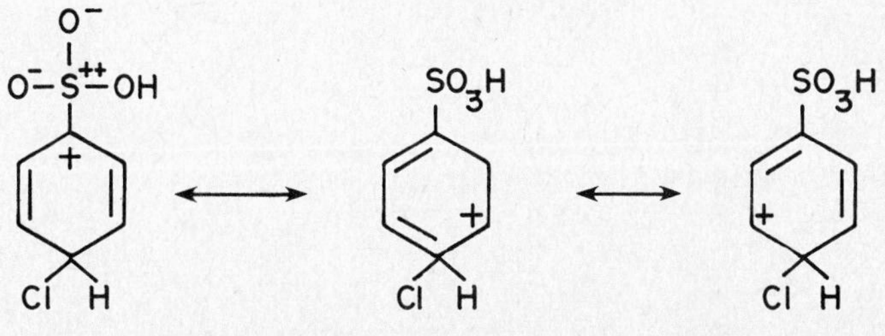

The sulfur atom in the sulfonic acid group carries a partial positive charge, because oxygen is more electronegative than sulfur. This partial charge is illustrated in the first contributor to the intermediates for *ortho* or *para* substitution. Note that two positive charges appear on adjacent atoms in these contributors, an unfavorable situation. No such contributor appears in the intermediate for *meta* substitution. Thus *meta* substitution predominates.

4.31 See Sec. 4.12 for a discussion of the orienting influence of substituents.

a.

Cl—⟨ ⟩—CH_3
(and <u>ortho</u>)

b.

NO_2

⟨ ⟩—SO_3H

c.

Cl—⟨ ⟩—Br
(and <u>ortho</u>)

d. Same as c.

e. Same as b.

f.

Br—⟨ ⟩—CH_2CH_3
(and <u>ortho</u>)

g.

Br—⟨ ⟩—I

4.32 If the halides were not identical, the following kind of exchange could occur (see eq. 4.15):

$$Cl_2 + FeBr_3 \rightleftharpoons \overset{\delta+}{Cl} \cdots Cl \underset{\underset{Br}{|}}{\overset{\overset{Br}{|}}{Fe}} {}^{\delta-}\!\!-Br \rightleftharpoons \overset{\delta+}{Cl} \cdots Br \underset{\underset{Br}{|}}{\overset{\overset{Cl}{|}}{Fe}} {}^{\delta-}\!\!-Br$$

$$\overset{\delta+}{Br} \cdots Cl \underset{\underset{Br}{|}}{\overset{\overset{Br}{|}}{Fe}} {}^{\delta-}\!\!-Cl \rightleftharpoons Cl\!-\!Br + FeBr_2Cl$$

In this way, electrophilic bromine (Br^+) could be formed. Consequently, a mixture of chlorinated and brominated aromatic products would be obtained.

4.33 a. Since the two substituents must end up in a *meta* relationship, the first one introduced into the benzene ring must be *meta*-directing. Therefore, nitrate first and then brominate.

b.

Alkyl groups are *ortho,para*-directing, but the ——SO_3H group is *meta*-directing.

c. First make the ethylbenzene, then nitrate it.

d.

87

e.

(+ some *ortho* isomer)

The nitro substituent must be introduced first, to block the *para* position from the bromination.

f. Compare with part a. The bromination must be performed first.

g.

If the chlorination were performed first, a considerable portion of product would have the chlorine *para* to the methyl group. Also, note that in the second step above, both substituents direct the chlorine to the desired position ($-CH_3$ is *o,p*-directing, $-NO_2$ is *m*-directing).

h.

4.34 D_2SO_4 is a strong acid and a source of the electrophile D^+ (analogous to H^+ from H_2SO_4).

Loss of H^+ from the intermediate benzenonium ion results in replacement of H by D. With a large excess of D_2SO_4, these equilibria are shifted to the right, eventually resulting in fully deuterated benzene, C_6D_6.

4.35 a. —S̈CH$_3$ *ortho,para*-directing because of the unshared electron pairs on sulfur

    b. $\overset{+}{N}(CH_3)_3$ *meta*-directing because of the positive charge on the nitrogen (electron-withdrawing)

    c. —Ö—C(=O)—CH$_3$ *ortho,para*-directing because of the unshared electron pairs on the oxygen

    d. —C(=O)—NH$_2$ *meta*-directing because of the partial positive charge on the carbonyl carbon, due to contributors such as

4.36 Nitro groups are ring-deactivating. Thus, as we substitute nitro groups for hydrogens, we make the ring less and less reactive toward further electrophilic substitution and, therefore, must increase the severity of the reaction conditions.

4.37   a.   anisole; the —OCH_3 group is ring-activating, whereas the —CO_2H group is ring-deactivating.

   b.   toluene; although both substituents (—CH_3 and —Br) are *ortho, para*-directing, the methyl group is ring-activating whereas the bromine is ring-deactivating.

4.38   3-Nitrobenzoic acid is better, because both substituents are *m*-directing:

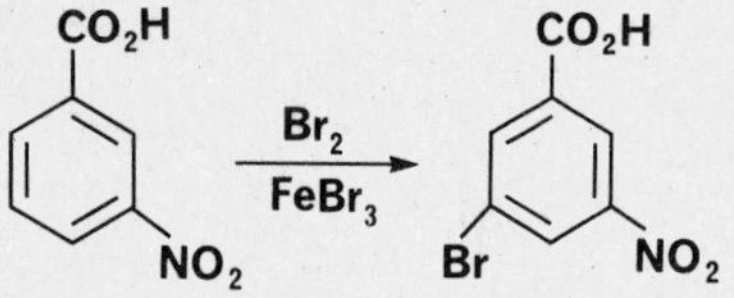

On the other hand, 3-bromobenzoic acid has an *o,p*-directing *and* an *m*-directing substituent, and on nitration would give a mixture of isomers:

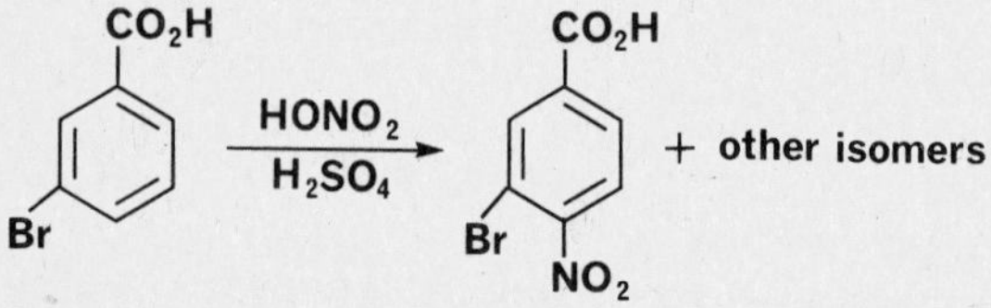

4.39

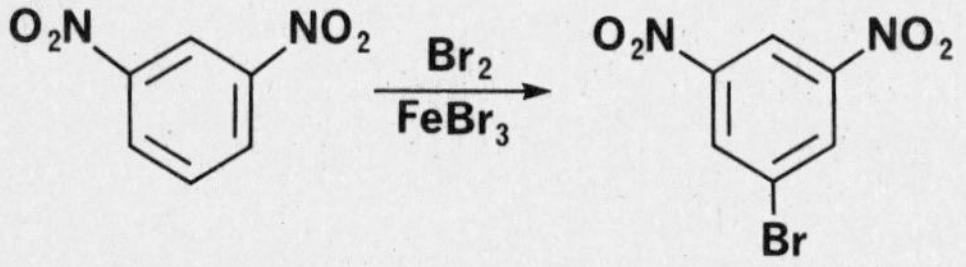

The *m*-directing effect of both nitro groups reinforces substitution in the position shown.

4.40   a.

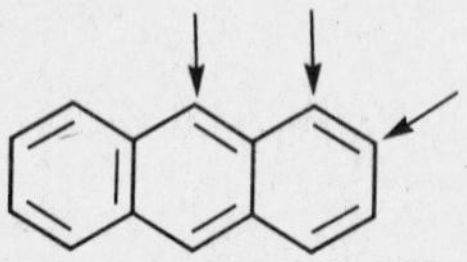

For the numbering of anthracene, see p. 127. Three different monosubstitution products are possible: at C-1 (equivalent to C-4, C-5, and C-8); at C-2 (equivalent to C-3, C-6, and C-7); and at C-9 (equivalent to C-10).

b.

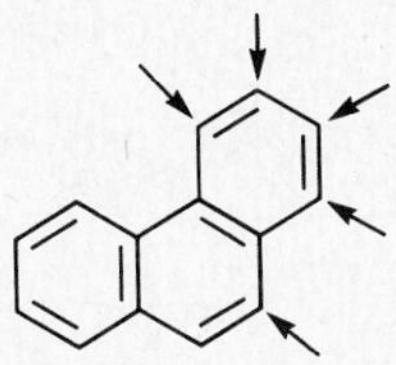

For the numbering of phenanthrene, see p. 127. Five different monosubstitution products are possible: at C-1 (equivalent to C-8); at C-2 (equivalent to C-7); at C-3 (equivalent to C-6); at C-4 (equivalent to C-5); and at C-9 (equivalent to C-10).

4.41

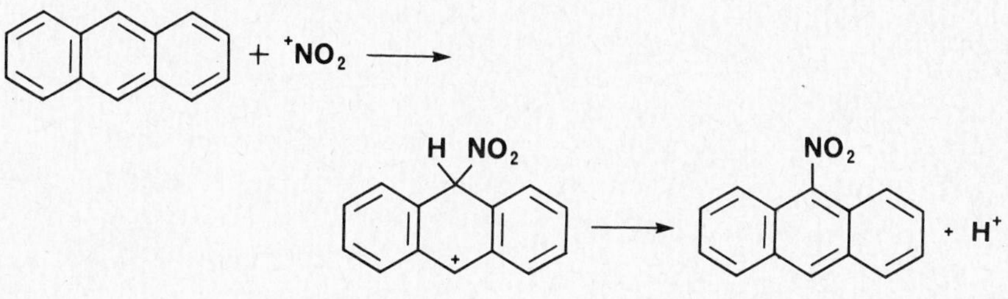

Nitration at C-9 is preferred over the other two possibilities (C-1 or C-2) because the intermediate benzenonium ion retains two benzenoid rings. Substitution at C-1 or C-2 gives an intermediate benzenonium ion with a naphthalene-like part, which has less resonance energy than two benzene rings.

4.42 Of these species, only the cyclopentadiene anion (part d) is aromatic. It has six π electrons in a monocyclic conjugated array:

As seen in the above resonance contributors, the negative charge is delocalized over all five ring carbons. Although we can write analogous resonance structures for the cation (part c) and the radical (part b), they are not aromatic because they contain only four and five π electrons, respectively. Cyclopentadiene itself (part a) has two conjugated double bonds, but the cyclic array is interrupted by a CH_2 group; cyclopentadiene is not aromatic.

4.43　Only the cycloheptatriene cation (part c) is aromatic. It has
six π electrons in a monocyclic array. In this ion, commonly
called the tropylium ion, the positive charge is equally distrib-
uted over all seven carbons. The radical (part b) and the anion
(part d) have seven and eight π electrons, respectively, in a cyclic
array and do *not* fulfill the Hückel rule for aromaticity. Cycloheptatriene (part a) has six π electrons, but the cyclic array is
interrupted by a CH$_2$ group; thus, cycloheptatriene is also *not*
aromatic.

　　　Hückel's prediction that the cyclopentadiene anion (Problem 4.42,
part d) and tropylium cation (Problem 4.43, part c) would be aromatic
and that the other ions or radicals of these types would not be was
one of the early triumphs of molecular orbital theory over resonance
theory. The prediction was verified experimentally, in the enhanced
stability of these species.

4.44　a.　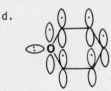　　The cyclic array is interrupted by the
　　　　　　　　　　　　　　　　　　　CH$_2$ group; the compound is *not* aromatic.

b.　The compound (isoxazole) is aromatic; the
nitrogen and each carbon contribute one
electron, and the oxygen contributes two
electrons to the six π system. In addition, there is an unshared electron pair
on the nitrogen and on the oxygen, in the
ring plane.

c.　The compound (oxazole) is aromatic, with
six π electrons in a cyclic array. The
orbital picture is similar to that for
isoxazole (part b).

d.　The compound (pyrylium ion) is aromatic;
its orbital picture is very much like that
of benzene. Each ring atom contributes
one electron to the six π system. The
unshared electron pair on oxygen lies in
an orbital in the ring plane. Use the
formula in Sec. 1.11 to verify that the
oxygen carries a formal +1 charge.

4.45

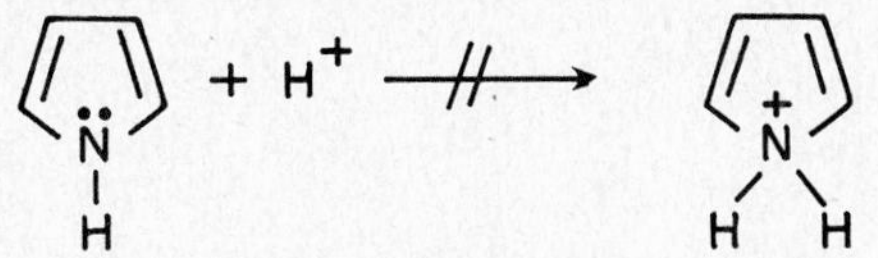

The unshared electron pair required to make six π electrons in pyrrole is no longer available if the nitrogen is protonated. Thus, the aromaticity of pyrrole would be destroyed by protonating the nitrogen, and the resonance energy associated with this aromaticity would be lost.

4.46

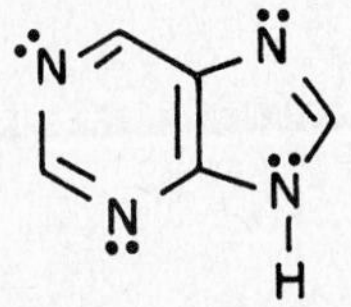

The N——H nitrogen contributes both electrons of the unshared pair to the π system. The other three nitrogens are doubly bonded; they contribute one electron each to the π system. They do *not* contribute either electron from the unshared pair. The nitrogen *least* likely to be protonated is the N——H nitrogen, because if it were protonated, the unshared electron pair would no longer be available to the aromatic π system.

Put another way, the N——H nitrogen is a pyrrole-type nitrogen and is not basic, whereas the other three nitrogens are pyridine-type nitrogens and, like the nitrogen in pyridine, are basic.

CHAPTER FIVE   STEREOISOMERISM

CHAPTER SUMMARY

Stereoisomers have the same atom connectivities but different arrangements of the atoms in space. They may be **chiral** or **achiral**. A stereoisomer
is chiral if its mirror image is not identical or superimposable on the
original molecule; it is achiral if the molecule and its mirror image are
identical. **Enantiomers** are a pair of molecules related as nonsuperimposable
mirror images. Any molecule with a **plane of symmetry** is achiral.

Chiral molecules are **optically active**; they rotate a beam of plane-
polarized light. They are dextrorotatory (+) or levorotatory (-), depending
on whether they rotate the beam to the right or left, respectively. The
rotations are measured with a polarimeter and are expressed as **specific
rotations,** defined as

$$[\alpha]_{\lambda}^{t} = \frac{\alpha}{l \times c} \quad \text{(solvent)}$$

where α = observed rotation, l = length of sample tube in decimeters,
c = concentration in g/mL, and the measurement conditions of temperature t,
wavelength of polarized light λ, and solvent are given. Achiral molecules
are **optically inactive**.

Pasteur showed that optical activity was related to molecular right- or
left-handedness (chirality). Later, van't Hoff and LeBel proposed that the
four valences of carbon are directed toward the corners of a tetrahedron.
If the four attached groups are different, two arrangements are possible
and are related as an object and its nonsuperimposable mirror image. Such
a carbon is a **chiral center** (or asymmetric carbon atom) and the two arrangements are called **enantiomers**. These molecules differ *only* in chiral (or
handed) properties, such as the *direction* of rotation of plane-polarized
light. They have identical **achiral properties**, such as melting and boiling
points.

Configuration refers to the arrangement of groups attached to a chiral
center. **Enantiomers** have opposite configurations. Configuration can be
designated by the *R-S* **convention**. Groups attached to the chiral center
are ranked in a priority order according to decreasing atomic number. When
the chiral center is viewed from the side *opposite* the lowest-priority group,
the center is said to be *R* if the other three groups, in decreasing priority

order, form a *clockwise* array. If the three-group array is counterclockwise, the configuration is *S*. A similar convention (*E-Z*) has been applied to alkene *cis-trans* isomers.

Diastereomers are stereoisomers that are not mirror images of one another. They may differ in all types of properties, and be chiral or achiral.

Compounds with n different chiral centers may exist in a maximum of 2^n forms. Of these, there will be $2^n/2$ pairs of enantiomers. Compounds from different enantiomeric pairs are diastereomers. If two (or more) of the chiral centers are identical, certain isomers will be achiral. A **meso form** is an optically inactive, achiral form of a compound with chiral centers. **Tartaric acid**, which has two identical chiral centers, exists in three forms, the *R,R* and *S,S* forms (a pair of enantiomers) and the achiral *meso* form.

A **racemic form** is a 50:50 mixture of enantiomers. It is optically inactive. A racemic mixture of configurational isomers cannot be separated (**resolved**) by ordinary chemical means (distillation, crystallization, chromatography) unless the reagent is chiral. One way to separate a pair of enantiomers is to first convert them to diastereomers by reaction with a chiral reagent, then separate the diastereomers and regenerate the (now separated) enantiomers.

LEARNING OBJECTIVES

1. Know the meaning of: plane-polarized light, polarimeter, optically active or optically inactive, observed rotation, specific rotation, dextrorotatory, levorotatory.

2. Know the meaning of: chiral, achiral, enantiomers, plane of symmetry, superimposable and nonsuperimposable mirror images, racemic mixture.

3. Know the meaning of: asymmetric carbon atom, chiral center, *R-S* convention, priority order, *E-Z* convention.

4. Know the meaning of: diastereomer, *meso* compound, lactic acid, tartaric acid, resolution.

5. Given the concentration of an optically active compound, length of the polarimeter tube, and observed rotation, calculate the specific rotation. Given any three of the four quantities mentioned, calculate the fourth.

6. Given a structural formula, draw it in three dimensions and locate any plane of symmetry.

7. Given the structure of a compound, determine if any chiral centers (asymmetric carbon atoms) are present.

8. Given the structure or name of a compound, tell whether it is capable of optical activity.

9. Know the rules for establishing priority orders of groups in the *R-S* convention.

10. Given a compound with a chiral center, assign the priority order of groups attached to it.

11. Given a chiral center in a molecule, assign the *R* or *S* configuration to it.

12. Draw the three-dimensional formula of a molecule with a particular configuration, *R* or *S*.

13. Given a pair of *cis-trans* isomers, assign the *E* or *Z* configuration.

14. Draw the formula of an alkene with a particular configuration, *E* or *Z*.

15. Given a structure with more than one chiral center, tell how many stereo-isomers are possible and draw the structure of each. Tell what relationship the stereoisomers have to each other (for example, enantiomeric, diastereomeric).

16. Tell whether or not a particular structure can exist as a *meso* form.

17. Given a structure with two or more identical chiral centers, draw the structure of the *meso* form.

18. Given a pair of stereoisomers, classify them as configurational or conformational, chiral or achiral, and enantiomers or diastereomers.

ANSWERS TO PROBLEMS

Problems Within the Chapter

5.1 a. chiral   b. achiral   c. achiral   d. chiral
     e. achiral   f. chiral   g. chiral   h. achiral

The achiral objects (tea cup, football, tennis racket, and pencil) can be used with equal ease by right- or left-handed persons. Their mirror images are superimposable on the objects themselves. On the other hand, a golf club must be either left- or right-handed and is chiral; a shoe will fit a left or a right foot; a corkscrew may have a right- or left-handed spiral. These objects, as well as a portrait, have mirror images that are *not* identical with the objects themselves, and thus they are chiral.

5.2   a.

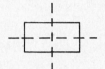

b.

rectangle
(two such
planes)

equilateral
triangle
(three such
planes)

c.   In a cube, symmetry planes may bisect opposite faces, or they may bisect the cube diagonally:

5.3   None of these chiral objects has a plane of symmetry.

5.4   b.

The plane of symmetry passes through the handle of the tea cup.

c.

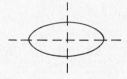

A football has an infinite number of symmetry planes that pass through the ends of the oblate spheroid; each of those planes can be perpendicularly bisected by a plane that is also a plane of symmetry.

e.

A tennis racket has a plane of symmetry that bisects the racket through the handle. The racket plane itself is also a plane of symmetry.

h. 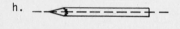 A pencil has an infinite number of symmetry planes that pass down the length, through its center.

These four objects are achiral. None of the other objects in Problem 5.1 has a plane of symmetry; thus they are all chiral.

5.5   eclipsed ethane:

The planes of symmetry are (a) the three planes that pass through any pair of eclipsed hydrogens and (b) the perpendicular bisector of the C—C bond. Ethane in this conformation is achiral.

5.6

```
Cl   Cl
  \ ¦ /
   C≑C
  / ¦ \
 H   H
```

cis -1,2-Dichloroethene has a plane of symmetry that bisects the double bond. The molecular plane is also a symmetry plane.

```
Cl    H
  \  /
   C=C
  /  \
 H    Cl
```

trans -1,2-Dichloroethene is planar; that plane is a symmetry plane.

Both molecules are *achiral* ; their mirror images are superimposable on the original structures.

5.7　$[\alpha]_D^{20} = \dfrac{+0.66}{0.5 \times \dfrac{1.5}{50}} = 44°$ (ethanol)

5.8　The chiral centers are marked with an asterisk. Note that each chiral center has four different groups attached.

a.　$CH_3CH_2 - \overset{\overset{\displaystyle H}{|}}{\underset{\underset{\displaystyle CH_3}{|}}{C}}\text{*} - CH_2CH_2CH_3$　　b.　$CH_3 - \overset{\overset{\displaystyle H}{|}}{\underset{\underset{\displaystyle Cl}{|}}{C}}\text{*} - \overset{\overset{\displaystyle H}{|}}{\underset{\underset{\displaystyle Cl}{|}}{C}}\text{*} - CH_3$

c.　

d.　$Br - \overset{\overset{\displaystyle H}{|}}{\underset{\underset{\displaystyle Cl}{|}}{C}}\text{*} - CH_3$

5.9　a.　$Br - \overset{\overset{\displaystyle H}{|}}{\underset{}{C}}\text{*}{-}CH_3$ （phenyl below）

chiral

b.　$Br - \overset{\overset{\displaystyle H}{|}}{\underset{\underset{\displaystyle H}{|}}{C}} - \overset{\overset{\displaystyle H}{|}}{\underset{\underset{\displaystyle H}{|}}{C}} - \text{(phenyl)}$

no chiral centers;
achiral

5.10

Note that if the right structure is rotated 180° about the carbon-phenyl bond, the methyl and phenyl groups can be superimposed on those of the left structure, but the positions of the hydrogen and bromine will be interchanged.

5.11 a. The two structures are identical. Two group interchanges (an
even number) convert one to the other:

$$CH_3 \xrightarrow[\substack{-CH_2CH_3}]{\substack{interchange \\ -Cl\ and}} CH_3 \xrightarrow[\substack{-CH_3}]{\substack{interchange \\ H\ and}} H$$

CH₃ · C//..H — CH₃CH₂ Cl → CH₃ · C//..H — Cl CH₂CH₃ → H · C//..CH₃ — Cl CH₂CH₃

b. Three group interchanges (an odd number) are required to convert
one structure to the other. Thus the structures are enantiomers.

CH₃ · C//..H — CH₃CH₂ Cl
interchange −CH₃ and −CH₂CH₃ →
CH₂CH₃ · C//..H — CH₃ Cl
interchange −Cl and −CH₂CH₃ →
Cl · C//..H — CH₃ CH₂CH₃
interchange −H and −CH₂CH₃ →
Cl · C//.. CH₂CH₃ — CH₃ H

The sequence of group interchanges can be varied in each example, but
regardless of the sequence the conclusion in each case will be the
same.

5.12 In each case, proceed from high to low priority.

a. —OH > —C(CH$_3$)$_3$ > —CH$_3$ > —H

b. —OCH$_3$ > —OH > —CH$_2$OH > —CH$_3$

c. —OH > —NHCH$_3$ > —CN > —CH$_2$NH$_2$

d. ⬡— > —C≡CH > —CH=CH$_2$ > —CH$_2$CH$_3$

The higher priority of ⬡— over —C≡CH is assigned because
of the connection of the ring carbon to *two* adjacent ring carbons,
whereas the acetylenic carbon is connected to only one other carbon
atom.

5.13 a.

CH=O
CH₃ · · · H · · OH

Counterclockwise, or *S*.

Priority order: OH > CH=O > CH$_3$ > H
Configuration is *S*.

b.

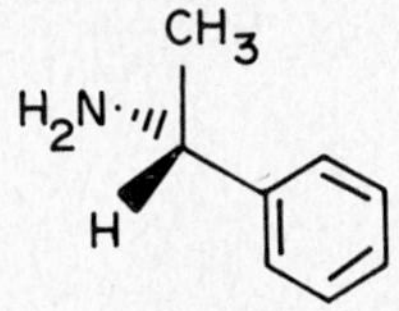

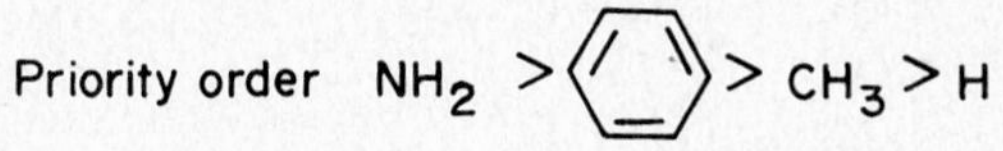

Priority order $NH_2 > \langle\!\bigcirc\!\rangle > CH_3 > H$

Configuration is R.

This can perhaps be seen more easily if the molecule is rotated 120° around the C—CH$_3$ bond so that the lowest priority group (H) points back:

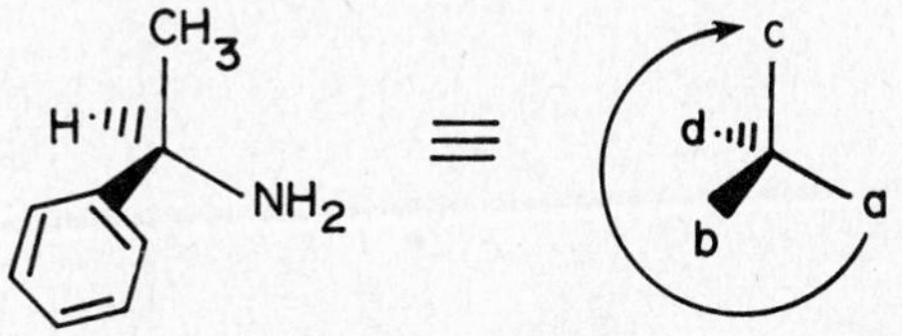

Clockwise, or R

5.14   a.   The priority order of groups around the chiral center is as follows:

$CH_3CH_2CH_2 > CH_3CH_2 > CH_3 > H$

First draw the lowest priority group pointing away from you:

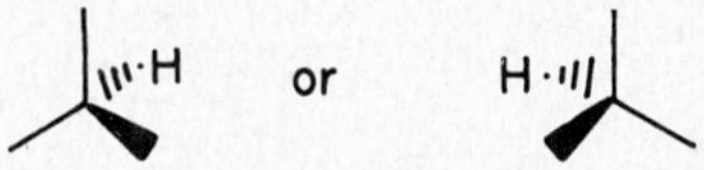

or

Then fill in the groups in priority order, clockwise (R):

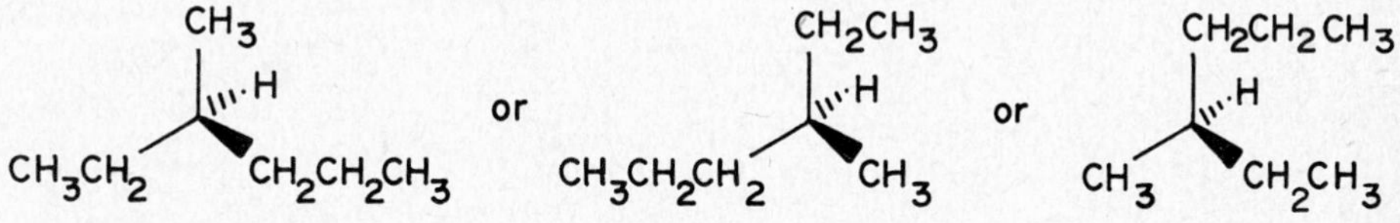

or   or

and similarly:

or   or

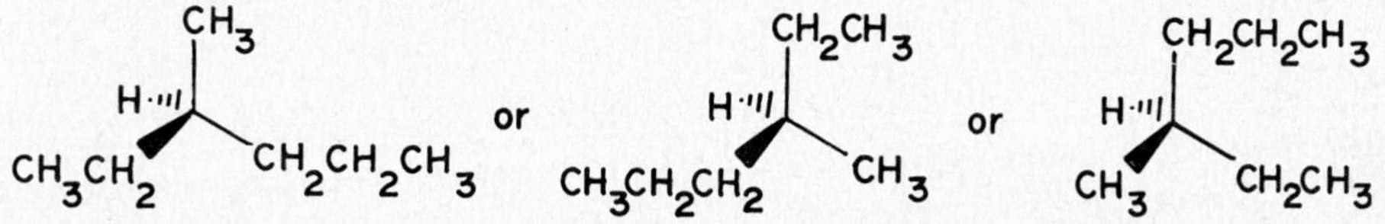

As you can see, there are many ways to write the correct answer. In subsequent problems, only one correct way will be shown. Work with models if you have difficulty.

b. The priority order of groups around the chiral center is as follows:

$$CH_2=CH- > CH_3CH_2- > CH_3- > H-$$

One correct way of writing the answer is shown below:

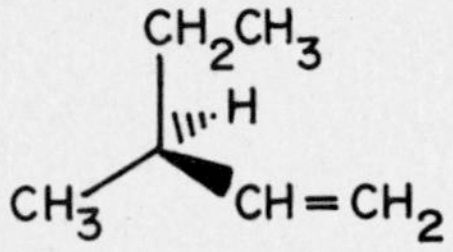

(S)-3-methyl-1-pentene

5.15 a. The priority order at each doubly bonded carbon is $CH_3 > H$ and $CH_3CH_2 > H$. The configuration is Z.

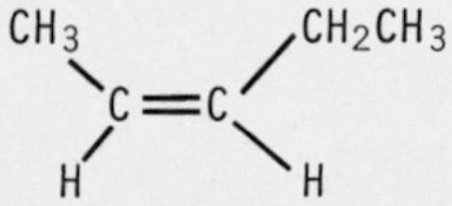

(Z)-2-pentene

b. The priority order is Br > Cl and F > H. The configuration is Z.

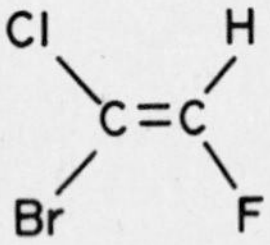

(Z)-1-bromo-1-chloro-2-fluoroethene

5.16 a.

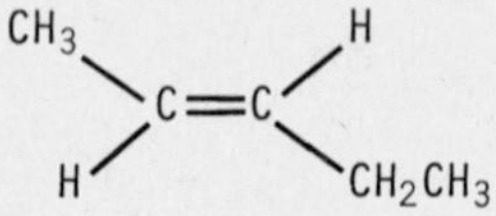

(E)-2-pentene

The two highest priority groups, CH_3 and CH_3CH_2, are *entgegen,* or opposite.

b.

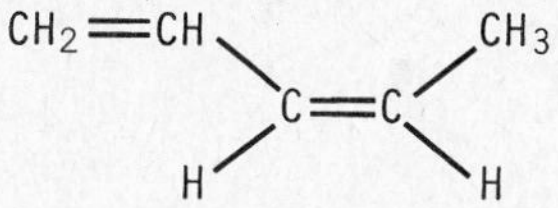

$$CH_2{=}CH \diagdown \quad \diagup CH_3$$
$$C{=}C$$
$$H \diagup \quad \diagdown H$$

(Z)-1,3-pentadiene

The priorities are $CH_2{=}CH{-} > H$ and $CH_3 > H$.

The two highest priority groups, $CH_2{=}CH{-}$ and $CH_3{-}$, are *zusammen* (together).

5.17   The $(2R,3R)$ and $(2S,3S)$ isomers are a pair of enantiomers. Their specific rotations will be equal in magnitude and opposite in sign.

The $(2R,3R)$ and $(2S,3R)$ isomers are a pair of diastereomers. Their specific rotations will be unequal in magnitude and may or may not differ in sign.

5.18   There are *four different* chiral centers, marked below with asterisks. There are therefore $2^4 = 16$ possible isomers.

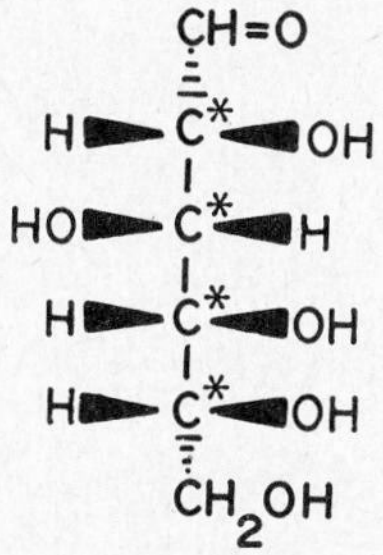

$$CH{=}O$$
$$H{-}C^*{-}OH$$
$$HO{-}C^*{-}H$$
$$H{-}C^*{-}OH$$
$$H{-}C^*{-}OH$$
$$CH_2OH$$

5.19

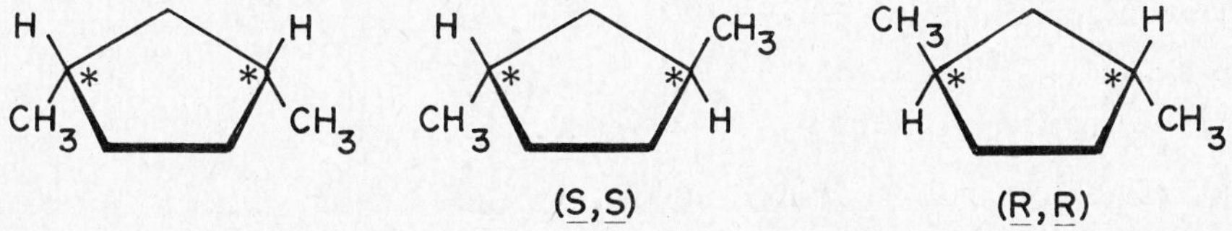

There are two chiral centers (marked with asterisks). The *cis* isomer is achiral, a *meso* form. The *trans* isomer is chiral and can exist as a pair of enantiomers, either (S,S) or (R,R).

5.20

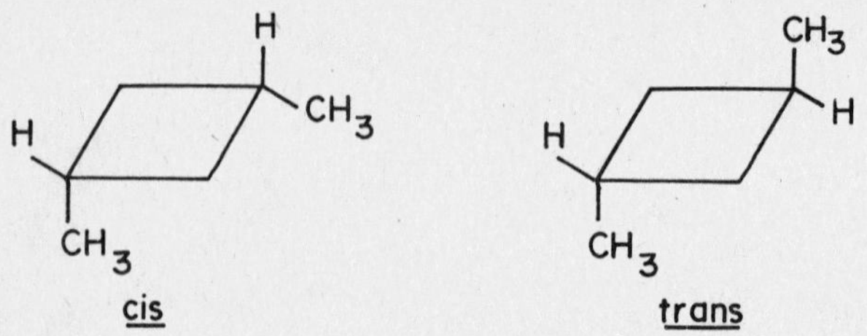

cis   trans

There are *no* chiral centers. Both molecules have planes of
symmetry. The *cis* isomer has two such planes, through opposite
corners of the ring. The *trans* isomer has one such plane, through
the opposite methyl-bearing corners. Both compounds are optically
inactive and achiral. They are *not meso* compounds, because there
are no chiral centers. To summarize, the two isomers are configurational, achiral diastereomers.

5.21 If we draw the structure in three dimensions, we see that either
hydrogen at C-2 can be replaced with equal probability:

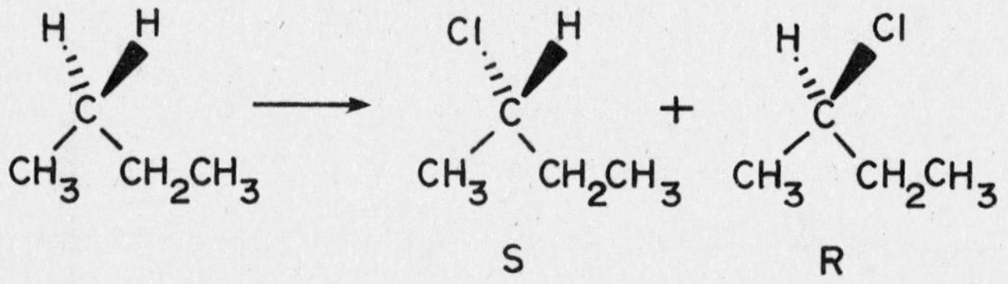

S   R

Thus a 50:50 mixture of the two enantiomers is obtained.

Additional Problems

5.22 Each of these definitions can be found explicitly or implicitly in
the following sections of the text:

a. 5.2   b. 5.2   c. 5.4   d. 5.5   e. 5.7
f. 5.3   g. 5.6   h. 5.11   i. 5.12   j. 5.14

5.23 In most cases one can examine the structure for a plane of symmetry--
if none is present, enantiomers are usually possible. But the deci-
sive test is to compare the molecule with its mirror image and determine whether they are superimposable.

5.24   a.

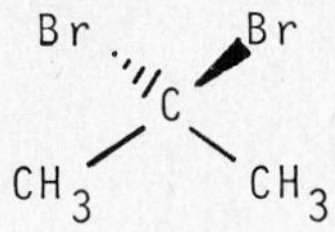

Br$\cdots$C$\cdots$Br
CH$_3$ CH$_3$

Optically inactive; the molecule has several planes of symmetry, for example, the one that passes through the three carbon atoms.

b.

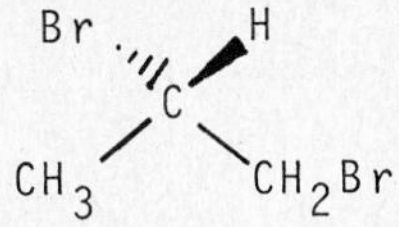

Br$\cdots$C$\cdots$H
CH$_3$ CH$_2$Br

Optically active; carbon-2 is a chiral center, with four different groups attached.

c. 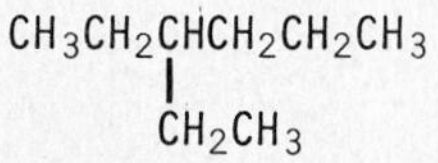 $CH_3CH_2CHCH_2CH_2CH_3$
              |
            CH_2CH_3

Optically inactive; none of the carbon atoms has four different groups attached.

d. 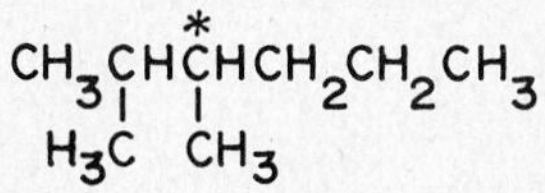 $CH_3CH\overset{*}{C}HCH_2CH_2CH_3$
          | |
        H_3C CH_3

Optically active; the carbon marked with an asterisk is a chiral center.

e.   H$\cdots$CH$_3$

5 ⟨ 1 ⟩ 2
    4   3

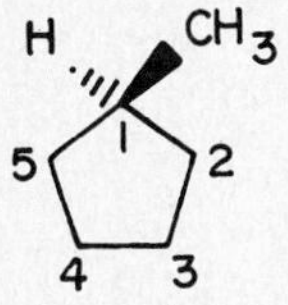

The substance is optically inactive; it has a plane of symmetry perpendicular to the five-membered ring, through carbon-1 and bisecting the bond between carbon-3 and carbon-4.

f. $CH_3-\overset{\overset{\displaystyle H}{|}}{\underset{\underset{\displaystyle D}{|}}{C}}{}^{*}-OH$

Optically active; the carbon marked with an asterisk is a chiral
center, with four different groups attached. Even isotopes of the
same element are sufficiently different to lead to optical activity.

5.25 a.

$C_6H_5-\overset{\overset{\displaystyle H}{|}}{\underset{\underset{\displaystyle OH}{|}}{C}}{}^{*}-CO_2H$

b.

$CH_2-\overset{\overset{\displaystyle H}{|}}{\underset{\underset{\displaystyle OH}{|}}{C}}{}^{*}-\overset{\overset{\displaystyle H}{|}}{\underset{\underset{\displaystyle HO}{|}}{C}}{}^{*}-CH=O$ OH

c.

$\overset{\overset{\displaystyle H}{|}}{\underset{\underset{\displaystyle OH}{|}}{C}}{}^{*}-CH_3$

d.

$CH_3-\overset{\overset{\displaystyle H}{|}}{\underset{\underset{\displaystyle Cl}{|}}{C}}{}^{*}-CCl_3$

e.

CH_3 — ⬡ — CH_3

no chiral centers

f.

HO CH_3

5.26 In each case the *observed* rotation would be doubled, but the specific
rotation would remain constant. For example, if c is doubled, α will
also double, but the fraction α/c in the formula for specific rotation
will remain constant.

5.27 Alter the concentration of the solution and measure the observed
rotation again. For example, we might increase the concentration by
10%. If the original rotation was +30° it would now be +33°. But
if the original rotation was -330° it would now be -363° (or -3°).
The new rotations, +33° and -3°, are easily distinguished.

5.28 The following are examples; there may be other possibilities.

a. $CH_3\overset{*}{C}HCH_2CH_3$
        |
        OH

b. $CH_3\overset{*}{C}HCH_2CH_2CH_3$
        |
        Br

c. $HOCH_2\overset{*}{C}HCH_2CH_3$
          |
          OH

d. $CH_3\overset{*}{C}HCH=CH_2$
        |
        CH_2CH_3

In each case the asymmetric carbon atom is marked with an asterisk.

5.29   All structures must contain only one double bond and no rings. We
       know this because if the monovalent chlorine were replaced by a
       hydrogen we would have C_5H_{10}, corresponding to the general formula
       C_nH_{2n} (a molecule with one double bond or one ring). Since the
       chloride is described as unsaturated, there can be no ring present.

       a. $CH_2\!=\!CHCH_2CH_2CH_2Cl$

       b. $CH_3CH\!=\!C\!-\!CH_2Cl$
                         |
                         CH_3

       c. $\overset{*}{CH_2}\!=\!CHCHCH_2CH_3$ or $CH_2\!=\!CHCH_2\overset{*}{CH}CH_3$
                         | |
                         Cl Cl

       d. $CH_3CH\!=\!\overset{*}{CH}CHCH_3$
                              |
                              Cl

5.30   The rules for priority order are given in Sec. 5.9.
       a. $HO\!-\! > CH_3CH_2\!-\! > CH_3\!-\! > H\!-\!$
       b. $Cl\!-\! > C_6H_5\!-\! > CH_3\!-\! > H\!-\!$
       c. $HO\!-\! > \!-\!CH_2Cl > \!-\!CH_2OH > \!-\!CH_3$
       d. $\!-\!CH\!=\!O > CH_2\!=\!CH\!-\! > CH_3CH_2CH_2\!-\! > CH_3CH_2\!-\!$

5.31   a. CH_3 b. CH_3

                          ⁣ıı·H   ıı·H
           CH_3CH_2 OH C_6H_5 Cl

       c. CH_2OH d. $CH_2CH_2CH_3$

                          ıı·CH_3 ıı·CH_2CH_3
           $ClCH_2$ OH $CH_2\!=\!CH$ $CH\!=\!O$

       There are many ways to write these structures. They have been drawn
       here by putting the lowest priority group receding away from the
       viewer, and the remaining three groups in a clockwise array, with the
       highest priority group at the lower right extending toward the viewer.

5.32 In each case, write down the groups in the proper priority order. Then view the chiral center from the face opposite the lowest priority group and determine whether the remaining array is clockwise (R) or counterclockwise (S). If you have difficulty, construct and examine molecular models.

a.

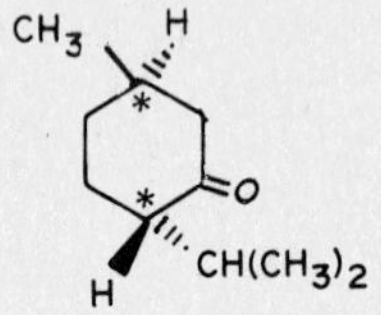

"top" chiral center

$$\overset{\overset{\text{O}}{\underset{\|}{}}}{-\text{CH}_2\text{C}} > -\text{CH}_2\text{CH}_2 > -\text{CH}_3 > -\text{H};$$
therefore R

"bottom" chiral center

$$\overset{\overset{\text{O}}{\underset{\|}{}}}{-\text{C}} > -\text{CH(CH}_3)_2 > -\text{CH}_2 > -\text{H};$$
therefore S

b.

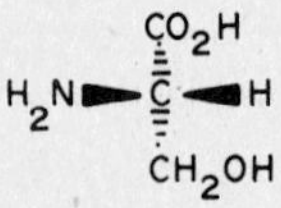

$-\text{NH}_2 > -\text{CO}_2\text{H} > -\text{CH}_2\text{OH} > -\text{H};$
therefore S

c.

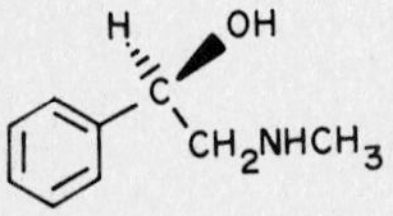

$-\text{OH} > -\text{CH}_2\text{NHCH}_3 > \bigcirc\hspace{-1.5em}\bigcirc > -\text{H};$
therefore R

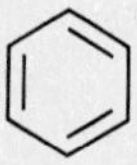

5.33

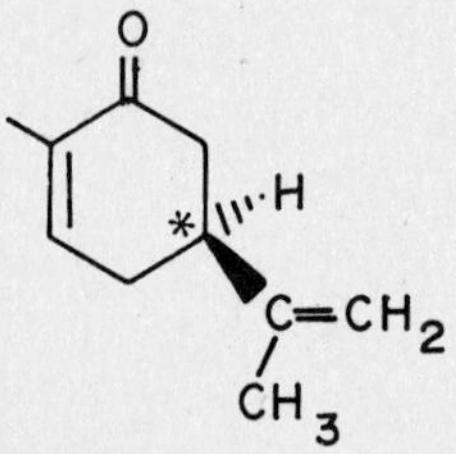

(+) − carvone

The chiral center is marked with an asterisk. The priority order of groups at this center is

and the configuration is S.

A word about the priority order may be helpful. The $-\overset{\underset{|}{CH_3}}{C}=CH_2$ group has three bonds from the attached carbon atom to the next atoms "out," and is therefore of the highest priority. The remaining groups both begin with $-CH_2$, so we must proceed further. One group is

$$-CH_2-\overset{\overset{O}{\|}}{C}-C$$

and the other is

$$-CH_2-\overset{\underset{|}{H}}{C}=C$$

Of these, the group with $C=O$ has the higher priority because oxygen has a higher atomic number than carbon.

5.34 a.

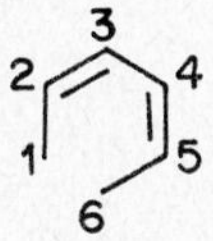

(Z,Z)-2,4-hexadiene or, more precisely, $(2Z,4Z)$-2,4-hexadiene

If you have difficulty, draw out the full structure:

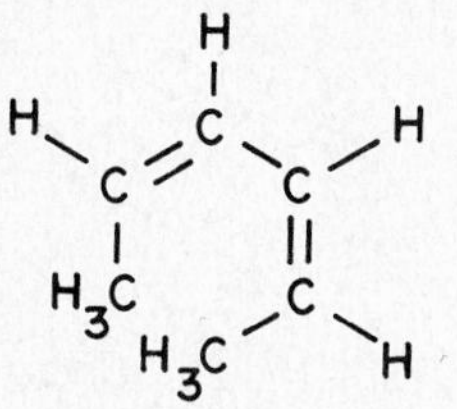

At the double bond between C-2 and C-3, the priority order is $CH_3 > H$ and $CH_3CH=CH- > H$. The two high-priority groups, CH_3 and $CH_3CH=CH-$, are Z or *zusammen*. The same is true at the double bond between C-4 and C-5.

b.

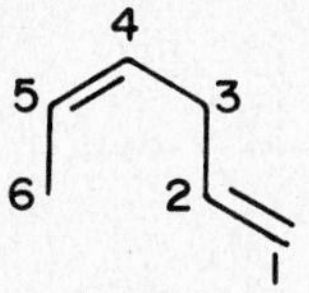

(Z)-1,4-hexadiene; there is no stereochemistry at the double bond joining C-1 and C-2 because both substituents at C-1 are identical ($=CH_2$).

c.

(E,E)-2,4-hexadiene

d.

There is no stereochemistry at either double bond; the IUPAC name is 1,5-hexadiene.

5.35 The structure is $CH_3CH{=}CH{-}\overset{*}{CH}{-}CH_3$. Four isomers are possible:
with Br below the starred carbon.

(R, E)   (S, E)

(R, Z)   (S, Z)

The upper and lower sets form two pairs of enantiomers.

5.36 a. $\overset{1}{CH_2}{=}\overset{2}{CH}{-}\overset{3}{CH}{-}\overset{4}{CH}{=}\overset{5}{CH_2}$
with CH_3 on carbon 3.

There are no chiral centers and no *cis-trans* possibilities at either double bond. Only one structure is possible.

b.

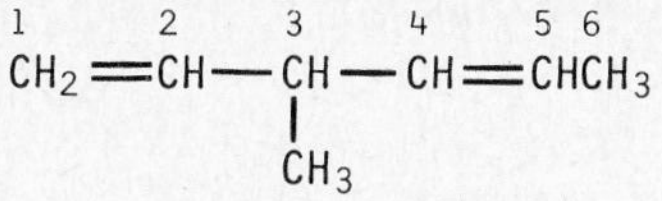

$$\overset{1}{CH_2}=\overset{2}{CH}-\overset{3}{CH}-\overset{4}{CH}=\overset{5}{CH}\overset{6}{CH_3}$$

(with CH_3 branch on C-3)

Carbon-3 is a chiral center, and *cis-trans* isomers are possible at the double bond between C-4 and C-5. Therefore, four structures are possible (*R* or *S* at C-3, and *E* or *Z* at the double bond).

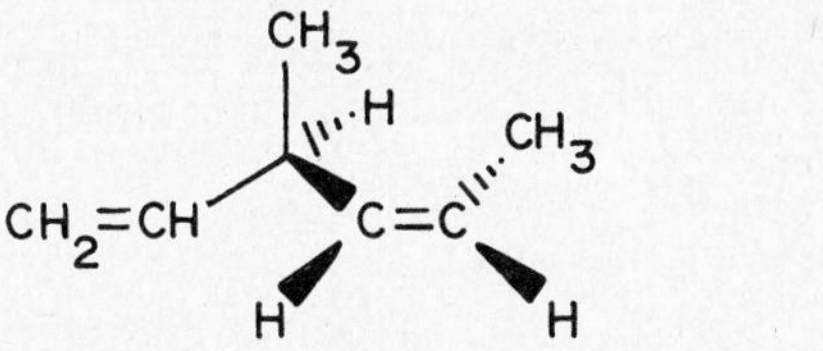

(<u>R</u>,<u>Z</u>)- 3-methyl-1,4-hexadiene

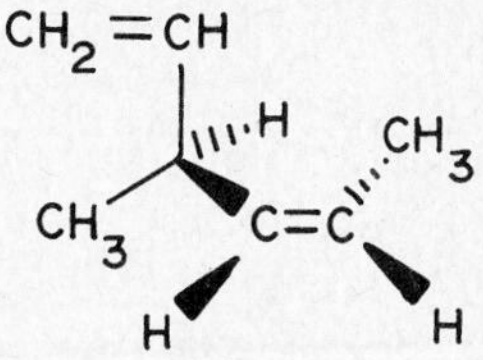

(<u>S</u>,<u>Z</u>)

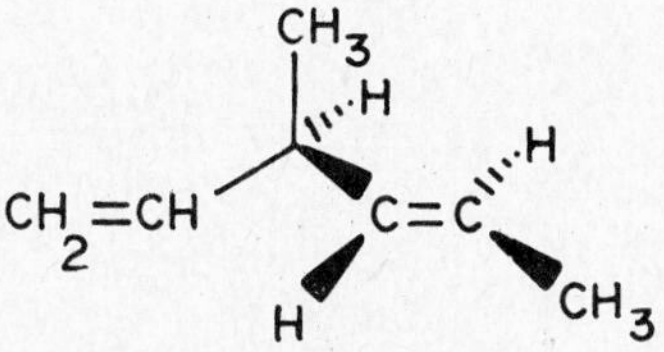

(<u>R</u>,<u>E</u>)

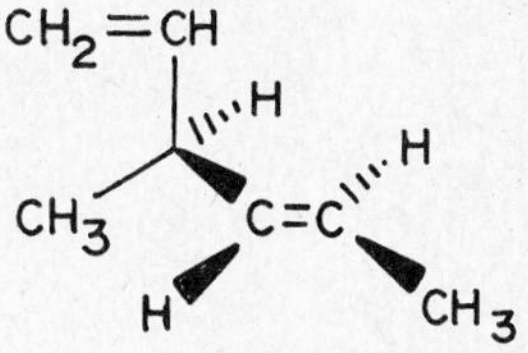

(<u>S</u>,<u>E</u>)

c.

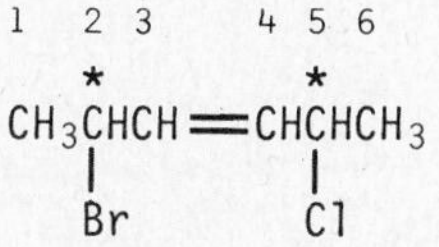

$$\overset{1}{CH_3}\overset{2}{\underset{*}{CH}}\overset{3}{CH}=\overset{4}{\underset{*}{CH}}\overset{5}{CH}\overset{6}{CH_3}$$

(with Br on C-2 and Cl on C-4)

There are two chiral centers, marked with asterisks. Each can be either *R* or *S*. Also, the double bond joining C-3 and C-4 can be *E* or *Z*. Thus eight isomers are possible:

R,Z,R	S,Z,R
R,Z,S	S,Z,S
R,E,R	S,E,R
R,E,S	S,E,S

The first of these is shown below:

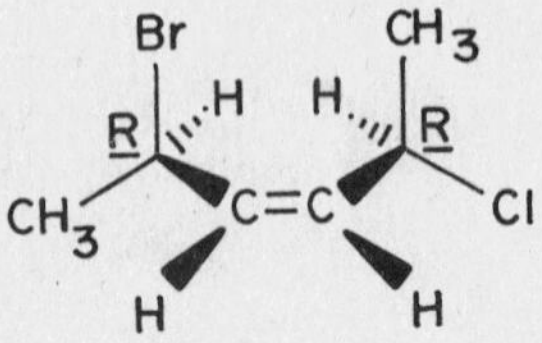

$$\underline{Z}$$

The other seven isomers can be drawn by interchanging one or more groups, using this structure as a guide. For example, the S,Z,R isomer is

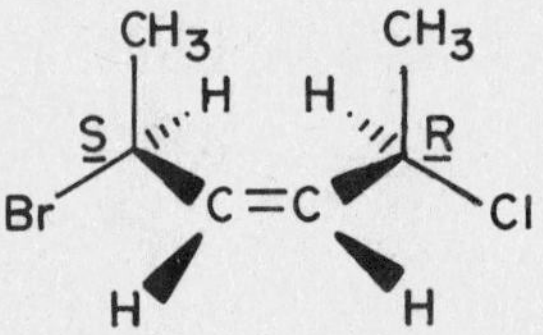

$$\underline{Z}$$

and so on.

d. 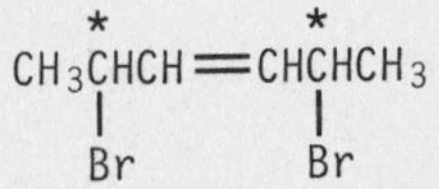 $CH_3\overset{*}{C}HCH=CH\overset{*}{C}HCH_3$
        | |
        Br   Br

Compare with part c. In this case both chiral centers are identical. Therefore, two *meso* forms are possible, and the total number of isomers is reduced to six:

R,Z,R
R,Z,S *(meso)*
R,E,R
R,E,S *(meso)*
S,Z,S
S,E,S

There are two sets of enantiomers:

R,Z,R *and* S,Z,S
R,E,R *and* S,E,S

And there are two optically inactive, *meso* forms: *R,Z,S* and *R,E,S*

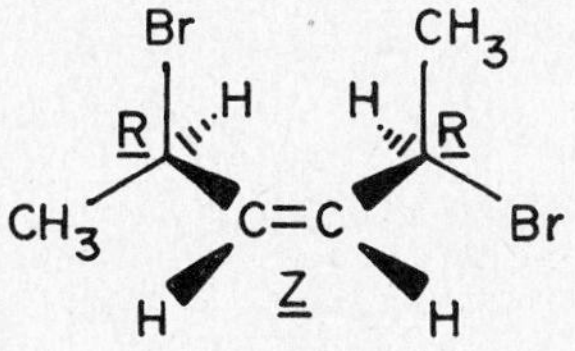

R,Z,R

The other five structures can be derived from this one by inter-changing groups. For example, the *R,Z,S meso* form is:

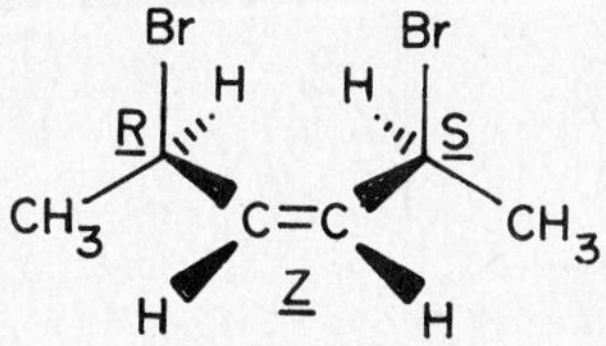

The plane of symmetry that causes this form to be achiral and optically inactive is the perpendicular bisector of the $C{=}C$ bond.

5.37   For three different asymmetric carbons, $2^3 = 8$. The possibilities are as follows:

$R - R - R$ $S - R - R$
$R - R - S$ $S - R - S$
$R - S - R$ $S - S - R$
$R - S - S$ $S - S - S$

For four different asymmetric carbons, $2^4 = 16$. The possibilities are as follows:

$R - R - R - R$ $R - S - R - R$ $S - R - R - R$ $S - S - R - R$
$R - R - R - S$ $R - S - R - S$ $S - R - R - S$ $S - S - R - S$
$R - R - S - R$ $R - S - S - R$ $S - R - S - R$ $S - S - S - R$
$R - R - S - S$ $R - S - S - S$ $S - R - S - S$ $S - S - S - S$

5.38 Suppose we consider the sawhorse formula for one of the two enantiomers, say, (R)-2-chlorobutane:

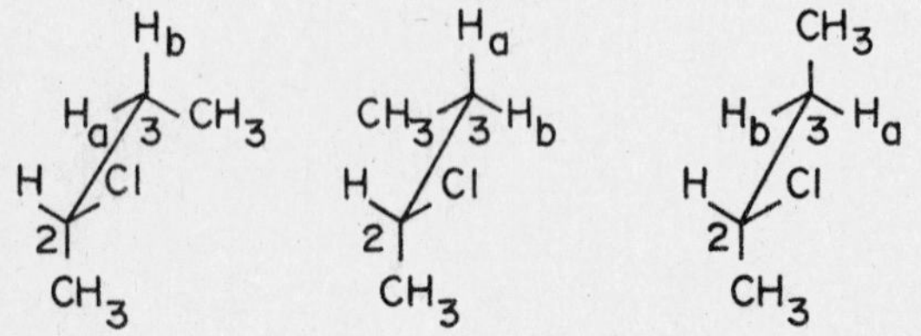

No matter which conformer we choose, the two hydrogens H_a and H_b (one of which is replaced by chlorine when the compound is chlorinated to give 2,3-dichlorobutane) are in different environments. For example, in the left conformer, H_a is flanked by H and CH_3 on C-2, whereas H_b is flanked by H and Cl. If we rotate to get the third conformer, H_b now is in the position occupied formerly by H_a (flanked by H and CH_3), but H_a *does not have the position occupied formerly by* H_b (it is flanked by CH_3 and Cl, not H and Cl). The positions of H_a and H_b cannot be interchanged. These hydrogens are said to be *diastereotopic*, since if one or the other were replaced by some group that would make C-3 asymmetric, the products would be diastereomers (*not* enantiomers).

Since H_a and H_b are diastereotopic (and not enantiotopic), there is no reason why they should react identically. Thus the ratio of *meso* to racemic product need not be 1:1. The argument applies regardless of which isomer of 2-chlorobutane we consider, and is therefore true of the racemic mixture as well.

5.39 Solubility in an achiral solvent such as water is an achiral property; it has no direction or handedness. Thus (-)-tartaric acid should have identical solubility to its enantiomer, (+)-tartaric acid. The solubility is 139 g/100 g of water. The (-) or (+) isomers are diastereomers of the *meso* form; therefore, they differ in solubility from the *meso* form.

5.40

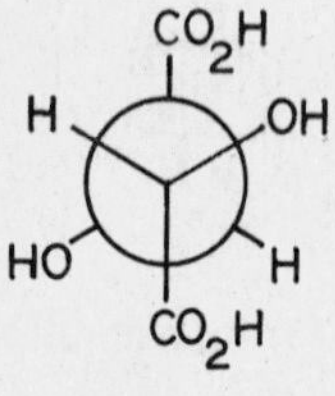

This is the *meso* form. As drawn, this conformation has a center of symmetry, the midpoint of the central C—C bond. The plane of symmetry is readily seen if we rotate the "rear" carbon 180°:

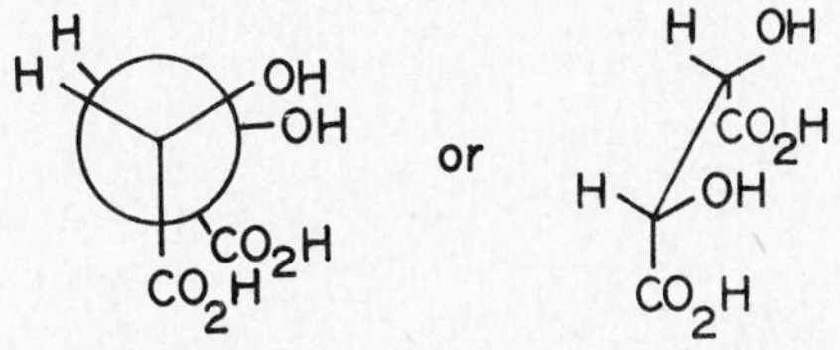

or

The remaining two structures correspond to the *S,S* and *R,R* isomers, respectively:

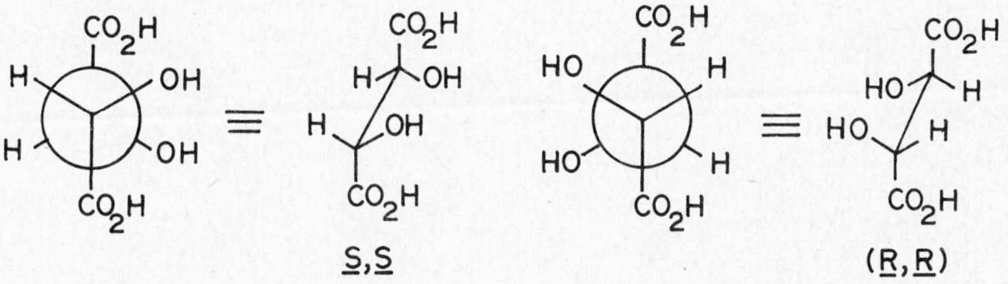

S,S   (R,R)

The priority order at each chiral center is $OH > CO_2H > CH(OH)CO_2H > H$.

5.41

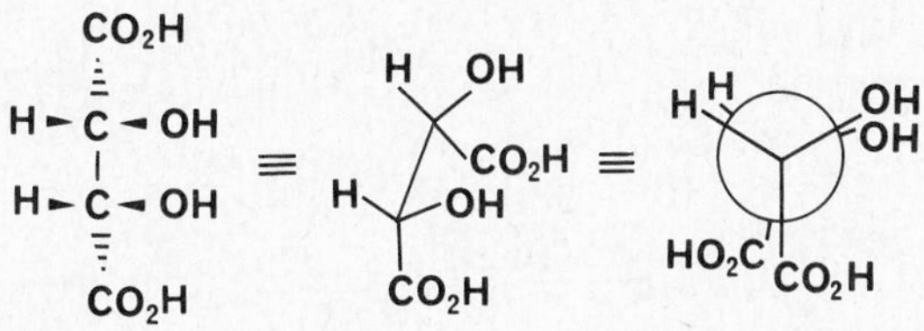

The chiral centers are eclipsed.

5.42 The conformation shown for *meso*-tartaric acid does *not* have a plane of symmetry. It does, however, have a center of symmetry (the mid-point of the C—C bond that joins the two chiral centers). This conformation of the *meso* form is achiral, as can be shown by the fact that the molecule and its mirror image are superimposable:

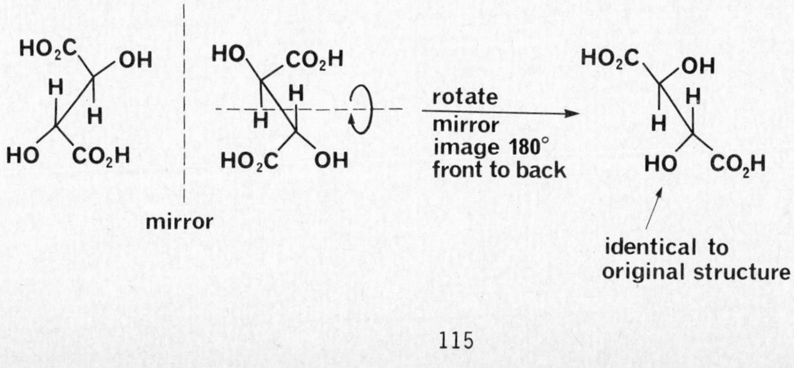

mirror

rotate
mirror
image 180°
front to back

identical to
original structure

5.43 The structures are conformational isomers; both are achiral; they are diastereomers (stereoisomers but not mirror images).

5.44 The structures are conformational isomers; both are chiral; they are enantiomers (they have nonsuperimposable mirror images). Although they are enantiomers, they interconvert readily through rotation about the C—C bond; they cannot, therefore, be separated from one another.

5.45 The compound has three different chiral centers, indicated by arrows in the formula below. There are therefore 2^3 or eight possible stereoisomers.

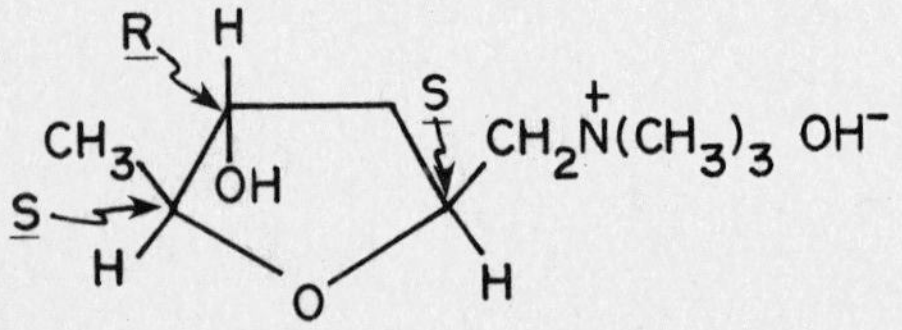

The configurational designation (R or S) at each center in the naturally occurring poison is shown.

5.46

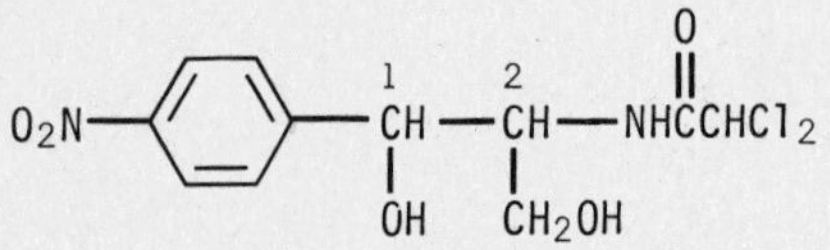

The priority order at C-1 is

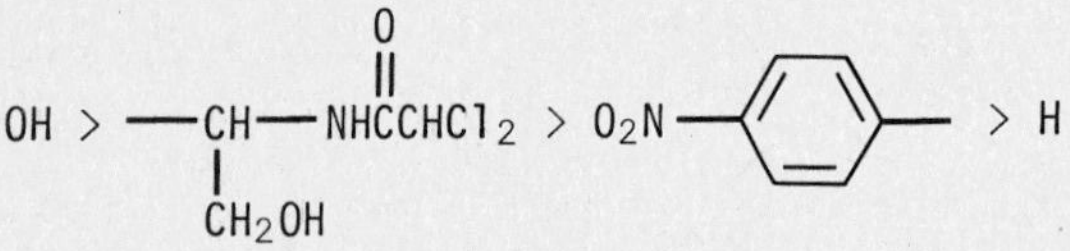

At C-2, the priority order is

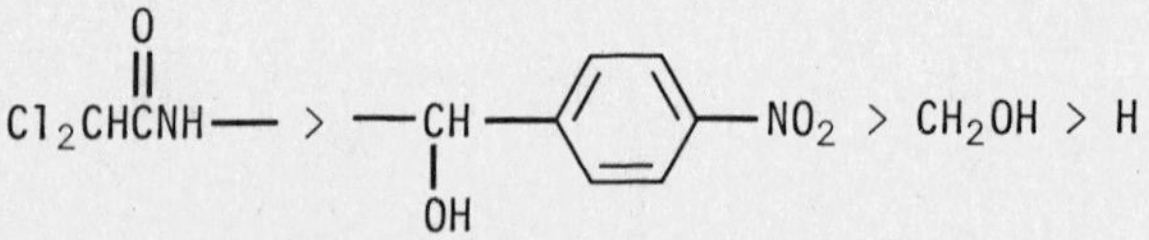

Using a sawhorse structure, we can write

$$CHCl_2CNH \overset{\overset{O}{\parallel}}{} \overset{\overset{H}{|}}{\underset{R}{C}} CH_2OH$$

$$HO \underset{H}{\overset{R}{C}} \text{—} NO_2$$

Using dashes and wedges, we have

$$HO \overset{R}{} \quad \overset{H}{\underset{R}{C}} CH_2OH$$
$$H\cdots\cdot \quad NHCCHCl_2$$
$$NO_2 \qquad \overset{\parallel}{O}$$

To obtain the diastereomer with the S configuration at C-2, simply interchange the positions of any two groups at that carbon:

$$CHCl_2CNH \overset{\overset{O}{\parallel}}{} \overset{\overset{CH_2OH}{|}}{\underset{S}{C}} H$$
$$HO \underset{H}{\overset{R}{C}} \text{—} NO_2$$

or

$$HO \quad HOH_2C \quad H$$
$$H\cdots\cdot \overset{R}{} \quad \underset{S}{} NHCCHCl_2$$
$$NO_2 \qquad \overset{\parallel}{O}$$

The enantiomer of the $(1R,2R)$ form is the $(1S,2S)$ form:

$$HOCH_2 \quad \overset{H}{\underset{S}{C}} H \qquad \overset{OH}{C} \cdots H$$
$$CHCl_2CNH \qquad \underset{S}{}$$
$$\overset{\parallel}{O}$$
$$NO_2$$

The enantiomer of the (1R,2S) form is the (1S,2R) form:

5.47   There are several theories about this question. It seems most likely that the origin of optical activity or chiral molecules is intimately connected with the origin of life itself. The stereospecificity that chirality allows seems essential for the control of living systems, but it is uncertain which came first, chirality or life.

    There are also mechanical theories. For example, there are many examples of chiral cavities in crystals of molecules which themselves are achiral. These chiral cavities may have acted to resolve a racemic mixture, by virtue of the fact that only one of the two enantiomers would fit the cavity. Mechanical theories of this type, however, seem less likely than biological theories.

    It is also possible that chirality evolved and that optical activity reached a maximum value for a particular substance only gradually.

CHAPTER SIX   ORGANIC HALOGEN COMPOUNDS;
SUBSTITUTION AND ELIMINATION REACTIONS

CHAPTER SUMMARY

Alkyl halides react with **nucleophiles**, reagents that can supply an electron pair to form a covalent bond, to give a product in which the nucleophile takes the place of the halogen. Table 6.1 gives sixteen examples of such **nucleophilic substitution reactions**, which can be used to convert alkyl halides to alcohols, ethers, esters, thiols, alkyl cyanides, or acetylenes.

Nucleophilic substitution may occur by two mechanisms. The S_N2 **mechanism** is a one-step process. Its rate depends on the concentrations of substrate and nucleophile. If the halogen-bearing carbon is chiral, substitution occurs with inversion of configuration. The reaction is fastest for primary halides and slowest for tertiary halides.

The S_N1 **mechanism** is a two-step process. In the first step, the alkyl halide ionizes to a carbocation and a halide ion. In the second, fast step, the carbocation combines with the nucleophile. The overall rate is independent of nucleophile concentration. If the halogen-bearing carbon is chiral, substitution occurs with racemization. The reaction is fastest for tertiary halides and slowest for primary halides. The two mechanisms are compared in Table 6.2.

Elimination reactions often compete with substitution. They involve elimination of the halogen and a hydrogen from an adjacent carbon to form an alkene. Like substitution, they occur by two main mechanisms. The **E2 mechanism** is a one-step process. The nucleophile acts as a base to remove the adjacent proton. The preferred form of the transition state is planar, with the hydrogen and the leaving group in an *anti* conformation.

The **E1 mechanism** has the same first step as the S_N1 mechanism. The resulting carbocation then loses a proton from a carbon atom adjacent to the positive carbon to form the alkene.

Substitution and elimination reactions occur in biological systems.

Several polyhalogen compounds have useful properties. Among them, carbon tetrachloride, chloroform, and methylene chloride are useful solvents. Other important polyhalogen compounds include $CBrClF_2$ and $CBrF_3$, used as fire

extinguishers, $CF_3CHClBr$ (called **halothane**), used as an anesthetic, tri-
and tetrachloroethenes, used as dry-cleaning solvents, and **chlorofluoro-
carbons** or **Freons** (CCl_3F, CCl_2F_2, $CClF_3$, $CHCl_2F$, $CHClF_2$), used as
refrigerants, aerosol propellants, and solvents. **Teflon** is a polymer
of tetrafluoroethene. It is used in nonstick coatings, "Gore-Tex" fabrics,
insulators, and many other things. Certain **perfluorochemicals** dissolve
high percentages of oxygen and can be used as artificial blood. Many
halogen-containing compounds are important pesticides.

REACTION SUMMARY

Nucleophilic Substitution

$$Nu: + R\!-\!X \longrightarrow R\!-\!\overset{+}{Nu} + X^-$$

or

$$Nu:^- + R\!-\!X \longrightarrow R\!-\!Nu + X^-$$

(See Table 6.1 for examples.)

Elimination

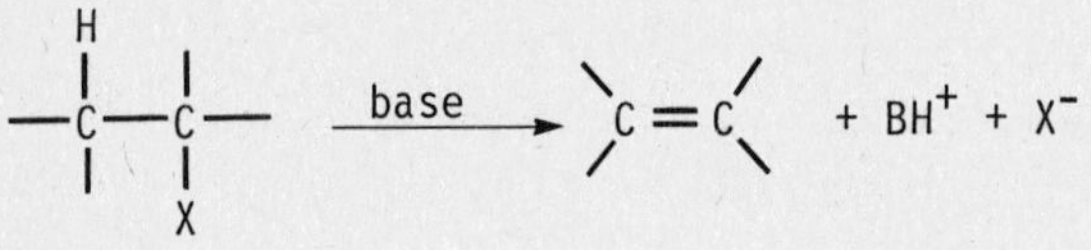

Preparation of Freons

$$CCl_4 \xrightarrow[SbF_5]{HF} CCl_3F, \; CCl_2F_2, \text{ etc.}$$

Preparation of Teflon

$$2CHClF_2 \xrightarrow{600\text{-}800°C} CF_2\!=\!CF_2 + 2HCl$$

$$CF_2\!=\!CF_2 \xrightarrow{peroxide} -\!\!(CF_2CF_2)\!\!-_n$$

MECHANISM SUMMARY

S_N2 (Bimolecular Nucleophilic Substitution)

$$Nu: + \quad \text{C—L} \longrightarrow \left[\overset{\delta+}{Nu} \cdots \text{C} \cdots \overset{\delta-}{L} \right] \longrightarrow \overset{+}{Nu}\text{—C} \quad + :L^-$$

S_N1 (Unimolecular Nucleophilic Substitution)

$$\text{C—L} \xrightarrow{\text{slow}} \overset{+}{\text{C}} + L^-$$

carbocation

$$\overset{+}{\text{C}} + Nu: \xrightarrow{\text{fast}} \text{C}\overset{+}{\diagdown}_{Nu} \quad or \quad \overset{+}{Nu}\diagup\text{C}$$

E2 (Bimolecular Elimination)

$$B: \quad H \quad \text{C—C} \quad X \longrightarrow BH^+ + \quad C=C \quad + X^-$$

E1 (Unimolecular Elimination)

$$-\overset{H}{\underset{|}{\text{C}}}-\text{C—L} \rightleftharpoons -\overset{H}{\underset{|}{\text{C}}}-\overset{+}{\text{C}} + L^-$$

$$\longrightarrow C=C + H^+$$

LEARNING OBJECTIVES

1. Know the meaning of: nucleophilic substitution reaction, nucleophile, substrate, leaving group.

2. Be familiar with the examples of nucleophilic substitution reactions listed in Table 6.1.

3. Know the meaning of: S_N2 mechanism, inversion of configuration, S_N1 mechanism, racemization, rate-determining step, E2 and E1 mechanism.

4. Know the formulas of carbon tetrachloride, chloroform, methylene chloride, Freons, Teflon, 2,4-D, dioxin.

5. Given the name of an alkyl halide or a polyhalogen compound, write its structural formula.

6. Given the structural formula of an alkyl halide, write a correct name for it.

7. Write the equation for the reaction of an alkyl halide with any of the nucleophiles listed in Table 6.1. Recognize the class of organic compound to which the product belongs.

8. Given the structure of an alkyl halide, predict whether it is most likely to react with nucleophiles by an S_N1 or an S_N2 mechanism.

9. Given the structure of an alkyl halide and a nucleophile, write the equations that illustrate the formation of both the substitution and elimination products, and be able to predict which path is likely to be favored.

10. Know the stereochemical outcome of S_N1 and S_N2 substitutions and of E1 and E2 eliminations.

11. Given an alkyl halide with a particular stereochemistry, a nucleophile, and reaction conditions, predict the stereochemistry of the product of nucleophilic substitution.

12. Combine nucleophilic substitutions with previously studied reactions to devise a multistep synthesis of a given product.

ANSWERS TO PROBLEMS

Problems Within the Chapter

6.1   a. $NaOH + CH_3CH_2CH_2Br \longrightarrow CH_3CH_2CH_2OH + Na^+Br^-$

(item 1, Table 6.1)

b. $(CH_3CH_2)_2N + CH_3CH_2Br \longrightarrow (CH_3CH_2)_4N^+ + Br^-$

(item 10, Table 6.1)

c. $KCN + \langle\!\!\!\bigcirc\!\!\!\rangle\!-CH_2Br \rightarrow \langle\!\!\!\bigcirc\!\!\!\rangle\!-CH_2CN + K^+Br^-$

(item 15, Table 6.1)

6.2   a. SH^- + $CH_3CH_2CH_2CH_2Br \longrightarrow CH_3CH_2CH_2CH_2SH + Br^-$

           nucleophile   substrate   leaving
                                                           group

      (item 11, Table 6.1)

  b. OH^- + $(CH_3)_2CHCH_2Br \longrightarrow (CH_3)_2CHCH_2OH + Br^-$

           nucleophile   substrate   leaving
                                                             group

      (item 1, Table 6.1)

  c. $CH_3CH_2CH_2\overset{..}{N}H_2$ + $CH_3CH_2CH_2Br \longrightarrow (CH_3CH_2CH_2)_2\overset{+}{N}H_2 + Br^-$

           nucleophile   substrate   leaving
                                                             group

      (item 8, Table 6.1)

      This reaction is followed by the acid-base equilibrium

$$CH_3CH_2CH_2NH_2 + (CH_3CH_2CH_2)_2\overset{+}{N}H_2 \rightleftharpoons CH_3CH_2CH_2\overset{+}{N}H_3 + (CH_3CH_2CH_2)_2NH$$

  d. $(CH_3CH_2)_2\overset{..}{S}:$ + $CH_3CH_2Br \longrightarrow (CH_3CH_2)_3\overset{+}{S} + Br^-$

           nucleophile   substrate   leaving
                                                           group

      (item 13, Table 6.1)

6.3   a. S_N2. The substrate is a secondary halide and may react by either S_N2 or S_N1; the nucleophile SH^- is a strong nucleophile, favoring S_N2.

  b. S_N1. The substrate is secondary and may react by either mechanism. The nucleophile (CH_3OH) is relatively weak and also polar, favoring the ionization mechanism.

  c. S_N2. The substrate is the same as in part b, but the nucleophile in this case is much stronger, favoring the displacement mechanism.

6.4   Each of these reactions involves a strong nucleophile, and usually proceeds by an S_N2 mechanism. Thus, the reactions work well only with primary and secondary halides. In eqs. 6.4 and 6.6, the substrate is a tertiary halide. Attack at the "rear" of the C—Br bond is sterically hindered, which slows down the rate of an S_N2 process drastically. This allows the competing elimination process to take over.

6.5 $CH_3-\overset{\overset{\displaystyle CH_3}{|}}{\underset{\underset{\displaystyle CH_3}{|}}{C}}-O^-K^+$ is a much bulkier alkoxide than $CH_3O^-Na^+$.

Thus it does not readily participate in S_N2 displacements. Instead, it acts as a base (instead of a nucleophile) in an E2 reaction:

$$CH_3CH_2-\overset{\overset{\displaystyle H}{|}}{\underset{\underset{\displaystyle H}{|}}{C}}-CH_2-Br \longrightarrow CH_3CH_2CH=CH_2 + (CH_3)_3COH + Br^-$$

$$CH_3-\overset{\overset{\displaystyle CH_3}{|}}{\underset{\underset{\displaystyle CH_3}{|}}{C}}-O^-$$

Attack at hydrogen is less hindered than attack at the "rear" of the C—Br bond.

6.6   For the *R* enantiomer, the equations are

and

Since the transition states leading to the two isomers of 2-butene are mirror images, their energies should be the same. The *trans*-to-*cis* ratios for the two enantiomers, *R* and *S*, will there-fore be identical.

6.7   Three products are possible:

Additional Problems

6.8   Each of these reactions involves displacement of a halogen by a nucleophile; review Sec. 6.3 and Table 6.1.

a. $CH_3CH_2CH_2CH_2Br$ + NaI $\xrightarrow{\text{acetone}}$ $CH_3CH_2CH_2CH_2I$ + $NaBr$

b. $CH_3CHCH_2CH_3$ + $Na^{+-}OC_2H_5$ $\longrightarrow$ $CH_3CHCH_2CH_3$ + Na^+Cl^-
     | |
   Cl OC_2H_5

c. $(CH_3)_3CBr$ + H_2O $\longrightarrow$ $(CH_3)_3COH$ + H_2O

The mechanism here is S_N1 (most of the other reactions in this problem occur by an S_N2 mechanism).

d. $Cl-\langle\bigcirc\rangle-CH_2Cl$ + $NaCN$ $\longrightarrow$ $Cl-\langle\bigcirc\rangle-CH_2CN$ + Na^+Cl^-

Substitution occurs only at the aliphatic (benzyl) carbon, and not on the aromatic ring.

e. $CH_3CH_2CH_2I$ + $Na^{+-}C\equiv CH$ $\longrightarrow$ $CH_3CH_2CH_2C\equiv CH$ + Na^+I^-

The use of acetylides as nucleophiles is a particularly important example of nucleophilic substitution because it results in a new carbon-carbon bond; thus, larger organic molecules can be assembled from smaller ones using this method. The same is true for cyanide ion as a nucleophile (part d).

f. CH_3CHCH_3 + $NaSH$ $\longrightarrow$ CH_3CHCH_3 + Na^+Cl^-
     | |
   Cl   SH

g. $CH_2=CHCH_2Cl$ + $2\ NH_3$ $\longrightarrow$ $CH_2=CHCH_2NH_2$ + $NH_4^+Cl^-$

h. $CH_2CH_2CH_2CH_2$ + $2\ NaCN$ $\longrightarrow$ $CH_2CH_2CH_2CH_2$ + $2\ Na^+Br^-$
   | | | |
  Br   Br   CN   CN

Displacement occurs at both possible positions.

i.   [cyclohexane ring with CH₃ and Br substituents] + CH_3OH $\longrightarrow$ [cyclohexane ring with CH₃ and OCH₃ substituents] + HBr

The starting halide is tertiary, and the mechanism is S_N1.

6.9   Use the equations in Table 6.1 as a guide.

   a. $CH_3CH_2CH_2Br$ + NH_3 (item 7)

   b. CH_3CH_2I + $CH_3CH_2S^-Na^+$ (item 12)

   c. $CH_3CH_2CH_2Br$ + $HC{\equiv}C^-Na^+$ (item 16)

   d. $(CH_3)_2CHBr$ + $(CH_3)_2CHO^-Na^+$ (item 2)

   e. $BrCH_2$—⟨benzene ring⟩—CH_2Br + Na^+CN^- (item 15)

   f. CH_3CH_2Br + ⟨benzene ring⟩—O^-Na^+ (item 2)

6.10   The configuration inverts if the reaction occurs by an S_N2 mechanism, but if the S_N1 mechanism prevails, considerable racemization occurs.

   a.   The nucleophile is methoxide ion, CH_3O^-; the alkyl halide is secondary, and the mechanism is S_N2.

$\underline{S}-2-\text{bromobutane}$ $\underline{R}-2-\text{methoxybutane}$ (inversion)

   b.   The alkyl halide is tertiary and the nucleophile is methanol (a weaker nucleophile than methoxide ion). The mechanism is S_N1 and the product is a mixture of R and S isomers.

$\underline{R}-3-\text{bromo}-3-\text{methylhexene}$

R   +   S

c. The alkyl halide is secondary, and the HS^- ion is a strong nucleophile; the mechanism is S_N2.

6.11 a. Sodium cyanide is a strong, anionic nucleophile. Thus the mechanism is S_N2 and the reactivity order of halides is primary > secondary > tertiary. Therefore,

$$(CH_3)_2CHCH_2Br > CH_3CHCH_2CH_3 >> (CH_3)_3CBr$$
$$\underset{Br}{|}$$

b. With 50% aqueous acetone, there is a weak nucleophile (H_2O) and a highly polar reaction medium favoring ionization, or the S_N1 mechanism. In this mechanism, the reactivity order of alkyl halides is tertiary > secondary > primary. Therefore,

$$(CH_3)_3CBr > CH_3CHCH_2CH_3 >> (CH_3)_2CHCH_2Br$$
$$\underset{Br}{|}$$

6.12 An S_N2 displacement can occur. Since the leaving group and the nucleophile are identical (iodide ion), there is no change in the gross structure of the product. However, the configuration inverts every time a displacement occurs.

Since the enantiomer is produced, the optical rotation of the solution decreases. Eventually, as the concentration of the S enantiomer builds up, it too reacts with iodide ion to form some R isomer. Eventually an equilibrium (50:50) or racemic mixture is formed, and the solution is optically inactive.

127

6.13   The first step in the hydrolysis of any one of these halides is the
        ionization to a t-butyl cation:

$$(CH_3)_3C-X \xrightarrow{H_2O} (CH_3)_3C^+ + X^-$$

(X = Cl, Br, or I)

The product-determining step involves the partition of this intermediate between two paths--reaction with water and loss of a proton:

$$(CH_3)_3COH \xleftarrow{H_2O} (CH_3)_3C^+ \xrightarrow{-H^+} CH_2=C(CH_3)_2$$

Since the halide ion is, to a first approximation, not involved in
these steps, this partition occurs in the same ratio regardless of
which alkyl halide is being hydrolyzed. This result provides experimental support for the S_N1 mechanism.

6.14   a.   The halide is tertiary, and the nucleophile is a relatively weak
             base. Hence the predominant mechanism is S_N1:

Some E1 reaction may occur in competition with S_N1, giving mainly
the product with the double bond in the ring:

major   minor

However, the main product will be the ether (S_N1).

        b.   The nucleophile in this case is stronger, but the S_N2 process is
             not possible because the alkyl halide is tertiary. This nucleo-
             phile is also a strong base. Therefore, an E2 reaction will be
             preferred.

or

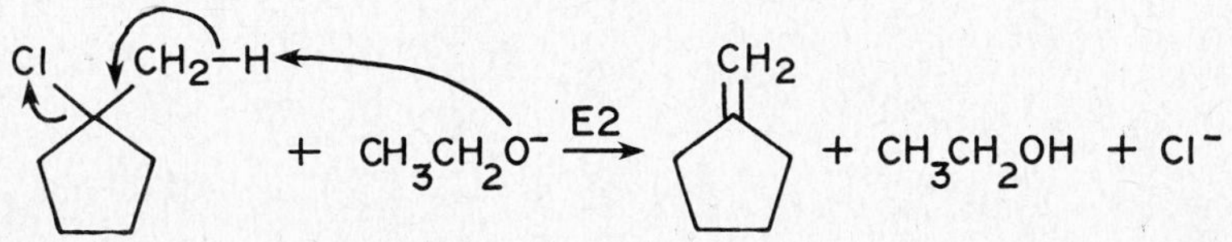

The predominant product is 1-methylcyclopentene, the more stable of the two possible alkenes.

6.15

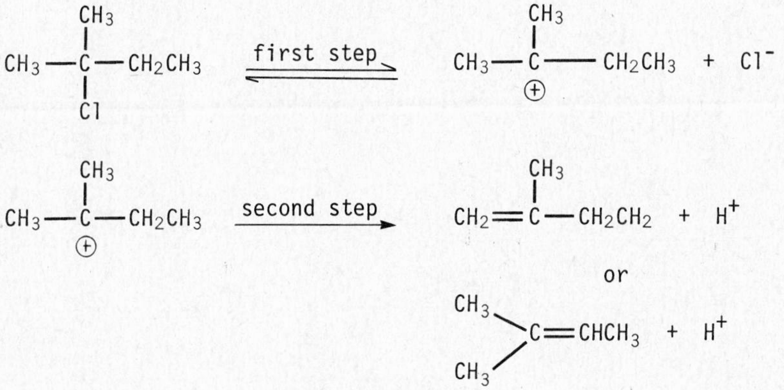

In the second step, the proton may be lost from either of the methyl carbons *or* from the methylene carbon, giving the two alkenes shown.

6.16 The first reaction involves a strong nucleophile (CH_3O^-), and the S_N2 mechanism is favored. Therefore, only one product is obtained. The second reaction involves a weak nucleophile (CH_3OH) that is also a fairly polar solvent, favoring the S_N1 mechanism:

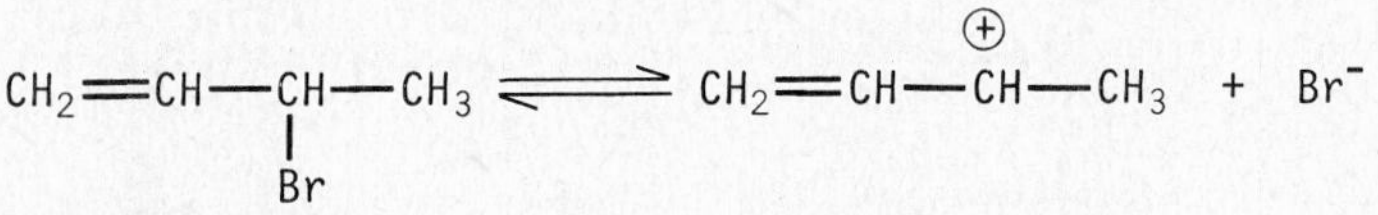

The carbocation is a resonance hybrid:

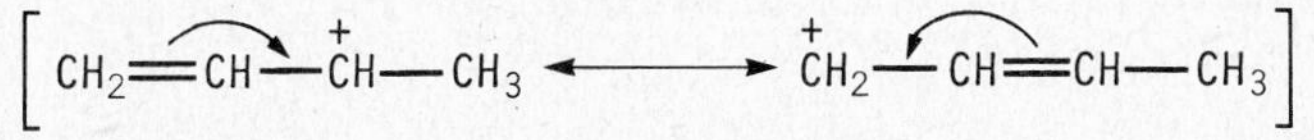

It can react with methanol at either positively charged carbon, giving the two observed products.

6.17 Cyclohexyl chlorides have the transoid coplanar geometry of the E2 transition state only when the chlorine is in an axial position:

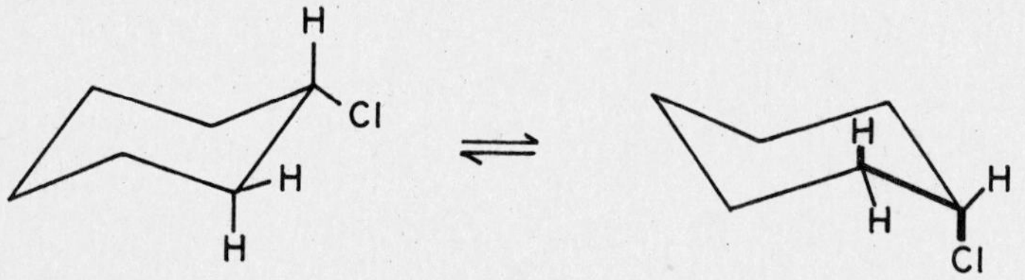

Consider menthyl chloride:

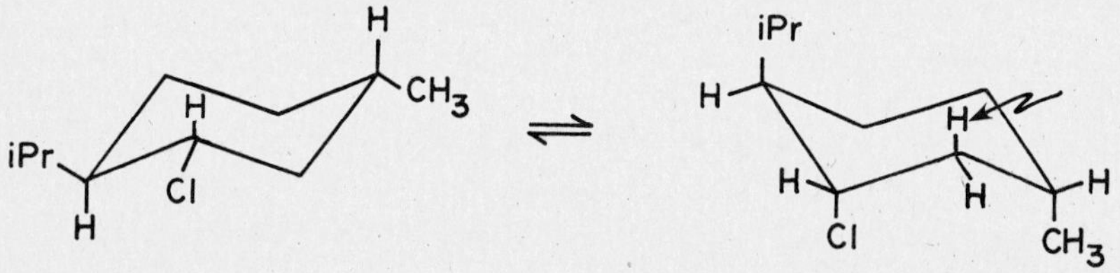

In the conformation on the right, with the chlorine axial, the only hydrogen on an adjacent carbon suitably located for E2 elimination is marked with an arrow. Therefore, the product is 2-menthene.

Now consider neomenthyl chloride:

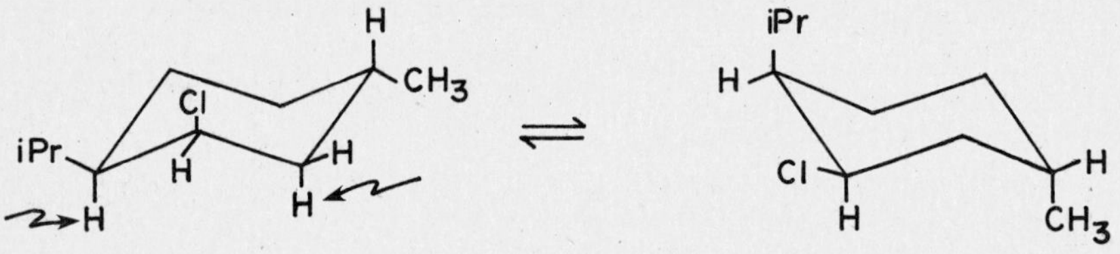

In this case, the conformation with the chlorine axial is on the left. Two hydrogens (marked with arrows) have the suitable geometry for E2 elimination, and both are eliminated to give 75% 3-menthene and 25% 2-menthene.

6.18 The polymerization occurs by a free-radical chain mechanism (review Sec. 3.18).

$$RO{-}OR \xrightarrow{\text{heat}} 2\ RO\cdot$$

$$RO\cdot\ +\ CF_2{=}CF_2 \longrightarrow ROCF_2CF_2\cdot$$

$$ROCF_2CF_2\cdot\ +\ CF_2{=}CF_2 \longrightarrow ROCF_2CF_2CF_2CF_2\cdot\ \text{and so on.}$$

The result is $-(CF_2CF_2-)_n$ terminated by $-OR$ or some other group.

6.19   a. $CH_3CH=CHCH_3$ $\xrightarrow{HBr}$ $CH_3CHCH_2CH_3$
                                                     $|$
                                                     Br

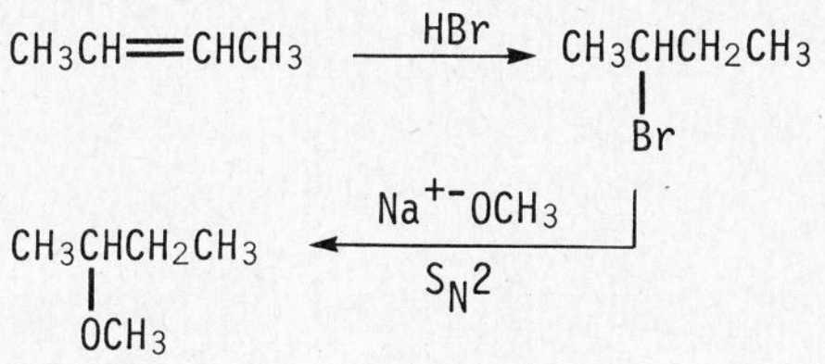

                           $Na^{+-}OCH_3$
$CH_3CHCH_2CH_3$ $\xleftarrow{\hspace{1.5cm}}$
                             S_N2
  $|$
  OCH_3

                    CH_3
                     $|$
b. $CH_3-C=CHCH_3$ $\xrightarrow{HBr}$

CH_3
$|$
$CH_3-C-CH_2CH_3$
$|$
Br

       CH_3
       $|$
$CH_3-C-CH_2CH_3$ $\xleftarrow[\;S_N1\;]{CH_3OH}$
       $|$
       OCH_3

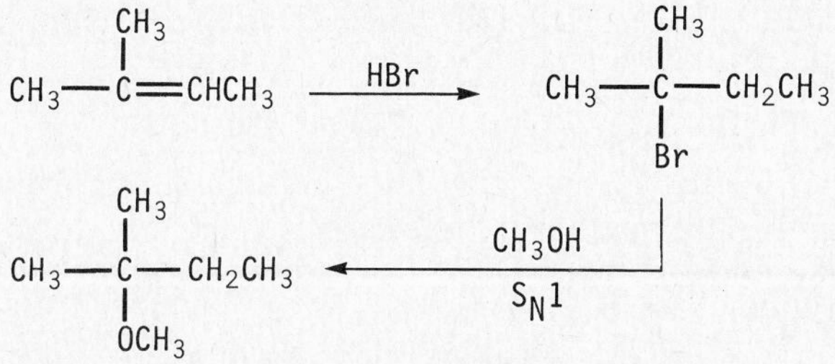

Here we use a weak electrophile, CH_3OH, because the substrate is a tertiary bromide and we want to favor the S_N1 mechanism. If we had used $Na^{+-}OCH_3$ instead, considerable elimination (E2) would have occurred. In part a, however, the alkyl halide was secondary, so the stronger nucleophile was required.

c.   [benzene ring]$-CH=CH_2$ $\xrightarrow{HBr}$ [benzene ring]$-CHCH_3$
                                               $|$
                                             Br

                       Na^+CN^-
[benzene ring]$-CHCH_3$ $\xleftarrow[\;S_N2\;]{\hspace{1.5cm}}$
             $|$
             CN

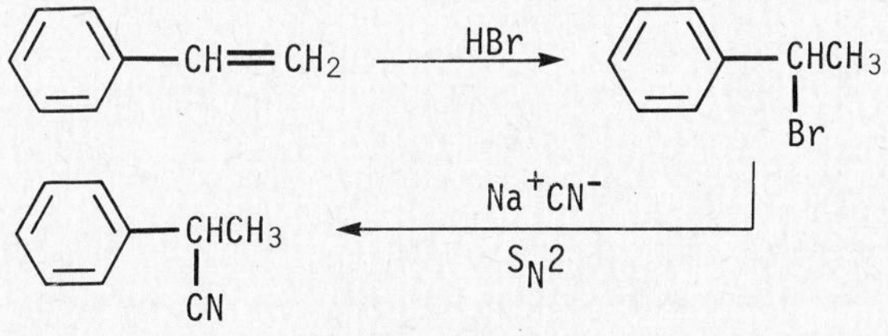

In the first step, addition occurs according to Markovnikov's rule; the second step occurs at the benzylic position by the S_N2 mechanism.

6.20   Work backwards from the desired product to a reasonable starting material:

a.   [benzene ring]$-CHCH_3$ $\xleftarrow{NH_3}$ [benzene ring]$-CHCH_3$
                    $|$ $|$
                    NH_2 Br

                                         HBr
[benzene ring]$-CH=CH_2$ $\xrightarrow{\hspace{1.5cm}}$

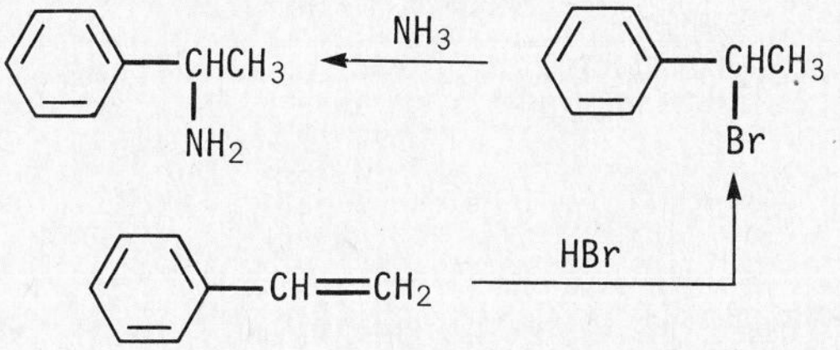

b.

$$CH_3CH_2CHCH_2CH_3 \xleftarrow{\quad Na^+SH^- \quad} CH_3CH_2CHCH_2CH_3$$

with SH (below first) and Br (below second).

$$CH_3CH=CHCH_2CH_3 \xrightarrow{\quad HBr \quad}$$

This synthesis will work, but it will *not* give a good yield, because the first step can give not only 3-bromopentane but its regioisomer, 2-bromobutane (both products are obtained via secondary carbocation intermediates, which have nearly equal stabilities).

6.21 a. $2\ CH_3OH\ +\ 2\ Na \longrightarrow 2\ CH_3O^-Na^+\ +\ H_2$

$$CH_3O^-Na^+\ +\ CH_3CH_2Br \xrightarrow{\quad S_N2 \quad} CH_3OCH_2CH_3\ +\ Na^+Br^-$$

This two-step synthesis of ethers is called the Williamson synthesis (see Sec. 8.6).

b. $(CH_3)_3CBr\ +\ CH_3OH \xrightarrow{\quad S_N1 \quad} (CH_3)_3COCH_3\ +\ HBr$

We select this combination of reagents, and *not* $(CH_3)_3COH\ +\ CH_3Br$, because methyl bromide, being primary, will not react by an S_N1 mechanism, and $(CH_3)_3COH$ is too weak a nucleophile to displace Br^- from CH_3Br in an S_N2 process. [$(CH_3)_3CO^-K^+$ would provide a strong enough nucleophile, and the reaction

$$(CH_3)_3CO^-K^+\ +\ CH_3Br \longrightarrow (CH_3)_3COCH_3\ +\ K^+Br^-$$

would provide an alternative synthesis of the desired product, but it uses an alkoxide, rather than an alcohol as the problem specifies.]

6.22 a. $CH_2=CH-CH=CH_2\ +\ HBr \xrightarrow{\quad 1,4 \quad} CH_3-CH=CH-CH_2Br$

$$CH_3CH=CH-CH_2CN \xleftarrow[S_N2]{\quad Na^+CN^- \quad}$$

b. $CH_2=CH-CH=CH_2\ +\ Br_2 \xrightarrow{\quad 1,4 \quad} CH_2-CH=CH-CH_2$

with Br below each terminal CH_2.

$$CH_2-CH=CH-CH_2 \xleftarrow[S_N2]{\quad 2\ Na^+CN^- \quad}$$

with CN below each terminal CH_2.

6.23 This type of reaction sequence is a useful method for constructing C—C bonds.

a. $CH_3C{\equiv}CH + Na^+NH_2^- \xrightarrow{NH_3} CH_3C{\equiv}C^-Na^+$

$CH_3C{\equiv}C-CH_2-\bigcirc \xleftarrow[S_N2]{} \bigcirc-CH_2Br$

b. $HC{\equiv}CH \xrightarrow[NH_3]{NaNH_2} HC{\equiv}C^-Na^+ \xrightarrow{CH_3Br} HC{\equiv}C-CH_3$

$CH_3CH_2C{\equiv}CCH_3 \xleftarrow{CH_3CH_2Br} Na^{+-}C{\equiv}C-CH_3 \xleftarrow[NH_3]{NaNH_2}$

The order in which the alkyl halides were used could be reversed, with the same overall result.

6.24 a. $CH_2{=}\underset{\underset{CH_3}{|}}{C}-CH_2CH_3 \xrightarrow{HBr} CH_3-\underset{\underset{CH_3}{|}}{\overset{\overset{Br}{|}}{C}}-CH_2CH_3$

$CH_3-\underset{\underset{CH_3}{|}}{C}{=}CH-CH_3 \xleftarrow{K^{+-}OC(CH_3)_3}$

The first step follows Markovnikov's rule. The second step, an E2 elimination, gives mainly the desired trisubstituted alkene, although some of the starting alkene may also be formed.

b.

In the first step, proton addition occurs mainly at the carbon adjacent to the methyl group because the resulting allylic cation is tertiary (at one end):

Elimination of HBr from this intermediate can yield the desired product.

6.25 a. $CH_2\!=\!CHCH_2Br \xrightarrow{Na^+OH^-} CH_2\!=\!CHCH_2OH \xrightarrow[Pt]{H_2} CH_3CH_2CH_2OH$

b. $HC\!\equiv\!CCH_3 \xrightarrow[NH_3]{NaNH_2} Na^+\ {}^-C\!\equiv\!CCH_3 \xrightarrow{CH_3I}$

134

CHAPTER SEVEN   ALCOHOLS, PHENOLS, AND THIOLS

CHAPTER SUMMARY

The functional group of alcohols and phenols is the **hydroxyl group**. In alcohols, this group is connected to an aliphatic carbon, whereas in phenols, it is attached to an aromatic ring.

In the IUPAC system of nomenclature, the suffix for alcohols is *-ol*. Alcohols are classified as **primary, secondary,** or **tertiary** depending on whether one, two, or three organic groups are attached to the hydroxyl-bearing carbon. The nomenclature of alcohols and phenols is summarized in Secs. 7.2-7.4.

Alcohols and phenols form **hydrogen bonds**. These bonds account for the relatively high boiling points of these substances and the water solubility of lower members of the series.

Brønsted-Lowry and Lewis definitions of acids and bases are reviewed in Sec. 7.6. Alcohols are comparable in acidity to water; phenols are 10^6 times more acidic. This increased acidity is due to charge delocalization (resonance) in phenoxide ions. Electron-withdrawing groups, such as ——F and ——NO_2, increase acidity, through either an **inductive** or a resonance effect, or both.

Alkoxides, the conjugate bases of alcohols, are prepared from alcohols by reaction with reactive metals or metal hydrides. They are used as organic bases. Because of the greater acidity of phenols, phenoxides can be obtained from phenols and aqueous base.

Alcohols and phenols are weak bases. They can be protonated on the oxygen by strong acids. This reaction is the first step in the acid-catalyzed dehydration of alcohols to alkenes and in the conversion of alcohols to alkyl halides by reaction with hydrogen halides. Alkyl halides can also be prepared from alcohols by reaction with **thionyl chloride** or **phosphorus halides**.

Inorganic esters (nitrates, nitrites, sulfates, phosphates) can be prepared from alcohols and the corresponding acids. Alkyl nitrates are explosives. Organic phosphates are important intermediates in cellular reactions.

Primary alcohols can be oxidized to **aldehydes**, whereas secondary alcohols give **ketones**.

Glycols have two or more hydroxyl groups on adjacent carbons. **Ethylene glycol, glycerol,** and **sorbitol** are examples of glycols that are commercially important.

Three important industrial alcohols are **methanol, ethanol,** and **2-propanol.**

Phenols readily undergo aromatic substitution since the hydroxyl group is ring-activating and *ortho-para*-directing. Phenols are easily oxidized to **quinones.** Phenols with bulky *ortho* substituents are commercial antioxidants.

Examples of biologically important alcohols are **geraniol, farnesol, cholesterol,** and **vitamin A.**

The functional group of **thiols** is the **sulfhydryl group,** ——SH. Thiols are also called **mercaptans** because of their reaction with mercury salts to form **mercaptides.** Thiols have intense, disagreeable odors. They are more acidic than alcohols and are easily oxidized to **disulfides.**

REACTION SUMMARY

Alkoxides from Alcohols

$$2 \text{ RO}\!-\!\text{H} + 2 \text{ Na} \longrightarrow 2 \text{ RO}^-\text{Na}^+ + \text{H}_2$$
$$\text{RO}\!-\!\text{H} + \text{NaH} \longrightarrow \text{RO}^-\text{Na}^+ + \text{H}_2$$

Phenoxides from Phenols

$$\text{ArO}\!-\!\text{H} + \text{Na}^+\text{OH}^- \longrightarrow \text{ArO}^-\text{Na}^+ + \text{H}_2\text{O}$$

Dehydration of Alcohols

Alkyl Halides from Alcohols

$$\text{R}\!-\!\text{OH} + \text{HX} \longrightarrow \text{R}\!-\!\text{X} + \text{H}_2\text{O}$$

$$\text{R}\!-\!\text{OH} + \text{SOCl}_2 \longrightarrow \text{R}\!-\!\text{Cl} + \text{HCl} + \text{SO}_2$$

$$3 \text{ R}\!-\!\text{OH} + \text{PX}_3 \longrightarrow 3 \text{ R}\!-\!\text{X} + \text{H}_3\text{PO}_3$$

Inorganic Esters from Alcohols

$ROH + HONO_2 \longrightarrow RONO_2 + H_2O$
  alkyl nitrate

$ROH + HONO \longrightarrow RONO + H_2O$
  alkyl nitrite

$ROH + HOSO_3H \longrightarrow ROSO_3H + H_2O$
  alkyl hydrogen sulfate

Oxidation of Alcohols

$RCH_2OH \xrightarrow{PCC} RCH{=}O$
primary   aldehyde

$R_2CHOH \xrightarrow[H^+]{Cr^{6+}} R_2C{=}O$
secondary   ketone

Aromatic Substitution in Phenols

Oxidation of Phenols to Quinones

Thiols

$RX + NaSH \longrightarrow RSH + Na^+X^-$ (preparation)

$RSH + NaOH \longrightarrow RS^-Na^+ + H_2O$ (acidity)

$2\ RSH \xrightleftharpoons[reduce]{oxidize} RS{-}SR$ (oxidation)
                    disulfide

LEARNING OBJECTIVES

1. Know the meaning of: alcohol, phenol, thiol, hydroxyl group, primary, secondary, and tertiary alcohol.

2. Know the meaning of: alkoxide, phenoxide, oxonium ion, alkyloxonium ion.

3. Know the meaning of: dehydration, thionyl chloride, inorganic ester, alkyl nitrate, alkyl hydrogen sulfate, alkyl phosphate, diphosphate, triphosphate.

4. Know the meaning of: chromic anhydride, PCC, aldehyde, ketone, glycol, glycerol, sorbitol, glyceryl trinitrate (nitroglycerine), quinone, antioxidant.

5. Be familiar with: NAD^+, geraniol, farnesol, squalene, cholesterol, vitamin A, lignin, urushiol.

6. Know the meaning of: thiol, mercaptan, mercaptide, sulfhydryl group, disulfide.

7. Given the structure of an alcohol, tell whether it is a primary, secondary, or tertiary alcohol.

8. Given the IUPAC name of an alcohol or phenol, draw its structure.

9. Given the structure of an alcohol or phenol, assign it a correct name.

10. Explain the significance of hydrogen bonding of an alcohol or phenol with regard to solubility in water and boiling point.

11. Given a small group of compounds, including alcohols, phenols, and hydrocarbons, arrange them in order of water solubility and construct a scheme for separating them based on acidity differences.

12. Given a group of compounds with similar molecular weights but differing potential for hydrogen bonding, arrange them in order of boiling point.

13. Draw the resonance contributors to phenoxide or substituted phenoxide ions, and discuss the acidity of the corresponding phenols.

14. Account for the acidity difference between alcohols and phenols.

15. Write equations for the reaction of a specific alcohol or phenol with sodium, sodium hydride, or with a strong aqueous base (NaOH, KOH).

16. Write the structures for all possible dehydration products of a given alcohol.

17. Write the steps in the mechanism for the dehydration of a given alcohol. Given alcohols of different classes, tell which dehydration mechanism is most likely, and what the relative dehydration rates will be.

18. Write equations for the reaction of a given alcohol with HCl, HBr, or HI, with cold, concentrated H_2SO_4 or HNO_3, with PCl_3 or PBr_3, with thionyl chloride ($SOCl_2$), or with an oxidant such as chromic anhydride or PCC.

19. Write the steps in the mechanism for conversion of an alcohol to an alkyl halide.

20. Write equations for the reaction of phenol with dilute aqueous nitric acid and with bromine water.

21. Contrast the acidity of alcohols and thiols; also contrast their reactivity toward oxidizing agents.

22. Write equations for the reaction of a given thiol with Hg^{2+}, base, or an oxidizing agent such as H_2O_2.

ANSWERS TO PROBLEMS

Problems Within the Chapter

7.1   a.   Number from the hydroxyl-bearing carbon:

$$\overset{3}{B}r\overset{}{C}H_2\overset{2}{C}H_2\overset{1}{C}H_2OH$$

$BrCH_2CH_2CH_2OH$ 3-bromo-1-propanol

b.   cyclopentanol

c. $CH_2{=\!\!=}CHCH_2CH_2OH$ 3-buten-1-ol

7.2   a. $CH_3CHCH_2CH_2CH_3$ b. CH_3CHOH c. $CH_3CHCH{=\!\!=}CHCH_3$
            | |
            OH   OH

7.3   Methanol is usually grouped with the primary alcohols:

Primary: CH_3OH, CH_3CH_2OH, $CH_3CH_2CH_2OH$, $CH_3CH_2CH_2CH_2OH$,
             $(CH_3)_2CHCH_2OH$ (all have a ⎯CH_2OH group)

Secondary: $(CH_3)_2CHOH$, $CH_3CHCH_2CH_3$ (both have a $\rangle CHOH$ group)
                                  |
                                  OH

Tertiary: $(CH_3)_3COH$

139

7.4 a.

b.

7.5 Use eq. 7.5.

$$pK_a = -\log (1.0 \times 10^{-16}) = 16.0$$

7.6 Acetic acid has the lower pK_a and is the stronger acid.

7.7 a. Lewis base; can donate its electron pair to an acid

b. Lewis acid; the boron has only six valence electrons around it and can accept two more:

c. Lewis acid; can accept electron pairs to neutralize the positive charge

d. Lewis base; because of the unshared electron pairs on the oxygen:

$$CH_3\overset{..}{\underset{..}{O}}CH_3$$

e. Lewis acid; this carbocation is isoelectronic with $(CH_3)_3B$ (part b)

f. Lewis base; because of the unshared electron pairs on the carbonyl oxygen

7.8 NH_4^+ is an acid; it can donate a proton to a base, as in the reverse of eq. 7.8.

7.9 A strong base is required:

$$NH_3 + :\overset{..}{N}H_2^- \rightleftharpoons :\overset{..}{N}H_2^- + NH_3$$
amide
ion

In this equilibrium, ammonia acts as an acid (proton donor) toward the very strong base, the amide ion. The reaction is exactly analogous to the reaction of water with hydroxide ion:

$$H_2O \; + \; :\ddot{O}H^- \; \rightleftharpoons \; H\ddot{O}:^- \; + \; H_2O$$

acid   base

7.10

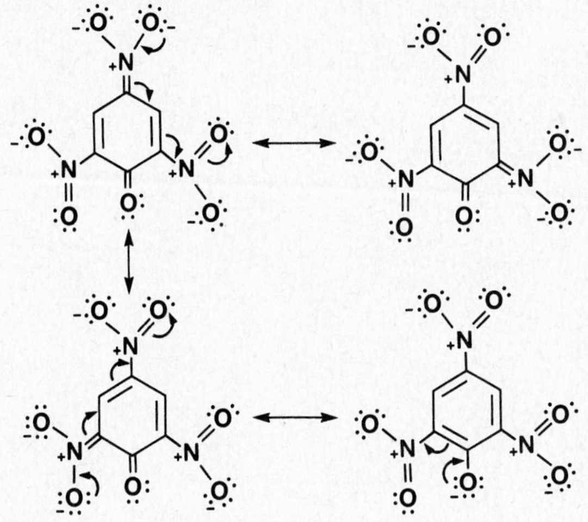

7.11 CH_3CH_2OH < $ClCH_2CH_2OH$ <   CH₃—⟨benzene ring⟩—OH

< ⟨benzene ring⟩—OH < Cl—⟨benzene ring⟩—OH

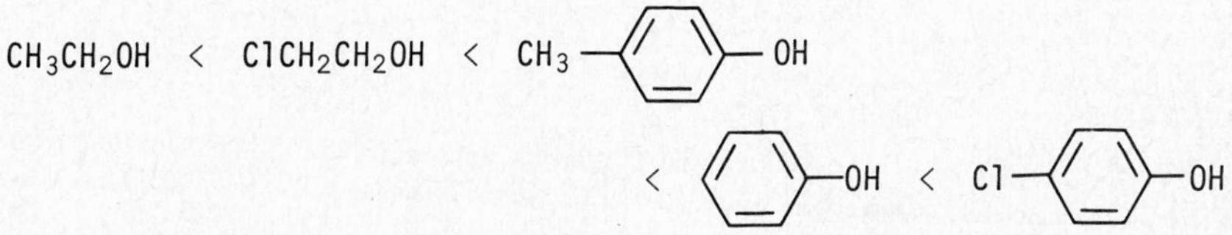

Both alcohols are weaker acids than the three phenols. Of the alcohols, 2-chloroethanol is the stronger acid because of the electronegativity of the chlorine substituent. Among the three phenols, acidity increases with increasing electronegativity of the *para* substituent: CH_3 < H < Cl.

7.12   Follow the pattern of eq. 7.10:

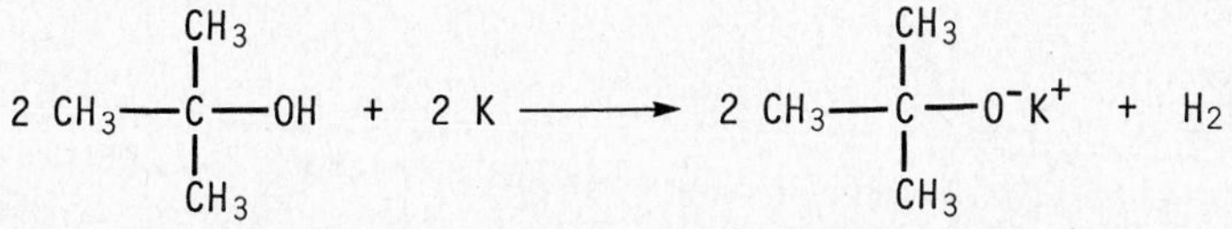

potassium *t*-butoxide

7.13 a. Follow the pattern of eq. 7.12:

$$CH_3-\langle\bigcirc\rangle-OH \ + \ NaOH \ \longrightarrow \ CH_3-\langle\bigcirc\rangle-O^-Na^+ \ + \ H_2O$$

p-cresol   sodium p-cresoxide

b. Alcohols do not react with aqueous base; the equilibrium favors
the starting material because hydroxide ion is a weaker base
than alkoxide ion:

$$\bigcirc\!\!\!<^H_{OH} \ + \ NaOH \ \rightleftharpoons \ \bigcirc\!\!\!<^H_{O^-Na^+} \ + \ H_2O$$

7.14 Write the structure of the alcohol, and consider products with a
double bond between the hydroxyl-bearing carbon and each adjacent
carbon that also has at least one hydrogen attached.

a.
$$\overset{\overset{\textstyle CH_3}{|}}{CH_3CH_2\overset{}{C}CH_2CH_2CH_3}\underset{OH}{}\ \xrightarrow[-H_2O]{H^+}$$

$$\overset{\overset{\textstyle CH_3}{|}}{CH_3CH=CCH_2CH_2CH_3} \ + \ \overset{\overset{\textstyle CH_2}{\|}}{CH_3CH_2CCH_2CH_2CH_3} \ + \ \overset{\overset{\textstyle CH_3}{|}}{CH_3CH_2C=CHCH_2CH_3}$$
$$(cis \text{ and } trans) \qquad\qquad\qquad\qquad\qquad\qquad\qquad\qquad (cis \text{ and } trans)$$

b.
$$\bigcirc\!\!\!<^{OH}_{CH_3} \ \xrightarrow[-H_2O]{H^+} \ \bigcirc\!\!=\!\!-CH_3 \ + \ \bigcirc\!\!=\!\!CH_2$$

The predominant product is generally the alkene with the most
substituted double bond. In part a, the first and third products have trisubstituted double bonds and should predominate
over the middle product, which is only disubstituted. To predict
more precisely is not possible. In part b, the product with the
double bond in the ring (1-methylcyclopentene) is trisubstituted
and will predominate.

7.15 The first step, which is rate-determining, is the formation of the *t*-butyl cation. The rate of this step does not depend on which acid is used:

$$(CH_3)_3COH \; + \; H^+ \; \rightleftharpoons \; (CH_3)_3C\overset{+}{\underset{\underset{H}{|}}{\overset{\cdot\cdot}{O}}}\text{---}H \; \xrightarrow[\substack{\text{slow} \\ \text{step}}]{S_N1} \; (CH_3)_3C^+ \; + \; H_2O$$
$$\text{\textit{t}-butyl cation}$$

Reaction of the carbocation with Cl^-, Br^-, or I^- is then fast.

7.16 Unlike the alcohol in Problem 7.15, the alcohol in this case is primary instead of tertiary. The rate-determining step is the S_N2 reaction:

$$CH_3CH_2CH_2CH_2OH \; + \; H^+ \; \rightleftharpoons \; CH_3CH_2CH_2CH_2\text{---}\overset{+}{\underset{\underset{H}{|}}{\overset{\cdot\cdot}{O}}}\text{---}H$$

$$X^- \; + \; \underset{CH_3CH_2CH_2}{\overset{\frown}{CH_2}}\text{---}\overset{+}{\underset{\underset{H}{|}}{O}}\text{---}H \; \xrightarrow[\substack{\text{slow} \\ \text{step}}]{S_N2} \; CH_3CH_2CH_2CH_2X \; + \; H_2O$$

The rate of this step varies with the nucleophilicity of X^-; this order of nucleophilicity is $I^- > Br^- > Cl^-$.

7.17 a. $CH_3(CH_2)_6CH_2OH \; + \; SOCl_2 \; \xrightarrow{\text{heat}} \; CH_3(CH_2)_6CH_2Cl \; + \; SO_2 \; + \; HCl$

b. $3 \; CH_3\underset{\underset{OH}{|}}{C}HCH_3 \; + \; PBr_3 \; \longrightarrow \; 3 \; CH_3\underset{\underset{Br}{|}}{C}HCH_3 \; + \; H_3PO_3$

7.18 Use eq. 7.28 as a guide.

$$(CH_3)_2CHCH_2CH_2OH \; + \; HONO \; \longrightarrow \; (CH_3)_2CHCH_2CH_2ONO \; + \; H_2O$$

7.19 a. $CH_3CH_2\underset{\underset{OH}{|}}{C}HCH_2CH_3 \; \xrightarrow[H^+]{CrO_3} \; CH_3CH_2\underset{\underset{O}{\|}}{C}CH_2CH_3$

The alcohol is secondary and gives a ketone.

b. $CH_3CH_2CH_2CH_2CH_2OH \; \xrightarrow[CH_2Cl_2, \; 25°C]{PCC} \; CH_3CH_2CH_2CH_2CH{=}O$

The alcohol is primary and gives an aldehyde.

7.20 a.

Phenol does not react.

b.

Phenol may undergo electrophilic aromatic substitution:

c.

(compare with eq. 7.26)

Phenol may react to form a phosphite ester:

But the phenolic OH group cannot be replaced by Cl in this way.

7.21 The phenoxide ion is negatively charged, and that charge can be delocalized to the *ortho* and *para* ring carbons. Therefore, attack by an electrophile at these positions is facilitated.

7.22 a.

$+$ HONO$_2$ $\longrightarrow$ [product with OH, NO$_2$, CH$_3$] $+$ H$_2$O

The hydroxyl group is more ring-activating than the methyl group
is. Thus substitution *ortho* to the hydroxyl group is preferred.

b.

$+$ Br$_2$ $\longrightarrow$ [product with OH, Cl, Br] $+$ HBr

The hydroxyl group is ring-activating, whereas the chlorine is
a ring-deactivating substituent. Therefore, substitution occurs
para to the hydroxyl group. Substitution *ortho* to the hydroxyl
group is less likely, since the product would have three adjacent
substituents, which would be quite crowded.

7.23 a.   b.

1,4-naphthoquinone   *ortho*-benzoquinone

The oxygens in the quinones are located on the same carbons as the
hydroxyl groups in the starting phenols.

7.24 a. $\overset{1}{C}H_3\overset{2}{C}H\overset{3}{C}H_2\overset{4}{C}H_3$ b. CH$_3$
              | $\diagdown$
              SH   CHSH
                                                   CH$_3$ $\diagup$

145

7.25   a.   Follow eq. 7.43 as a guide.

$$CH_3CH_2SH \ + \ KOH \ \longrightarrow \ CH_3CH_2S^-K^+ \ + \ H_2O$$

       b.   Follow eq. 7.41 as a guide.

$$2 \ CH_3CH_2SH \ + \ HgCl_2 \ \longrightarrow \ (CH_3CH_2S)_2Hg \ + \ 2 \ HCl$$

Additional Problems

7.26   a.

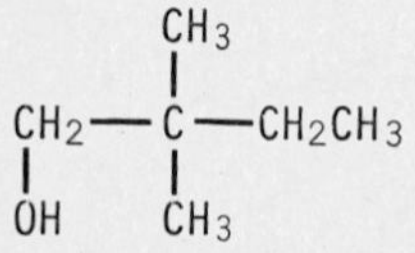

       b.

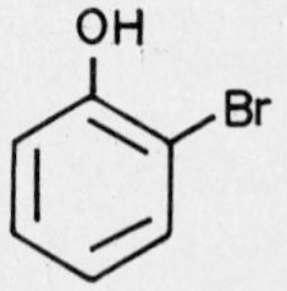

       c. 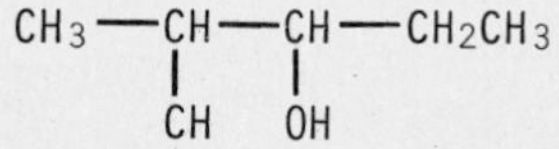 $CH_3 - CH - CH - CH_2CH_3$

       d.

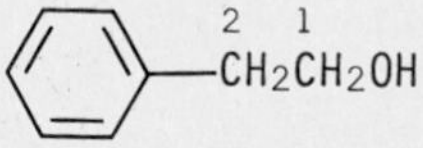

       e. $CH_3CH_2OSO_3H$

       f.

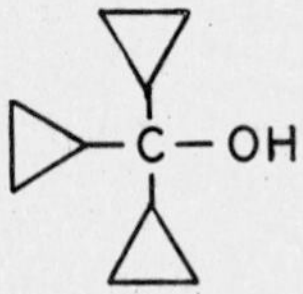

       g. $CH_3CH_2O^-Na^+$

       h.

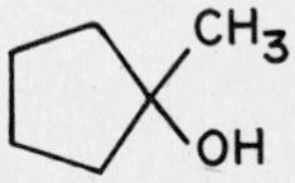

       i.

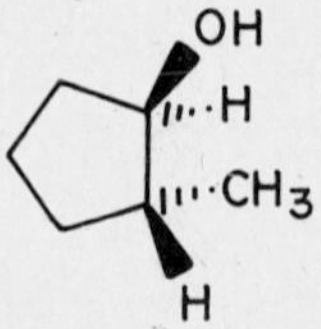

       j.

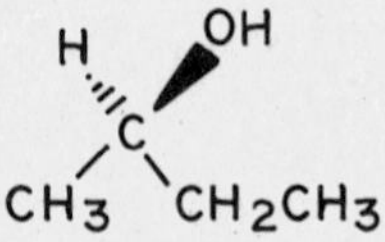

7.27   a.   primary   d.   primary   f.   tertiary
       h.   tertiary   i.   secondary

7.28   a.   3,3-dimethyl-2-butanol   b.   3-bromo-2-methyl-2-butanol
         c.   2,4-dichlorophenol   d.   cyclopropanol
         e.   *m*-bromophenol   f.   diphenylmethanol
         g.   2-buten-1-ol   h.   2-propanethiol (or isopropyl mercaptan)

         i.   1,2,3,4-butanetetraol   j.   potassium *n*-propoxide
         k.   *cis*-3-methylcyclobutanol   l.   cyclopropanethiol

7.29   a.   The hydroxyl should get the lower number; 3,3-dimethyl-2-butanol.
         b.   2-Methyl-1-butanol; the longest chain was not selected.
         c.   2-Propen-1-ol (or allyl alcohol); the hydroxyl group should get the lower number.
         d.   3-Chlorocyclohexanol; number the ring from the hydroxyl-bearing carbon, in a direction that gives substituents the lowest possible numbers.
         e.   2-Bromo-*p*-cresol or 2-bromo-4-methylphenol; give substituents the lowest possible numbers.
         f.   1,2-Propanediol; the chain was numbered in the wrong direction.

7.30   a.   Ethyl chloride < 1-hexanol < ethanol. Both alcohols can hydrogen-bond with water and will be more soluble than the alkyl chloride. The lower-molecular-weight alcohol will be more soluble (it has a shorter hydrophobic carbon chain).
         b.   1-Pentanol < 1,5-pentanediol < 1,2,3,4,5-pentanepentaol. All three compounds have the same number of carbon atoms. Water solubility will therefore increase with increasing numbers of hydroxyl groups (that is, as the ratio of hydroxyl groups to carbon atoms increases).

7.31   a.

$$R-\ddot{O}-R \;+\; H^+ \;\rightleftharpoons\; R-\overset{\displaystyle H}{\underset{\displaystyle \cdot\cdot}{O^+}}-R$$

         b.

$$R-\overset{\displaystyle }{\underset{\displaystyle R}{\ddot{N}}}-R \;+\; H^+ \;\rightleftharpoons\; R-\overset{\displaystyle H}{\underset{\displaystyle R}{N^+}}-R$$

         c.

$$\overset{\displaystyle R}{\underset{\displaystyle R}{>}}C=\ddot{O}: \;+\; H^+ \;\rightleftharpoons\; \overset{\displaystyle R}{\underset{\displaystyle R}{>}}C=\overset{\cdot\cdot}{O^+}-H$$

7.32

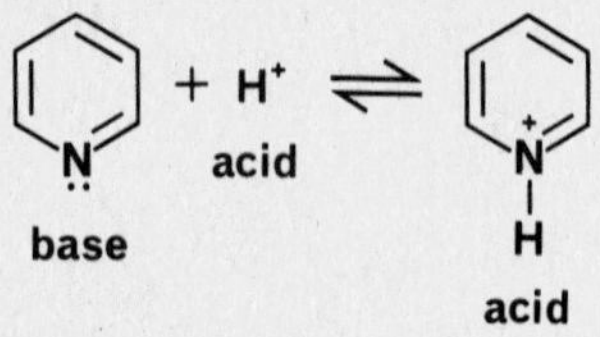

base   acid   H

acid

The chloride ion appears on both sides of the equation and plays
no role (except to balance the positive charges).

7.33   a.

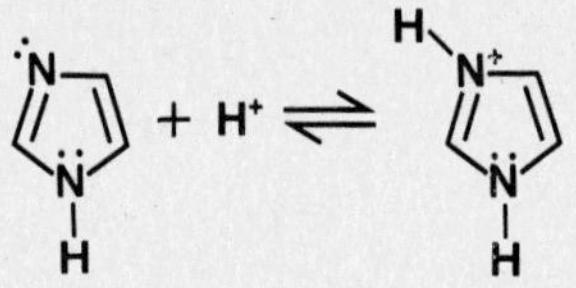

The nitrogen *without* an attached hydrogen is the more basic
nitrogen (see Problem 4.19).

The positive charge in the resulting imidazolium ion is
delocalized over both nitrogens:

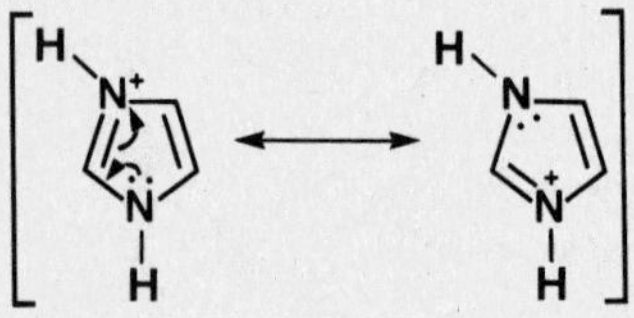

b.

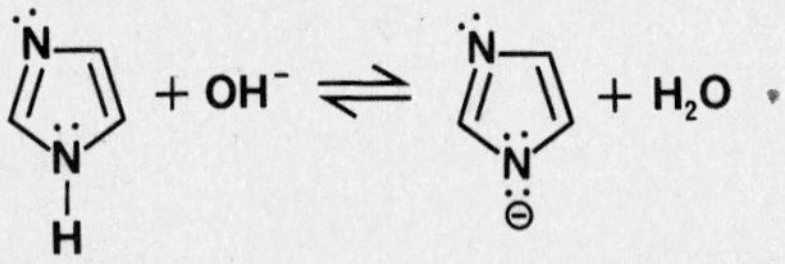

The N—H proton is acidic. The negative charge in the resulting
anion can be delocalized over both nitrogens:

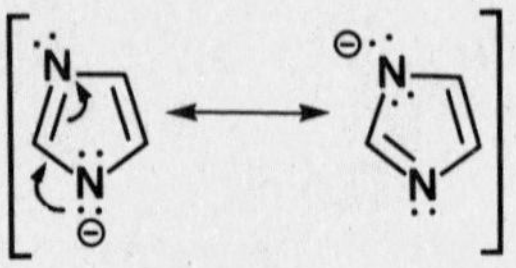

7.34

The two alcohols are less acidic than the two phenols. The electron-withdrawing chlorine substituent makes 2-chlorocyclohexanol a stronger acid than cyclohexanol. The electron-withdrawing cyano substituent makes p-cyanophenol a stronger acid than phenol. The negative charge in the p-cyanophenoxide ion can be delocalized to the nitrogen:

7.35 Since t-butyl alcohol is approximately 100 times *weaker* than ethanol as an acid ($pK_a = 18$ and $pK_a =$ approximately 16, respectively), it follows that if we consider the conjugate bases, t-butoxide ion is a stronger base than ethoxide ion.

7.36 p-Cresol reacts with aqueous base, and the product is water-soluble:

soluble in water layer

The cyclohexanol is not sufficiently acidic to react with sodium hydroxide. When the layers are separated, the cyclohexanol will be in the organic layer and the sodium p-cresoxide in the water layer. Acidification of the water layer allows the p-cresol to be recovered:

The p-cresol can then be extracted by an organic solvent.

7.37 In the laboratory, separations of this type are usually performed by dissolving the mixture in a low-boiling, inert organic solvent such as ether or methylene chloride. The solution is then extracted with an aqueous (neutral, acidic, or alkaline) solution that extracts one of the two components. The layers are then separated, and the organic layer is evaporated to recover the compound that was not extracted into the aqueous base. The aqueous layer is then treated in some way to recover the extracted compound.

a.

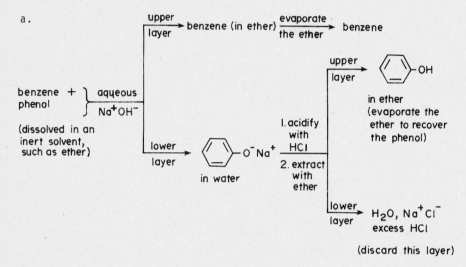

b. The same procedure used in part a works here, since phenol is extracted by base, whereas 1-hexanol, a much weaker acid, is not.
c. An ether solution of the two alcohols can be extracted with water. 1-Propanol is soluble, whereas 1-heptanol, with a much longer carbon chain, is not.

7.38 a. Use eq. 7.10 as a guide.

$$2 \ CH_3CHCH_2CH_3 \ + \ 2 \ K \ \longrightarrow \ 2 \ CH_3CHCH_2CH_3 \ + \ H_2$$
$$\qquad\qquad | \qquad\qquad\qquad\qquad\qquad\qquad | $$
$$\qquad\quad OH \qquad\qquad\qquad\qquad\qquad\quad O^-K^+$$
$$\qquad\qquad\qquad\qquad\qquad\quad \text{potassium 2-butoxide}$$

b. Use eq. 7.11 as a guide.

$$CH_3CHCH_3 \ + \ NaH \ \longrightarrow \ CH_3CHCH_3 \ + \ H_2$$
$$\qquad | \qquad\qquad\qquad\qquad\qquad\quad | $$
$$\quad OH \qquad\qquad\qquad\qquad\qquad O^-Na^+$$
$$\qquad\qquad\qquad\qquad\qquad \text{sodium isopropoxide}$$
$$\qquad\qquad\qquad\qquad\qquad \text{(or 2-propoxide)}$$

c. Use eq. 7.12 as a guide.

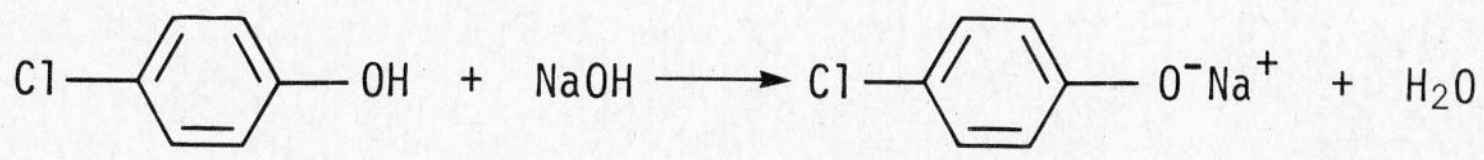

$$Cl-\underset{}{\bigcirc}-OH \ + \ NaOH \longrightarrow Cl-\underset{}{\bigcirc}-O^-Na^+ \ + \ H_2O$$

sodium *p*-chlorophenoxide

7.39 If you have any difficulty with this problem, review Sec. 7.9.

a.

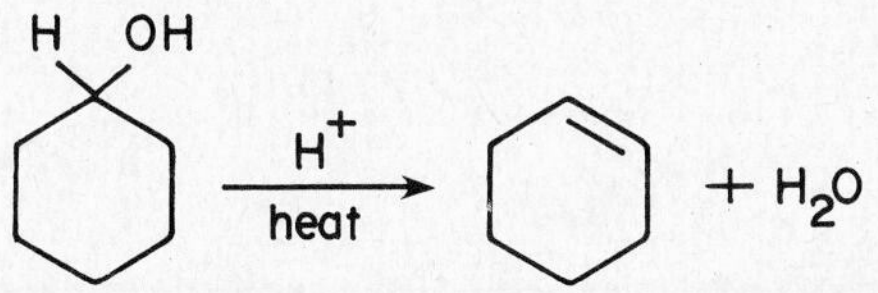

b. $CH_3CHCH_2CH_3 \xrightarrow[\text{heat}]{H^+} CH_2{=}CHCH_2CH_3$ and $CH_3CH{=}CHCH_3 \ + \ H_2O$

OH   (*cis* and *trans*)

Of the three possible alkenes, *trans*-2-butene is the most stable and predominates.

c.

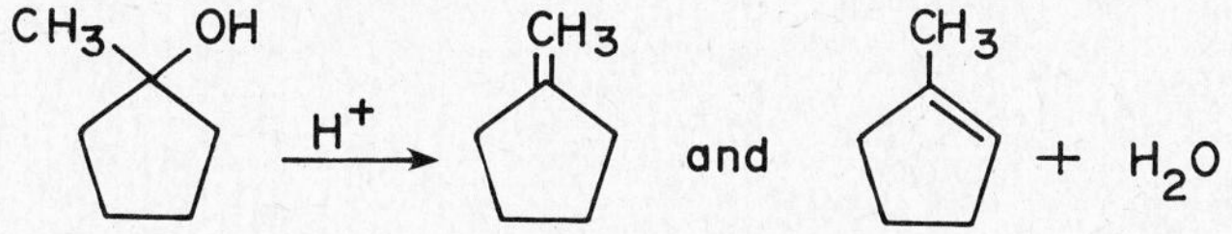

The predominant product is 1-methylcyclopentene.

d.

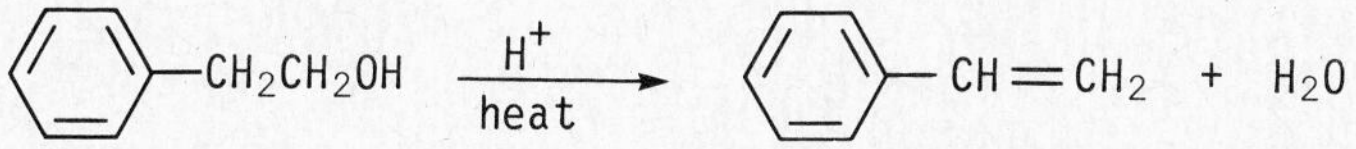

7.40 In the reaction

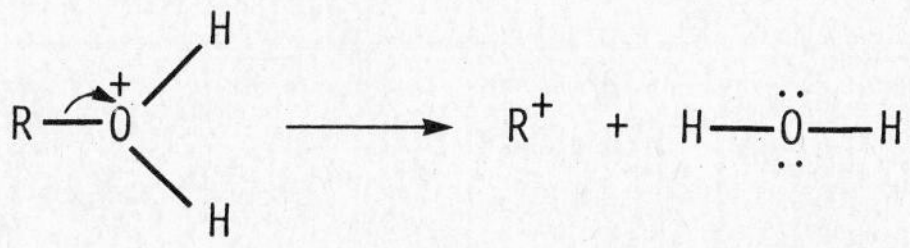

electrons flow toward the positive oxygen, and positive charge passes from the oxygen to carbon (R group).

In the reaction

$$R\text{—}\overset{..}{\underset{..}{O}}\text{—}H \longrightarrow R^+ + {}^-\text{:}\overset{..}{\underset{..}{O}}\text{—}H$$

(a) the oxygen is not charged and therefore is less electron-demanding, and (b) two oppositely charged species, R^+ and OH^-, must be separated. The second reaction thus requires much more energy than the first.

7.41 To begin, protonation of the oxygen and loss of water yield a tertiary carbocation:

$$CH_3\text{—}\underset{\underset{CH_3}{|}}{\overset{\overset{:\overset{..}{O}H}{|}}{C}}\text{—}CH_2CH_3 \underset{}{\overset{H^+}{\rightleftharpoons}} CH_3\text{—}\underset{\underset{CH_3}{|}}{\overset{\overset{\overset{H\;\;\;H}{\diagdown\overset{+}{O}\diagup}}{|}}{C}}\text{—}CH_2CH_3 \xrightarrow[\substack{rds \\ (S_N1)}]{slow} CH_3\text{—}\underset{\underset{CH_3}{|}}{\overset{+}{C}}\text{—}CH_2CH_3 + H_2O$$

The carbocation can then lose a proton from a carbon adjacent to the one that bears the positive charge, to give either product:

$$CH_3\text{—}\underset{\underset{CH_3}{|}}{\overset{+}{C}}\text{—}CH_2CH_3 \xrightarrow{-H^+} CH_2\text{=}\underset{\underset{CH_3}{|}}{C}\text{—}CH_2CH_3 \text{ and } CH_3\text{—}\underset{\underset{CH_3}{|}}{C}\text{=}CHCH_3$$

7.42 The by-product in eq. 7.23 is isobutylene, formed by an E1 process which competes with the main S_N1 reaction:

$$(CH_3)_3C\overset{..}{\underset{..}{O}}H \xrightarrow{H^+} (CH_3)_3C\text{—}\underset{\underset{H}{|}}{\overset{+}{O}}\text{—}H \xrightarrow[step]{slow} (CH_3)_3C^+ + H_2O$$

$$(CH_3)_3C^+ \xrightarrow[step]{fast} \begin{cases} \xrightarrow[S_N1]{Cl^-} (CH_3)_3CCl \qquad 80\% \\ \\ \xrightarrow[E1]{-H^+} CH_2\text{=}\underset{\underset{CH_3}{|}}{C}\text{—}CH_3 \quad 20\% \end{cases}$$

The reaction in eq. 7.24, on the other hand, involves a primary alcohol and proceeds by an S_N2 process:

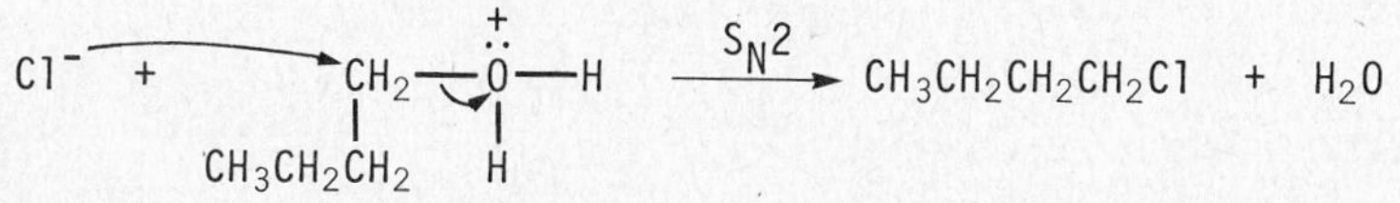

Cl⁻ + CH_2—$\overset{+}{\overset{..}{O}}$—H $\xrightarrow{S_N2}$ $CH_3CH_2CH_2CH_2Cl$ + H_2O
       $CH_3CH_2CH_2$ H

Chloride ion is a *very* weak base; thus the E2 process cannot compete with the S_N2 and the yield of the substitution product is nearly 100%.

7.43   a. 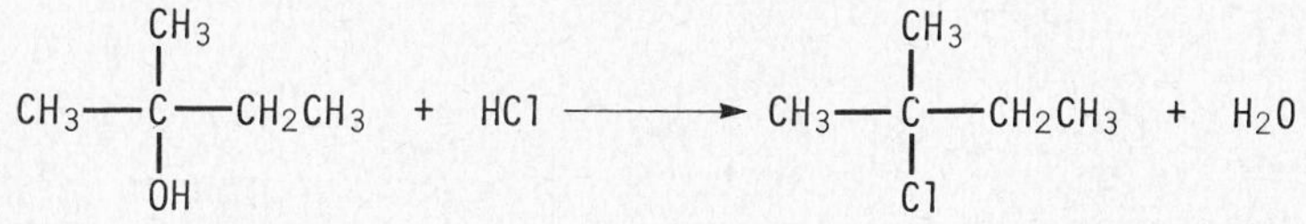 CH_3—$\underset{\underset{OH}{|}}{\overset{\overset{CH_3}{|}}{C}}$—$CH_2CH_3$ + HCl $\longrightarrow$ CH_3—$\underset{\underset{Cl}{|}}{\overset{\overset{CH_3}{|}}{C}}$—$CH_2CH_3$ + H_2O

b. $2\ CH_3CH_2CH_2CH_2CH_2OH$ + 2 Na $\longrightarrow$

$2\ CH_3CH_2CH_2CH_2CH_2O^-Na^+$ + H_2

sodium 1-pentoxide

c. 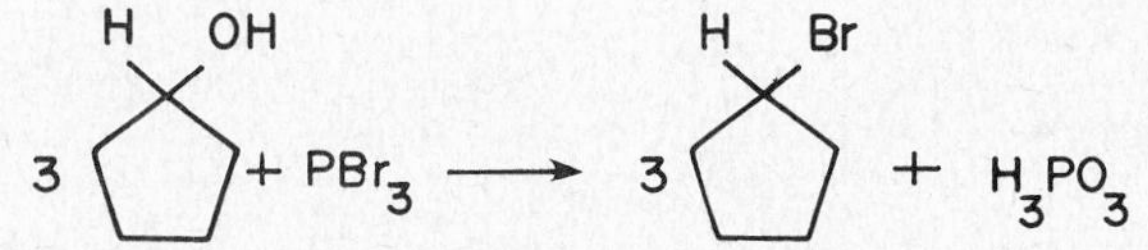   3 [cyclopentane ring with H OH] + PBr₃ $\longrightarrow$ 3 [cyclopentane ring with H Br] + H_3PO_3

d. 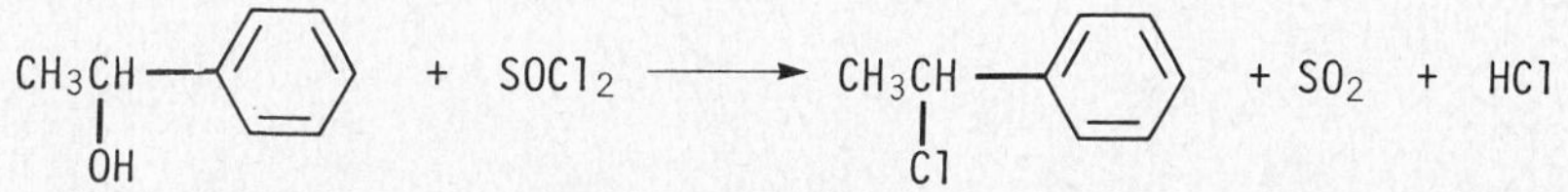 $CH_3\underset{\underset{OH}{|}}{CH}$—[benzene ring] + $SOCl_2$ $\longrightarrow$ $CH_3\underset{\underset{Cl}{|}}{CH}$—[benzene ring] + SO_2 + HCl

e. $CH_3CH_2CH_2CH_2OH$ + $HOSO_3H$ $\xrightarrow{cold}$

$CH_3CH_2CH_2CH_2OSO_3H$ + H_2O

f. 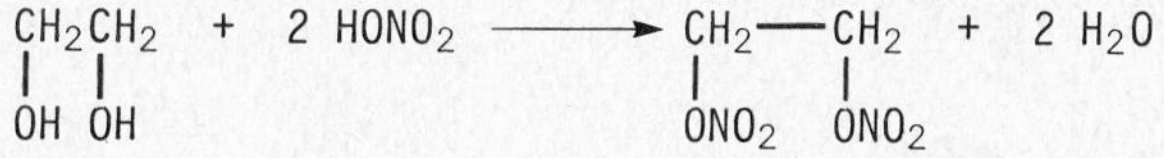 $\underset{\underset{OH}{|}}{CH_2}\underset{\underset{OH}{|}}{CH_2}$ + 2 $HONO_2$ $\longrightarrow$ $\underset{\underset{ONO_2}{|}}{CH_2}$—$\underset{\underset{ONO_2}{|}}{CH_2}$ + 2 H_2O

g. $CH_3CH_2CH_2CH_2CH_2OH$ + NaOH $\longrightarrow$ no reaction

h. $CH_3(CH_2)_6CH_2OH$ + HBr $\xrightarrow{ZnBr_2}$ $CH_3(CH_2)_6CH_2Br$ + H_2O

i. 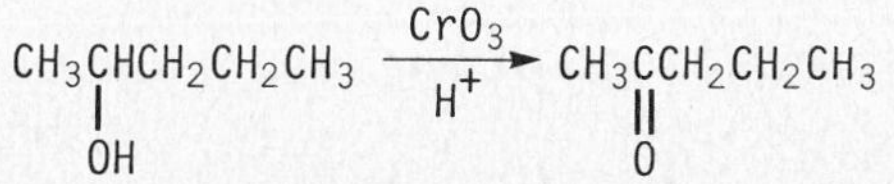 $CH_3\underset{\underset{OH}{|}}{CH}CH_2CH_2CH_3$ $\xrightarrow[H^+]{CrO_3}$ $CH_3\overset{\overset{O}{||}}{C}CH_2CH_2CH_3$

j.

$$\text{C}_6\text{H}_5\text{—CH}_2\text{OH} + \text{CH}_3\overset{\overset{\text{O}}{\|}}{\text{C}}\text{—OH} \xrightarrow{\text{H}^+} \text{C}_6\text{H}_5\text{—CH}_2\text{O}\overset{\overset{\text{O}}{\|}}{\text{C}}\text{CH}_3 + \text{H}_2\text{O}$$

This last reaction is similar to the formation of inorganic esters discussed in Sec. 7.12. Organic esters are discussed more fully in Chapter 10.

7.44 The mechanism involves protonation of the hydroxyl group and loss of water to form a carbocation:

$$\text{CH}_2\text{=CH—CH—CH}_3 \underset{\text{—H}_2\text{O}}{\overset{\text{H}^+}{\longrightarrow}} \text{CH}_2\text{=CH—}\overset{+}{\text{CH}}\text{—CH}_3$$

(with OH below the CH)

The carbocation is allylic and stabilized by resonance. This allylic ion, can react with the nucleophile Cl⁻ at either end, giving the observed products:

$$\text{CH}_2\text{=CH—}\overset{+}{\text{CH}}\text{CH}_3 \updownarrow \overset{+}{\text{CH}}_2\text{—CH=CHCH}_3 \quad \xrightarrow{\text{Cl}^-} \quad \begin{array}{l} \text{CH}_2\text{=CH—CHCH}_3 \\ \quad\quad\quad\quad | \\ \quad\quad\quad\quad \text{Cl} \\ \text{3-chloro-1-butene} \\ \text{or} \\ \text{CH}_2\text{—CH=CHCH}_3 \\ | \\ \text{Cl} \\ \text{1-chloro-2-butene} \end{array}$$

7.45 If necessary, review Secs. 3.10 through 3.14.

$$\text{CH}_3\text{CHCH}_2\text{CH}_3$$
$$\quad\;\;|$$
$$\quad\;\;\text{OH}$$

can be prepared by acid-catalyzed hydration of $\text{CH}_2\text{=CHCH}_2\text{CH}_3$ or $\text{CH}_3\text{CH=CHCH}_3$.

$$\quad\;\;\text{CH}_3$$
$$\quad\;\;\;|$$
$$\text{CH}_3\text{CCH}_3$$
$$\quad\;\;\;|$$
$$\quad\;\;\text{OH}$$

can be prepared by acid-catalyzed hydration of $\text{CH}_2\text{=C—CH}_3$
$$\quad\quad\quad\quad\quad\quad\quad\quad\quad\quad\quad\quad\quad\quad | $$
$$\quad\quad\quad\quad\quad\quad\quad\quad\quad\quad\quad\quad\;\text{CH}_3$$

Neither $CH_3CH_2CH_2CH_2OH$ nor $(CH_3)_2CHCH_2OH$ can be obtained by this route [though they can be obtained by the (more expensive) hydroboration-oxidation sequence, Sec. 3.20].

7.46 a.

b. $CH_3CH_2CH_2CH_2Br \xrightarrow{\text{aq. NaOH}} CH_3CH_2CH_2CH_2OH \xrightarrow[CH_2Cl_2, \ 25°C]{PCC}$

$$CH_3CH_2CH_2CH{=}O$$

c. $CH_3CH_2CH_2CH_2OH \xrightarrow[ZnBr_2]{HBr} CH_3CH_2CH_2CH_2Br \xrightarrow{Na^+SH^-} CH_3CH_2CH_2CH_2SH$

In problems of this type, working backwards from the final product sometimes helps. Ask yourself: What reactions have I studied that give this type of product? If there is more than one, see which one requires a precursor that is easily obtained from the compound given as the starting point for the synthesis.

7.47 The secondary alcohol should be oxidized to a ketone:

That is the only part of the molecule that is altered.

7.48 The overall equation is as follows:

155

The mechanism involves protonation of the isobutylene according to Markovnikov's rule:

$$CH_2{=}\underset{\underset{\displaystyle CH_3}{|}}{C}{-}CH_3 \ + \ H^+ \ \longrightarrow \ CH_3{-}\underset{\underset{\displaystyle CH_3}{|}}{\overset{+}{C}}{-}CH_3$$

The tertiary carbocation then acts as an electrophile toward the *p*-cresol:

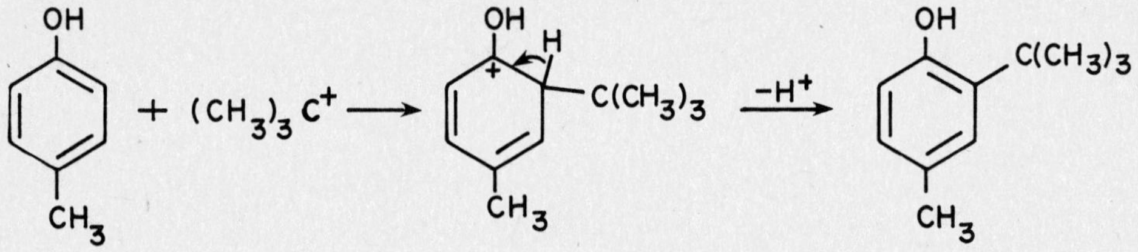

The steps are repeated to give the final product.

7.49   squalene

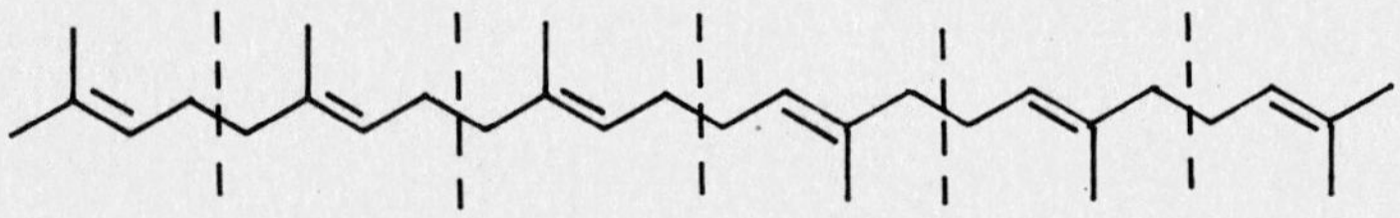

vitamin A

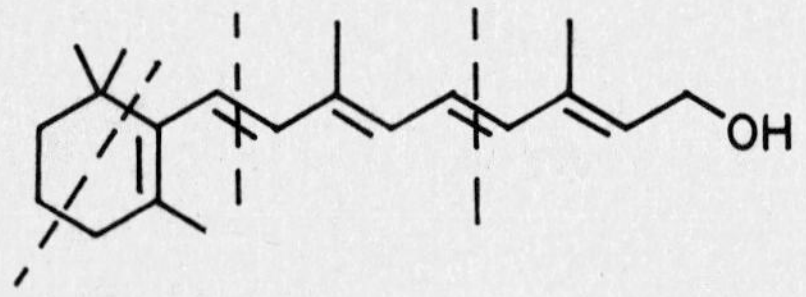

β-carotene

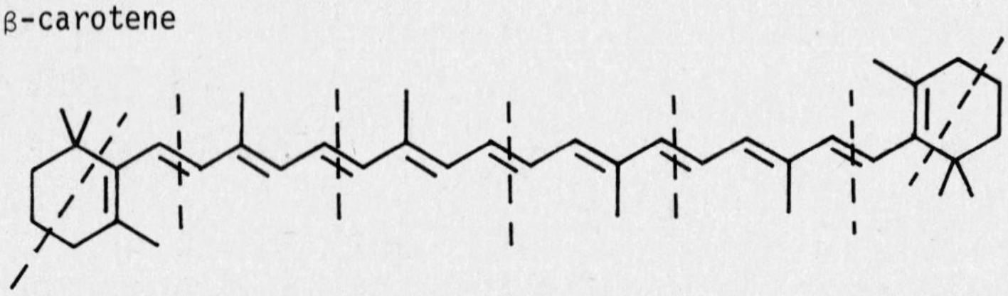

7.50
$$\overset{4}{C}H_3\overset{3}{C}H\!=\!\overset{2}{C}H\overset{1}{C}H_2SH$$

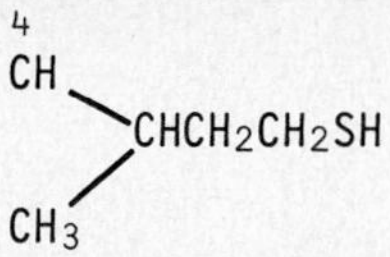

2-buten-1-thiol

$$\overset{4}{C}H\diagdown_{CH_3}^{}CHCH_2CH_2SH$$

3-methyl-1-butanethiol

7.51 $2\ CH_3SH \xrightarrow{\ H_2O_2\ } CH_3S\!-\!SCH_3$ (compare with eq. 7.44)

7.52 $CH_3CHCH_2CH_2OH \xrightarrow[\substack{or \\ PBr_3}]{HBr} CH_3CHCH_2CH_2Br$
            | |
          CH_3 CH_3

$\downarrow$ NaSH

$(CH_3)_2CHCH_2CH_2 \xleftarrow{\ H_2O_2\ } CH_3CHCH_2CH_2SH$
            | |
            S CH_3
            |
            S
            |
$(CH_3)_2CHCH_2CH_2$

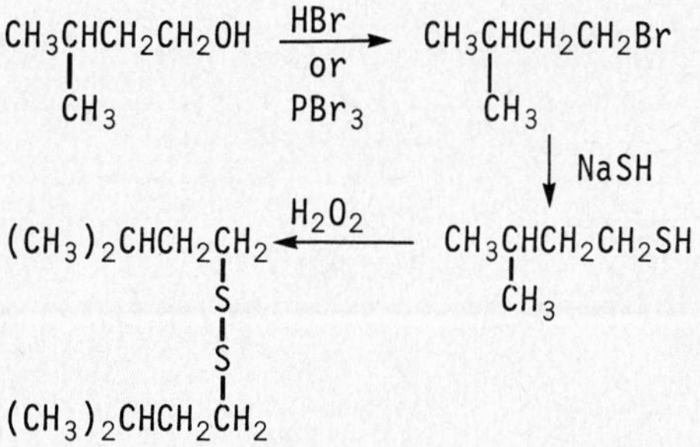

Work backwards: from the disulfide to the thiol (eq. 7.44), to the alkyl halide (eq. 7.42), to the alcohol (eq. 7.22 or eq. 7.26).

CHAPTER EIGHT   ETHERS, EPOXIDES, AND SULFIDES

CHAPTER SUMMARY

Ethers have two organic groups, either alkyl or aryl, connected to a single oxygen atom (R——O——R'). In common names, the two organic groups are named and followed by the word *ether,* as in *ethyl methyl ether,* $CH_3CH_2OCH_3$. In the IUPAC system, the smaller **alkoxy group** is named as a substituent on the longer carbon chain; for the preceding example, the IUPAC name is *methoxyethane.*

Ethers have much lower boiling points than the alcohols with which they are isomeric because ethers cannot form intermolecular hydrogen bonds with themselves. They do, however, act as Lewis bases to form hydrogen bonds with compounds containing an ——OH group (alcohols or water).

Ethers are excellent solvents for organic compounds. Their relative inertness makes them good solvents in which to carry out organic reactions.

Alkyl or aryl halides react with magnesium metal in **diethyl ether** or **tetrahydrofuran (THF)** to form **Grignard reagents, R——MgX**. Ethers stabilize these reagents by coordinating with the magnesium. Grignard reagents react with water, and the ——MgX is replaced by ——H, or if D_2O is used, by ——D.

Diethyl ether is prepared commercially by intermolecular dehydration of ethanol with sulfuric acid. The **Williamson ether synthesis,** another route to ethers, involves preparation of an alkoxide from an alcohol and a reactive metal, followed by an S_N2 displacement between the alkoxide and an alkyl halide.

Ethers can be cleaved at the C——O bond by strong protonic (HBr) or Lewis (BBr_3) acids; the products are alcohols and/or alkyl halides.

Epoxides (oxiranes) are three-membered cyclic ethers. The simplest and commercially most important example is **ethylene oxide**, manufactured from ethylene, air, and a silver catalyst. In the laboratory, epoxides are most commonly prepared from alkenes and organic peracids.

Epoxides react with nucleophiles to give products in which the ring has opened. For example, acid-catalyzed hydration of ethylene oxide gives **ethylene glycol**. Other nucleophiles (such as alcohols and ammonia) add similarly to epoxides, as do Grignard reagents. The latter is a useful two-carbon chain-lengthening reaction.

Epoxy resins are polymers prepared by similar ring-opening reactions, using as raw materials the epoxide **epichlorhydrin** and the nucleophile **bisphenol-A.**

Cyclic ethers with larger rings than epoxides include **furan, tetra-hydrofuran** (THF), **tetrahydropyran** (THP), and **dioxane.** Large-ring cyclic polyethers, called **crown ethers,** can selectively bind metallic ions, depending on the ring size.

Thioethers (**sulfides**) are the sulfur analogs of ethers. They can be prepared from thiols, alkyl halides, and base. Sulfides can be oxidized to **sulfoxides** and **sulfones.** They also react with alkyl halides to give **sulfonium salts,** some of which act as alkylating agents in biological processes.

REACTION SUMMARY

Grignard Reagents

$$R\text{---}X \ + \ Mg \ \xrightarrow[\text{THF}]{\text{ether or}} \ R\text{---}MgX \quad (\text{preparation})$$

$$RMgX \ + \ H\text{---}OH \ \longrightarrow \ RH \ + \ Mg(OH)X$$
$$RMgX \ + \ D\text{---}OD \ \longrightarrow \ RD \ + \ Mg(OD)X$$
$$\left.\right\} \text{hydrolysis}$$

Ether Preparation

$$2 \ ROH \ \xrightarrow[140°C]{H_2SO_4} \ ROR \ + \ H_2O$$

(best for primary alcohols; gives symmetric ethers)

$$2 \ ROH \ + \ 2 \ Na \ \longrightarrow \ 2 \ RO^-Na^+ \ + \ H_2$$
$$RO^-Na^+ \ + \ R'X \ \longrightarrow \ ROR' \ + \ Na^+X^-$$
$$\left.\right\} \text{Williamson synthesis}$$

(best for R' = primary)

Ether Cleavage

$$R\text{---}O\text{---}R \ + \ HBr \ \longrightarrow \ ROH \ + \ RBr$$

$$\xrightarrow{HBr} \ RBr \ + \ H_2O$$

$$3 \ R\text{---}O\text{---}R \ + \ 2 \ BBr_3 \ \longrightarrow \ 6 \ RBr \ + \ B_2O_3$$

$$\xrightarrow{3 \ H_2O} \ 2 \ H_3BO_3$$

Ethylene Oxide

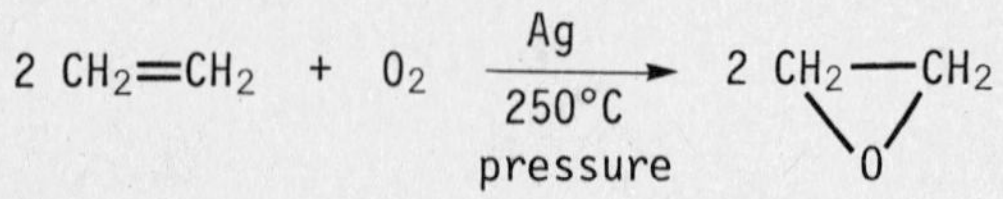

$$2 \ CH_2{=}CH_2 \ + \ O_2 \ \xrightarrow[\substack{250°C \\ pressure}]{Ag} \ 2 \ CH_2{-}CH_2 \ \underset{O}{\diagdown}$$

Other Epoxides

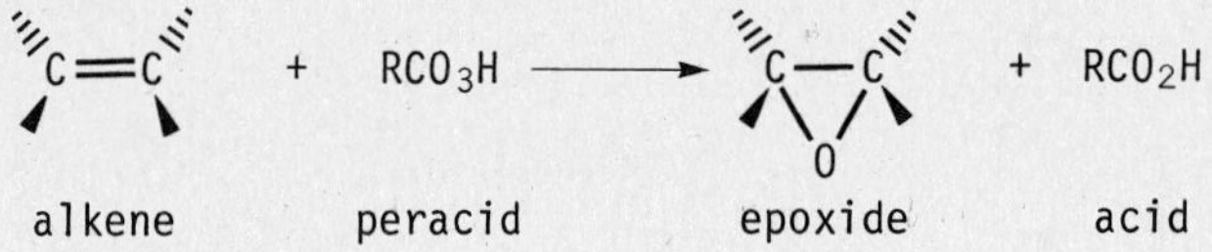

$$\text{C}{=}\text{C} \ + \ RCO_3H \ \longrightarrow \ \underset{O}{C{-}C} \ + \ RCO_2H$$

alkene peracid epoxide acid

Epoxide Ring Openings

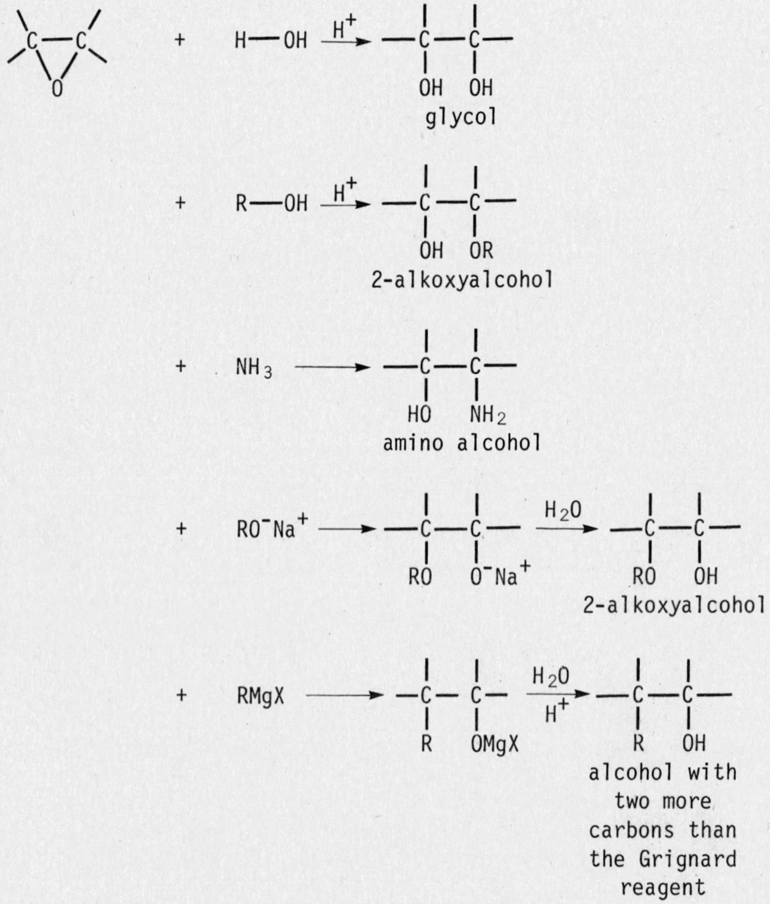

$$\underset{O}{C{-}C} \ + \ H{-}OH \ \xrightarrow{H^+} \ \underset{\substack{| \quad | \\ OH \ OH}}{-C{-}C-}$$

glycol

$$+ \ R{-}OH \ \xrightarrow{H^+} \ \underset{\substack{| \quad | \\ OH \ OR}}{-C{-}C-}$$

2-alkoxyalcohol

$$+ \ NH_3 \ \longrightarrow \ \underset{\substack{| \quad | \\ HO \ \ NH_2}}{-C{-}C-}$$

amino alcohol

$$+ \ RO^-Na^+ \ \longrightarrow \ \underset{\substack{| \quad | \\ RO \ \ O^-Na^+}}{-C{-}C-} \ \xrightarrow{H_2O} \ \underset{\substack{| \quad | \\ RO \ \ OH}}{-C{-}C-}$$

2-alkoxyalcohol

$$+ \ RMgX \ \longrightarrow \ \underset{\substack{| \ \ | \\ R \ \ OMgX}}{-C{-}C-} \ \xrightarrow[H^+]{H_2O} \ \underset{\substack{| \quad | \\ R \ \ OH}}{-C{-}C-}$$

alcohol with
two more
carbons than
the Grignard
reagent

Sulfides

$$RSH \xrightarrow{NaOH} RS^-Na^+ \xrightarrow{R'X} RSR' \qquad (preparation)$$

$$RSR \xrightarrow[25°C]{H_2O_2} \overset{O}{\underset{}{\overset{\|}{RSR}}} \xrightarrow[100°C]{H_2O_2} \overset{O}{\underset{\overset{\|}{O}}{\overset{\|}{RSR}}} \qquad (oxidation)$$

sulfoxide   sulfone

$$RSR + R'\!-\!X \longrightarrow R\!-\!\overset{\overset{\textstyle R'}{|}}{S^+}\!\!-\!R \;\; X^- \qquad (alkylation)$$

sulfonium salt

LEARNING OBJECTIVES

1. Know the meaning of: ether, alkoxy group, Grignard reagent, organometallic compound, ether cleavage.

2. Know the meaning of: epoxide, oxirane, organic peracid, nucleophilic
   addition to epoxides, diethylene glycol, ethanolamine, epoxy resin,
   epichlorhydrin, bisphenol-A.

3. Know the meaning of: cyclic ether, tetrahydrofuran, furan, tetrahydropyran, dioxane, crown ethers.

4. Know the meaning of: sulfide, thioether, sulfoxide, sulfone, sulfonium
   ion.

5. Given the name of an ether, thioether, or epoxide, write its structure,
   and vice versa.

6. Given the molecular formula, draw the structures of isomeric ethers
   and alcohols.

7. Compare the boiling points and solubilities in water of isomeric ethers
   and alcohols.

8. Write an equation for the preparation of a given Grignard reagent, and
   be able to name it.

9. Write an equation for the reaction of a given Grignard reagent and H_2O
   or D_2O.

10. Write an equation, using the appropriate Grignard reagent, for the preparation of a specific deuterium-labeled hydrocarbon.

11. Write equations for the preparation of a symmetric and an unsymmetric ether.

12. Write the equation for the cleavage of an ether by a strong acid (HBr, HI, H_2SO_4) or a Lewis acid (BBr_3).

13. Write the steps in the mechanism for cleavage of an ether.

14. Write an equation for the preparation of an epoxide from the corresponding alkene.

15. Write equations for the reaction of ethylene oxide or other epoxides with nucleophiles such as H^+ and H_2O, H^+ and alcohols, ammonia and amines, or a Grignard reagent.

16. Write the steps in the mechanism for ring-opening reactions of ethylene oxide and other epoxides.

17. Write an equation for the preparation of a given sulfide and of the corresponding sulfoxide, sulfone, or sulfonium salt.

ANSWERS TO PROBLEMS

Problems Within the Chapter

8.1 a. isopropyl methyl ether, or 2-methoxypropane
     b. phenyl *n*-propyl ether or 1-propoxybenzene
     c. 2-methoxy-3-methylbutane (note that the alkoxy substituent has priority in numbering)

8.2 a.   b. $CH_3CHCH_2CH_2CH_2CH_2CH_3$
                                               |
                                            OCH_2CH_3

8.3 $CH_2CH_2CH_2CH_2$ 1,4-butanediol
     | |
     OH   OH

     $CH_2CH_2CH_2OH$ 3-methoxy-1-propanol
     |
     OCH_3

     $CH_3OCH_2CH_2OCH_3$ 1,2-dimethoxyethane

The compounds are listed in order of decreasing boiling point. The fewer hydroxyl groups, the fewer possibilities there are for inter- molecular hydrogen bonding, and the lower the boiling point.

8.4 $(CH_3)_2CHOH$ $\xrightarrow[\text{or } PBr_3]{HBr}$ $(CH_3)_2CHBr$ $\xrightarrow[\text{ether}]{Mg}$

$(CH_3)_2CHMgBr$ $\xrightarrow{D_2O}$ $(CH_3)_2CHD$

We must first convert the alcohol to an alkyl halide before we can make the Grignard reagent.

8.5 Follow eq. 8.6, then eq. 8.5 as guides.

$CH_3CH_2CH_2Br$ $\xrightarrow[\text{ether}]{Li}$ $CH_3CH_2CH_2Li$ $\xrightarrow{D_2O}$ $CH_3CH_2CH_2D$

The resulting propane is labeled with one deuterium atom on one of the terminal carbon atoms.

8.6 Use eq. 8.7 as a model.

$2\ CH_3CH_2CH_2OH$ $\xrightarrow[140°C]{H_2SO_4}$ $CH_3CH_2CH_2OCH_2CH_2CH_3$ $+$ H_2O

8.7 a. $2\ CH_3OH$ $+$ $2\ Na$ $\longrightarrow$ $2\ CH_3O^-Na^+$ $+$ H_2

$CH_3O^-Na^+$ + [benzene ring]—CH_2Br $\longrightarrow$ [benzene ring]—CH_2OCH_3 + Na^+Br^-

or

2 [benzene ring]—CH_2OH + $2\ Na$ $\longrightarrow$ 2 [benzene ring]—$CH_2O^-Na^+$ + H_2

[benzene ring]—$CH_2O^-Na^+$ + CH_3Br $\longrightarrow$ [benzene ring]—CH_2OCH_3 + Na^+Br^-

Although both methods will work, the first is probably preferable, because S_N2 displacements occur quite easily for benzyl halides, and also because of the commercial availability of sodium meth- oxide.

b. $2\ (CH_3)_3COH$ $+$ $2\ K$ $\longrightarrow$ $2\ (CH_3)_3CO^-K^+$ $+$ H_2

$(CH_3)_3CO^-K^+$ $+$ CH_3Br $\longrightarrow$ $(CH_3)_3COCH_3$ $+$ K^+Br^-

We cannot use the alternative combination because the second step would fail; tertiary halides do *not* undergo S_N2 reactions:

$$CH_3O^-K^+ \ + \ (CH_3)_3CBr \ \longrightarrow\!\!\!/\!\!\!\longrightarrow \ CH_3OC(CH_3)_3 \ + \ K^+Br^-$$

8.8   Using HBr

Using BBr₃

8.9   Follow eq. 8.13 as a guide.

a.

b.

Note that the oxygen adds in such a way that the original geometry (*cis* or *trans*) in the alkene is maintained in the resulting epoxide.

8.10   In the first step, the epoxide oxygen is protonated by the acid catalyst:

In the second step, water acts as a nucleophile in an S_N2-type displacement:

The product is *trans*-1,2-cyclohexanediol:

8.11 $CH_2\!\!-\!\!CH_2$ (O) + $CH_3CH_2O^-Na^+$ $\xrightarrow{\text{EtOH}}$ $CH_2CH_2OCH_2CH_3$ | OH

The mechanism involves an S_N2 displacement:

Protonation gives the observed product.

8.12 Since two carbons, including the one bearing the hydroxyl group, come from the ethylene oxide (see eq. 8.17), we need a Grignard reagent with only three carbon atoms:

$$CH_3CH_2CH_2MgBr + CH_2\text{—}CH_2 \longrightarrow CH_3CH_2CH_2CH_2CH_2OMgBr$$

$$\Big\downarrow H^+$$

$$CH_3CH_2CH_2CH_2CH_2OH$$

8.13 The reaction in eq. 8.18 involves first the formation of the thiolate ion:

$$CH_3SH + NaOH \longrightarrow CH_3S^-Na^+ + H_2O$$

The second step is an S_N2 displacement:

$$CH_3S^-Na^+ + CH_3Br \longrightarrow CH_3SCH_3 + Na^+Br^-$$

Additional Problems

8.14 a. $CH_3CH_2CH_2OCH_2CH_2CH_3$
     c. $CH_3CH_2CHCH_2CH_2CH_3$
                  |
                  OCH_3

     b. $(CH_3)_3C\text{—}O\text{—}CH_3$
     d. $CH_2\text{=}CH\text{—}CH_2\text{—}O\text{—}CH_2\text{—}CH\text{=}CH_2$

     e. $Br\text{—}\langle\bigcirc\rangle\text{—}OCH_2CH_3$

     f.

     g. $CH_3OCH_2CH_2OCH_3$
     h. $CH_3CH_2\text{—}S\text{—}CH_2CH_3$

     i. $CH_3CH\text{—}CH_2$
             $\diagdown O\diagup$

     j. $CH_3CH_2CH\text{—}CH_2$
                 $\diagdown O\diagup$

8.15 a. diisopropyl ether
     c. propylene oxide (or methyloxirane)
     e. methoxycyclohexane (or cyclohexyl methyl ether)

     b. isobutyl methyl ether
     d. *p*-bromoanisole (or *p*-bromophenyl methyl ether)
     f. *t*-butyl phenyl ether

g. 2-ethoxypentane

h. 2-methoxyethanol

i. 1,2-epoxybutane (or
    ethyloxirane, or 1-butene
    epoxide)

j. methyl n-propyl sulfide

8.16   Be systematic.

$CH_3CH_2CH_2CH_2OH$
1-butanol

$CH_3OCH_2CH_2CH_3$
methyl n-propyl ether

$CH_3CH_2CHCH_3$
|
OH
2-butanol

$CH_3OCH(CH_3)_2$

methyl isopropyl ether

CH_3CHCH_2OH
|
CH_3
2-methyl-1-propanol

$CH_3CH_2OCH_2CH_3$

diethyl ether

$(CH_3)_3COH$
2-methyl-2-propanol

8.17   The actual boiling points are as follows:

1-pentanol	$CH_3CH_2CH_2CH_2CH_2OH$	137°C
1,2-dimethoxyethane	$CH_3OCH_2CH_2OCH_3$	83°C
hexane	$CH_3CH_2CH_2CH_2CH_2CH_3$	69°C
ethyl n-propyl ether	$CH_3CH_2OCH_2CH_2CH_3$	64°C

1-Pentanol is the only one of these compounds capable of forming
hydrogen bonds with itself. Thus it has the highest boiling point.
Judging from the table in Sec. 8.3, we might expect hexane to have
a slightly higher boiling point than a corresponding monoether,
although the boiling points should be quite close. 1,2-Dimethoxy-
ethane, with four polar C——O bonds, is expected to associate more
than the monoether (with only two C——O bonds); therefore we expect
it to have a boiling point that is appreciably higher than that of
ethyl n-propyl ether.
     Regarding water solubility, 1,2-dimethoxyethane has two oxygens
that can hydrogen-bond with water; the pentanol and the other ether
have only one oxygen, and the hexane has none. We expect
1,2-dimethoxyethane to be the most soluble in water of these
four compounds, and it is. In fact, the dimethoxyethane is completely soluble in water; 1-pentanol and ethyl n-propyl ether are
only slightly soluble in water, and hexane is essentially insoluble
in water.

8.18 a. $CH_3CH_2CH_2CH_2Br \xrightarrow[\text{ether}]{\text{Mg}} CH_3CH_2CH_2CH_2MgBr \xrightarrow{D_2O} CH_3CH_2CH_2CH_2D$

b. $CH_3OCH_2CH_2CH_2Br \xrightarrow[\text{ether}]{\text{Mg}} CH_3OCH_2CH_2CH_2MgBr \xrightarrow{D_2O} CH_3OCH_2CH_2CH_2D$

Note that the ether functionality can be tolerated in making a Grignard reagent (part b).

8.19 a. Since both alkyl groups are identical and primary, the dehydration route using sulfuric acid is preferred because it is least expensive and gives a good yield (see eq. 8.7).

$$2\ CH_3CH_2CH_2CH_2OH \xrightarrow[\text{ether}]{H_2SO_4} CH_3CH_2CH_2CH_2OCH_2CH_2CH_2CH_3$$

b. The Williamson method is preferred. The sodium phenoxide can be prepared using NaOH (instead of Na) because of the acidity of phenols:

8.20

Both alcohols will be protonated by the acid catalyst, but only the tertiary alcohol (t-butyl alcohol) can form a carbocation; the cation is then trapped by the methanol, which is only a weak nucleophile but is present in large excess. Thus the various equilibria are driven in the forward direction, as shown.

8.21   The second step, that is:

$$\langle\!\!\bigcirc\!\!\rangle\!-O^-Na^+ \;+\; \langle\!\!\bigcirc\!\!\rangle\!-Br \;\xrightarrow{\;\;/\!\!\!/\;\;}\; \langle\!\!\bigcirc\!\!\rangle\!-O\!-\!\langle\!\!\bigcirc\!\!\rangle \;+\; Na^+Br^-$$

fails because S_N2 displacements cannot be carried out on aryl halides.

8.22   The oxygen of the ether can be protonated, and the resulting highly polar dialkyloxonium ion is soluble in sulfuric acid:

$$R\!-\!\overset{..}{\underset{..}{O}}\!-\!R \;+\; H_2SO_4 \;\longrightarrow\; R\!-\!\overset{\overset{\displaystyle H}{|}}{\underset{..}{O}}{}^{\!+}\!-\!R \;+\; HSO_4^-$$

dialkyloxonium ion

Alkanes have no unshared electron pairs and are not protonated by sulfuric acid, and thus remain insoluble in it.

8.23   a.   No reaction; ethers (except for epoxides) are inert toward base.
        b. $CH_3OCH_2CH_2CH_3 \;+\; 2\,HBr \longrightarrow CH_3Br \;+\; CH_3CH_2CH_2Br \;+\; H_2O$
        c.   No reaction; ethers can be distinguished from alcohols by their inertness toward sodium metal.

        d. $CH_3CH_2\overset{..}{\underset{..}{O}}CH_2CH_3 \;+\; H_2SO_4 \xrightarrow{\text{cold}} CH_3CH_2\overset{\overset{\displaystyle +}{\,}}{\underset{\underset{\displaystyle H}{|}}{\overset{..}{O}}}CH_2CH_3 \;+\; HSO_4^-$

        The ether acts as a base and dissolves in the strong acid.

        e. $\langle\!\!\bigcirc\!\!\rangle\!-OCH_2CH_3 \xrightarrow{\;BBr_3\;} \langle\!\!\bigcirc\!\!\rangle\!-OH \;+\; CH_3CH_2Br$

        Compare with eq. 8.11.

8.24 $\begin{array}{c} H_2C\!-\!CH_2 \\[-2pt] \diagup\qquad\diagdown \\ H_2C\underset{O}{\diagdown\quad\diagup}CH_2 \end{array} \xrightarrow{\;2\,HBr\;} BrCH_2CH_2CH_2CH_2Br \;+\; H_2O$

tetrahydrofuran

8.25 $CH_2\!\!=\!\!CHCH_2CH_3 \xrightarrow[\substack{\text{(refer to}\\ \text{eq. 8.13)}}]{CH_3CO_3H} CH_2\!\!-\!\!CHCH_2CH_3$

$CH_2\!\!-\!\!CHCH_2CH_3 \xleftarrow[\substack{\text{(refer to}\\ \text{eq. 8.14)}}]{H_2O,\ H^+}$
| |
OH   OH

8.26   a. $CH_2\!\!-\!\!CH_2 + HBr \longrightarrow CH_2\!\!-\!\!CH_2$
                                                    | |
                                                   OH   Br

        b. $CH_2\!\!-\!\!CH_2 + 2HBr \longrightarrow CH_2\!\!-\!\!CH_2 + H_2O$
                                                      | |
                                                     Br   Br

The 2-bromoethanol formed in part a reacts as an alcohol with the second mole of HBr, to produce the dibromide.

        c. $CH_2\!\!-\!\!CH_2 + HO\!\!-\!\!\bigcirc \xrightarrow{H^+} CH_2CH_2\!\!-\!\!O\!\!-\!\!\bigcirc$
                                                          |
                                                         OH

                                              2-phenoxyethanol

8.27   See eq. 8.15 for comparison.

$CH_2\!\!-\!\!CH_2 + CH_3CH_2OH \xrightarrow{H^+} HOCH_2CH_2OCH_2CH_3$
                                              ethyl cellosolve

$CH_2\!\!-\!\!CH_2 + HOCH_2CH_2OCH_2CH_3 \xrightarrow{H^+} HOCH_2CH_2OCH_2CH_2OCH_2CH_3$
                                              ethyl carbitol

8.28

$\bigcirc\!\!-\!\!Br \xrightarrow[\text{ether}]{Mg} \bigcirc\!\!-\!\!MgBr \xrightarrow{\substack{CH_2\!\!-\!\!CH_2 \\ O}}$

$\bigcirc\!\!-\!\!CH_2CH_2OMgBr \xrightarrow[H_2O]{H^+} \bigcirc\!\!-\!\!CH_2CH_2OH$

                                              2-phenylethanol
                                              (oil of roses)

Compare with eq. 8.1 and eq. 8.17.

8.29 After protonation of the oxygen, the epoxide ring opens in an S_N1 manner to give the tertiary carbocation; this ion then reacts with a molecule of methanol (a weak nucleophile).

The regioisomer
$$CH_3\overset{\overset{\displaystyle CH_3}{|}}{\underset{\underset{\displaystyle OH}{|}}{C}}-CH_2OCH_3$$
is *not* formed, because methanol is a weak nucleophile and S_N2 attack on the protonated oxirane at the primary carbon cannot compete with the fast S_N1 process.

8.30
$$CH_2{=\!=}CH_2 + HO-Cl \longrightarrow \overset{\displaystyle }{\underset{\underset{\displaystyle OH}{|}}{CH_2}}-\overset{\displaystyle }{\underset{\underset{\displaystyle Cl}{|}}{CH_2}}$$

This step occurs by an electrophilic addition mechanism:

This step occurs by an intramolecular S_N2 displacement mechanism:

8.31

$$HO-\langle\bigcirc\rangle-\overset{\overset{CH_3}{|}}{\underset{\underset{CH_3}{|}}{C}}-\langle\bigcirc\rangle-OH \;+\; 2\,NaOH \;\rightleftharpoons\; Na^+\,{}^-O-\langle\bigcirc\rangle-\overset{\overset{CH_3}{|}}{\underset{\underset{CH_3}{|}}{C}}-\langle\bigcirc\rangle-O^-\,Na^+$$

$$ClCH_2CH-CH_2 \;+\; {}^-O-\langle\bigcirc\rangle-\overset{\overset{CH_3}{|}}{\underset{\underset{CH_3}{|}}{C}}-\langle\bigcirc\rangle-O^- \;+\; CH_2-CHCH_2Cl \xrightarrow{\;S_N2\;}$$

$$Cl^- CH_2CH-CH_2O-\langle\bigcirc\rangle-\overset{\overset{CH_3}{|}}{\underset{\underset{CH_3}{|}}{C}}-\langle\bigcirc\rangle-OCH_2-CHCH_2-Cl \xrightarrow[-2\,Cl^-]{\;S_N2\;}$$

$$CH_2-CHCH_2O-\langle\bigcirc\rangle-\overset{\overset{CH_3}{|}}{\underset{\underset{CH_3}{|}}{C}}-\langle\bigcirc\rangle-OCH_2CH-CH_2 \xrightarrow{\;2\text{ bisphenol-A}\;}$$

$${}^-O-\langle\bigcirc\rangle-\overset{\overset{CH_3}{|}}{\underset{\underset{CH_3}{|}}{C}}-\langle\bigcirc\rangle-OCH_2\underset{\underset{OH}{|}}{C}HCH_2O-\langle\bigcirc\rangle-\overset{\overset{CH_3}{|}}{\underset{\underset{CH_3}{|}}{C}}-\langle\bigcirc\rangle-OCH_2\underset{\underset{OH}{|}}{C}HCH_2-O-\langle\bigcirc\rangle-\overset{\overset{CH_3}{|}}{\underset{\underset{CH_3}{|}}{C}}-\langle\bigcirc\rangle-O^-$$

$$\xrightarrow{\;2\text{ epichlorhydrin}\;} \quad \text{and so on}$$

8.32 First the ethylene oxide is protonated by the acid catalyst:

$$CH_2-CH_2 \;+\; H^+ \;\rightleftharpoons\; CH_2-CH_2$$

The alcohol or glycol then act as nucleophiles in an S_N2 displacement, which occurs quite easily because the epoxide ring opens in the process, thus relieving the strain associated with the small ring:

$$CH_2-CH_2 \;+\; R-\ddot{O}H \;\longrightarrow\; \underset{\underset{OH}{|}}{CH_2}-CH_2-\overset{+}{\underset{\underset{H}{|}}{O}}-R$$

For eq. 8.15,
R = $-CH_3$ or $-CH_2CH_2OH$

$$\underset{\underset{OH}{|}}{CH_2}-CH_2-OR \;+\; H^+$$

8.33 a. Add a little of each compound to concentrated sulfuric acid, in separate test tubes. The ether is protonated and dissolves, whereas the hydrocarbon, being inert and less dense than sulfuric acid, simply floats on top.

 b. Add a little bromine in carbon tetrachloride to each ether. The allyl phenyl ether, being unsaturated, quickly decolorizes the bromine, whereas the ethyl phenyl ether does not.

$$CH_2\!\!=\!\!CHCH_2OC_6H_5 \ + \ Br_2 \ \longrightarrow \ \underset{\underset{Br}{|}}{CH_2}\!\!-\!\!\underset{\underset{Br}{|}}{CHCH_2OC_6H_5}$$

$$CH_3CH_2OC_6H_5 \ + \ Br_2 \ \longrightarrow \ \text{no reaction}$$

 c. Add a small piece of sodium to each compound. The alcohol liberates a gas (hydrogen), whereas no gas bubbles are apparent in the ether.

$$2 \ \underset{\underset{OH}{|}}{CH_3CHCH_2CH_3} \ + \ 2 \ Na \ \longrightarrow \ 2 \ \underset{\underset{O^-Na^+}{|}}{CH_3CHCH_2CH_3} \ + \ H_2$$

$$CH_3OCH_2CH_2CH_3 \ + \ Na \ \longrightarrow \ \text{no reaction}$$

 d. Add each compound to a little 10% aqueous sodium hydroxide. The phenol dissolves, whereas the ether is inert toward the base.

8.34 Since the product has only two carbons and the starting material has four carbons, two groups of two carbons must be separated by an ether oxygen.

$$-C-C-O-C-C-$$

The remaining two oxygens ($C_4H_{10}O_3$) must be at the ends of the chain. The desired structure is

$$HO-CH_2-CH_2-O-CH_2-CH_2-OH$$

and the equation for the reaction with HBr is

$$HOCH_2CH_2OCH_2CH_2OH \;+\; 4\,HBr \longrightarrow 2\,BrCH_2CH_2Br \;+\; 3\,H_2O$$

In this step the ether is cleaved, and the alcohol functions are also converted to alkyl halides.

8.35 The overall equation is

$$2\ HOCH_2CH_2OH \xrightarrow[\text{heat}]{H^+} \quad\begin{array}{c}O\\ \bigcirc \\ O\end{array}\quad +\ 2H_2O$$

The reaction is similar to the preparation of diethyl ether from ethanol (eq. 8.7).

Protonation of the hydroxyl group is followed by a nucleophilic (S_N2) displacement:

$$H\ddot{O}-CH_2CH_2-\ddot{O}H \;+\; H^+ \;\rightleftharpoons\; HO-CH_2CH_2-\overset{+}{\underset{H}{O}}-H$$

$$HO-CH_2CH_2-\ddot{O}H \;+\; \begin{array}{c}CH_2-\overset{+}{\underset{\;}{O}}-H\\ |\\ CH_2OH\end{array} \longrightarrow \begin{array}{c}H\\ |\\ HOCH_2CH_2\overset{+}{\underset{\;}{O}}-CH_2\\ |\\ CH_2OH\end{array}$$

$$\Big\Updownarrow -H^+$$

$$HOCH_2CH_2OCH_2CH_2OH$$

The process is now repeated, except that the nucleophilic displacement is *intra*molecular:

$$\text{(structure)} + H^+ \rightleftharpoons \text{(structure)}$$

$\downarrow S_N2$ (intramolecular)

$$\text{(structure)} \overset{-H^+}{\rightleftharpoons} \text{(structure)} + H_2O$$

8.36 The unshared electron pair on oxygen in furan is delocalized into the aromatic π system (review Sec. 4.16). Therefore, this oxygen is less capable of forming hydrogen bonds with water than is the oxygen in tetrahydrofuran, whose unshared pairs are localized and available for hydrogen bonding.

8.37 $CH_3 — \overset{\cdot\cdot}{S} — CH_3$ no formal charge

$$\left[\begin{array}{ccc} \overset{:O:}{\underset{}{\|}} & & \overset{:\overset{..}{O}:^-}{\underset{}{|}} \\ CH_3 — \overset{\cdot\cdot}{S} — CH_3 & \longleftrightarrow & CH_3 — S^+ — CH_3 \end{array} \right]$$

The sulfur-oxygen bond is highly polarized, with a positive charge on sulfur and a negative charge on oxygen. Polarity causes the molecules of dimethyl sulfoxide to associate. Here are two ways this can happen:

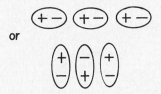

or

Energy (heat) is required to break up this molecular association. Therefore, the boiling point of dimethyl sulfoxide is much higher than that of dimethyl sulfide.

8.38　a.　$\langle\!\!\bigcirc\!\!\rangle$—$CH_2OH$ $\xrightarrow{\ HBr\ }$ $\langle\!\!\bigcirc\!\!\rangle$—$CH_2Br$ $\xrightarrow{\ NaSH\ }$

$\langle\!\!\bigcirc\!\!\rangle$—$CH_2SH$ $\xrightarrow[\ 2.\ CH_3CH_2Br\ ]{1.\ NaOH}$ $\langle\!\!\bigcirc\!\!\rangle$—$CH_2SCH_2CH_3$

b.　$CH_2\!\!=\!\!CHCH_2Br$ $\xrightarrow{\ NaSH\ }$ $CH_2\!\!=\!\!CHCH_2SH$ $\xrightarrow[\ 2.\ CH_2=CHCH_2Br\ ]{1.\ NaOH}$

$CH_2\!\!=\!\!CHCH_2SCH_2CH\!\!=\!\!CH_2$

CHAPTER NINE   ALDEHYDES AND KETONES

CHAPTER SUMMARY

The **carbonyl group**, $\diagdown$C$=$O, is present in both **aldehydes** (RCH$=$O)
and **ketones** (R$_2$C$=$O). The IUPAC ending for naming aldehydes is -*al*, and
numbering begins with the carbonyl carbon. The ending for the names of
ketones is -*one*, and the longest chain is numbered as usual. Common names
are also widely used. Nomenclature is outlined in Sec. 9.2.

Formaldehyde, acetaldehyde, and **acetone** are important commercial
chemicals, synthesized by special methods. In the laboratory, aldehydes
and ketones are most commonly prepared by oxidizing alcohols, but they can
also be prepared by ozonizing alkenes and by hydrating alkynes. Aldehydes
and ketones occur widely in nature (see Fig. 9.1).

The carbonyl group is planar, with the sp^2 carbon trigonal. The C$=$O
bond is polarized, with C positive and O negative. Many carbonyl reactions
are initiated by nucleophilic addition to the positive carbon and completed
by addition of a proton to the oxygen.

With acid catalysis, alcohols add to the carbonyl group of aldehydes
to give **hemiacetals** [RCH(OH)OR']; further reaction with excess alcohol gives
acetals [RCH(OR')$_2$]. Ketones react similarly to give **hemiketals** or **ketals**.
The reactions are reversible; that is, acetals or ketals can be readily
hydrolyzed (by H$_2$O, H$^+$) to their alcohol and carbonyl components.

Water adds similarly to the carbonyl group of certain aldehydes (for
example, formaldehyde and chloral) to give hydrates.

Grignard reagents add to carbonyl compounds. The products, after
hydrolysis, are alcohols whose structure depends on that of the starting
carbonyl compound. Formaldehyde gives *primary* alcohols, other aldehydes
give *secondary* alcohols, and ketones give *tertiary* alcohols.

Hydrogen cyanide adds to carbonyl compounds as a carbon nucleophile
to give **cyanohydrins** [R$_2$C(OH)CN].

Nitrogen nucleophiles add to the carbonyl group. Often, addition is followed by elimination of water to give a product with a $\diagup\!\!\!\diagdown C\!=\!NR$ group in place of the $\diagup\!\!\!\diagdown C\!=\!O$ group. For example, primary amines give **imines** (Schiff's bases); **hydroxylamine** (NH_2OH) gives **oximes** $\diagup\!\!\!\diagdown C\!=\!NOH$; and **hydrazine** ($NH_2NH_2$) gives **hydrazones** $\diagup\!\!\!\diagdown C\!=\!NNH_2$.

Aldehydes and ketones are easily reduced to primary or secondary alcohols, respectively. Useful reagents for this purpose are **lithium aluminum hydride** ($LiAlH_4$), **sodium borohydride** ($NaBH_4$), or hydrogen and a metal catalyst.

Aldehydes are more easily oxidized than ketones. The **Tollens' silver mirror test** and the **Fehling's** and **Benedict's tests** are positive for aldehydes, negative for ketones.

Aldehydes or ketones with an α-hydrogen exist as an equilibrium mixture of **keto** $\diagup\!\!\!\diagdown\!\overset{\alpha}{C}H\!-\!C\!=\!O$ and **enol** $\diagup\!\!\!\diagdown C\!=\!C\!-\!OH$ **tautomers.** The keto form usually predominates. An α-hydrogen is weakly acidic and can be removed by a base to produce a resonance-stabilized **enolate anion.** Deuterium exchange of α-hydrogens provides experimental evidence for enols as reaction intermediates.

In the **aldol condensation,** an enolate anion acts as a carbon nucleophile and adds to a carbonyl group to form a new carbon-carbon bond. Thus, the α-carbon of one aldehyde molecule becomes bonded to the carbonyl carbon of another aldehyde molecule to form an aldol (a 3-hydroxyaldehyde). In the **mixed aldol condensation,** the reactant with an α-hydrogen supplies the enolate anion, and the other reactant, usually without an α-hydrogen, supplies the carbonyl group to which the enolate ion adds. The aldol reaction is used commercially and also occurs in nature.

Quinones are cyclic conjugated diketones. They are colored compounds used as dyes. They also play important roles in reversible biological oxidation-reduction (electron-transfer) reactions.

REACTION SUMMARY

Preparation of Aldehydes and Ketones

$$\underset{C}{\overset{H}{>}}\underset{OH}{\overset{H}{<}}C \xrightarrow{\text{PCC}} \underset{C}{\overset{H}{>}}C{=}0 \qquad \text{(from primary alcohols)}$$

aldehyde

$$\underset{C}{\overset{C}{>}}\underset{OH}{\overset{H}{<}}C \xrightarrow{Cr^{6+}} \underset{C}{\overset{C}{>}}C{=}0 \qquad \text{(from secondary alcohols)}$$

ketone

$$\underset{R}{\overset{H}{>}}C{=}C\underset{R}{\overset{R}{<}} \xrightarrow{O_3} \underset{R}{\overset{H}{>}}C{=}0 \;+\; 0{=}C\underset{R}{\overset{R}{<}} \qquad \text{(from alkenes)}$$

$$R-C{\equiv}C-H \xrightarrow[\text{Hg}^{2+}]{\text{H}_3O^+} R-\overset{\overset{\displaystyle O}{\|}}{C}-CH_3 \qquad \text{(from alkynes)}$$

Hemiacetals and Acetals

$$ROH \;+\; R'-\overset{\overset{\displaystyle O}{\|}}{CH} \underset{}{\overset{H^+}{\rightleftharpoons}} R'-CH\overset{OH}{\underset{OR}{<}} \xrightarrow[\text{}]{ROH,\ H^+} R'-CH\overset{OR}{\underset{OR}{<}} \;+\; H_2O$$

$$\qquad\qquad\qquad\qquad\qquad \text{hemiacetal} \qquad\qquad\qquad \text{acetal}$$

Grignard Reagents

$$RMgX \;+\; CH_2{=}0 \longrightarrow RCH_2OMgX \xrightarrow{H_3O^+} RCH_2OH$$

$$\qquad\qquad \text{formaldehyde} \qquad\qquad\qquad\qquad \text{primary alcohol}$$

$$RMgX \;+\; R'CH{=}0 \longrightarrow \underset{R'}{\overset{R}{>}}CHOMgX \xrightarrow{H_3O^+} \underset{R'}{\overset{R}{>}}CHOH$$

$$\qquad\qquad \text{other aldehydes} \qquad\qquad\qquad \text{secondary alcohol}$$

$$RMgX \;+\; \underset{R''}{\overset{R'}{>}}C{=}0 \longrightarrow R'-\overset{\overset{\displaystyle R}{|}}{\underset{\underset{\displaystyle R''}{|}}{C}}-OMgX \xrightarrow{H_3O^+} R'-\overset{\overset{\displaystyle R}{|}}{\underset{\underset{\displaystyle R''}{|}}{C}}-OH$$

$$\qquad\qquad \text{ketone} \qquad\qquad\qquad\qquad\qquad\qquad \text{tertiary alcohol}$$

Hydrogen Cyanide

$$HCN \; + \; \underset{/}{\overset{\backslash}{C}}{=}O \; \rightleftharpoons \; \underset{/}{\overset{\backslash}{C}} \underset{\underset{N}{\overset{|}{C}}}{\overset{OH}{\Big|}}$$

cyanohydrin

Nitrogen Nucleophiles

$$-\ddot{N}H_2 \; + \; \underset{/}{\overset{\backslash}{C}}{=}O \; \rightleftharpoons \; \underset{/}{\overset{\backslash}{C}} \underset{NH-}{\overset{OH}{\Big|}} \; \xrightarrow{-H_2O} \; \underset{/}{\overset{\backslash}{C}}{=}N$$

Reduction

$$\underset{/}{\overset{\backslash}{C}}{=}O \; \xrightarrow[\text{or } H_2, \text{ catalyst, heat}]{\text{LiAlH}_4 \text{ or NaBH}_4} \; \underset{/}{\overset{\backslash}{C}} \underset{OH}{\overset{H}{\Big\backslash}}$$

Oxidation

$$RCH{=}O \; + \; 2 \; Ag(NH_3)_2^+ \; + \; 3 \; OH^- \; \xrightarrow{\underset{\text{test}}{\text{Tollens'}}}$$

aldehyde

$$R{-}\overset{\overset{O}{\|}}{C}{-}O^- \; + \; 2 \; Ag \; + \; 4 \; NH_3 \; + \; 2 \; H_2O$$

acid anion   silver
mirror

$$2 \; RCH{=}O \; + \; O_2 \longrightarrow 2 \; RCO_2H$$

Tautomerism

$$-\overset{\overset{H}{|}}{\underset{|}{C}}{-}\overset{\overset{O}{\|}}{C}{-} \; \rightleftharpoons \; -\overset{}{C}{=}\overset{\overset{H}{|}}{C}{-}$$

keto form   enol form

$$\xrightarrow{\text{base}} \left[-\overset{\overset{\displaystyle ..}{}}{\underset{|}{C}}{-}\overset{\overset{O}{\|}}{C}{-} \longleftrightarrow -\overset{}{C}{=}\overset{\overset{O^-}{|}}{C}{-} \right]$$

enolate anion

180

REACTION SUMMARY

Deuterium Exchange

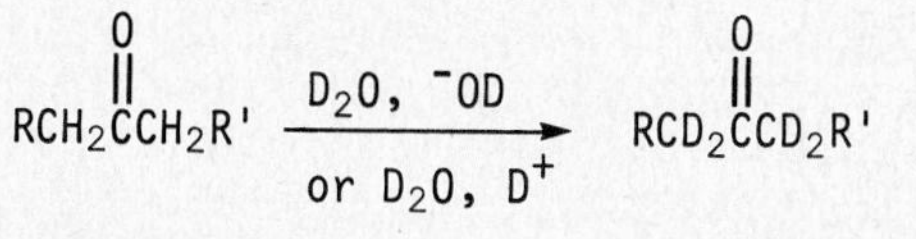

$$\underset{\displaystyle \mathrm{RCH_2CCH_2R'}}{\overset{\displaystyle \mathrm{O}}{\|}} \xrightarrow[\text{or } D_2O,\ D^+]{D_2O,\ ^-OD} \underset{\displaystyle \mathrm{RCD_2CCD_2R'}}{\overset{\displaystyle \mathrm{O}}{\|}} \quad \text{(only } \alpha\text{-hydrogens exchange)}$$

Aldol Condensation

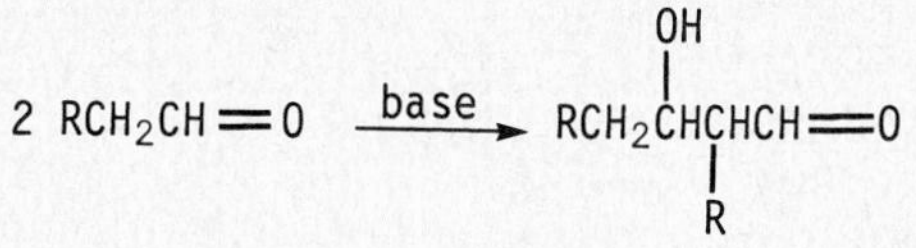

$$2\ \mathrm{RCH_2CH{=}O} \xrightarrow{\text{base}} \underset{\displaystyle \mathrm{R}}{\mathrm{RCH_2\overset{\displaystyle \mathrm{OH}}{\underset{|}{CH}}CHCH{=}O}}$$

Mixed Aldol Condensation

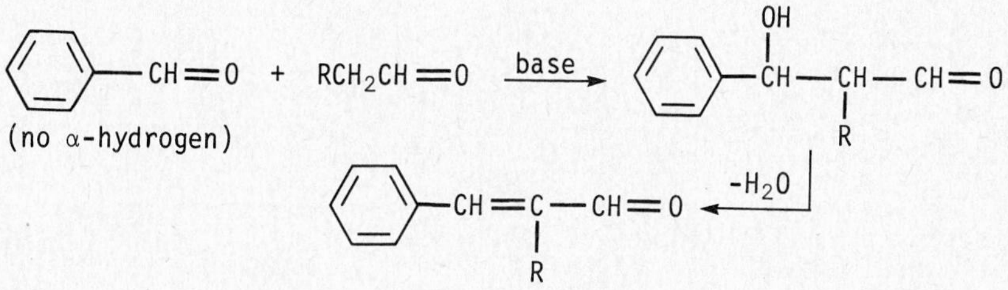

(no α-hydrogen)

MECHANISM SUMMARY

Nucleophilic Addition

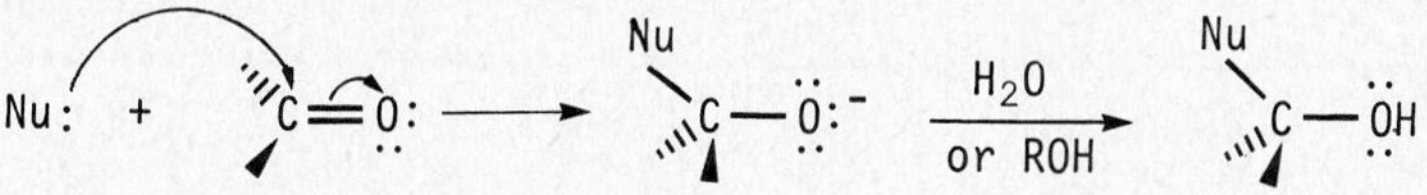

LEARNING OBJECTIVES

1. Know the meaning of: aldehyde, ketone, carbonyl group, formaldehyde, acetaldehyde, benzaldehyde, acetone, salicylaldehyde, acetophenone, benzophenone, carbaldehyde group.

2. Know the meaning of: nucleophilic addition, hemiacetal and acetal, hemiketal and ketal, aldehyde hydrate, cyanohydrin.

3. Know the meaning of: imine, Schiff's base, hydroxylamine, oxime, hydrazine, hydrazone, phenylhydrazine, phenylhydrazone.

4. Know the meaning of: lithium aluminum hydride, sodium borohydride, Tollens' reagent, silver mirror test, Fehling's and Benedict's test.

5. Know the meaning of: keto form, enol form, tautomers, tautomerism, enolate anion, α-hydrogen and α-carbon, aldol condensation, mixed aldol condensation.

6. Given the structure of an aldehyde or ketone, state its IUPAC name.

7. Given the IUPAC name of an aldehyde or ketone, write its structure.

8. Write the resonance contributors to the carbonyl group.

9. Given the structure or name of an aldehyde or ketone, write an equation for its reaction with the following nucleophiles: alcohol or thiol, cyanide ion, Grignard reagent or acetylide, hydroxylamine, hydrazine, phenylhydrazine, 2,4-dinitrophenylhydrazine, primary amine, lithium aluminum hydride, and sodium borohydride.

10. Explain the mechanism of acid catalysis of nucleophilic additions to the carbonyl group.

11. Write the steps in the mechanism of acetal formation and hydrolysis; draw the structures of resonance contributors to intermediates in the mechanism.

12. Given a carbonyl compound and a Grignard reagent, write the structure of the alcohol that is formed when they react.

13. Given the structure of a primary, secondary, or tertiary alcohol, deduce what combination of aldehyde or ketone and Grignard reagent can be used for its synthesis.

14. Given the structure of an aldehyde or ketone, write the formula of the alcohol that is obtained from it by reduction.

15. Given the structure of an aldehyde, write the structure of the acid that is formed from it by oxidation.

16. Know which tests can distinguish an aldehyde from a ketone.

17. Given the structure of an aldehyde or ketone, write the structure of the corresponding enol and enolate anion.

18. Identify the α-hydrogens in an aldehyde or ketone, and be able to recognize that these hydrogens can be exchanged readily for deuterium.

19. Write the structure of the aldol formed by the self-condensation of an aldehyde of given structure.

20. Given two reacting carbonyl compounds, write the structure of the mixed aldol obtained from them.

21. Write the steps in the mechanism of the aldol condensation.

ANSWERS TO PROBLEMS

Problems Within the Chapter

9.1 a. $CH_3CH_2CH_2CH_2CH{=}0$ b. 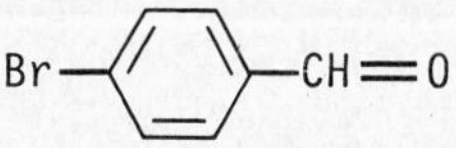 $Br{-}\langle\bigcirc\rangle{-}CH{=}0$

c. 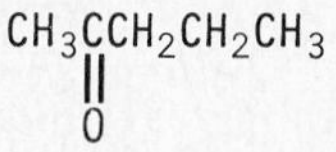 $CH_3CCH_2CH_2CH_3$ d. 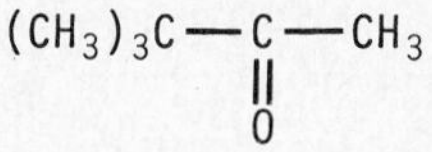 $(CH_3)_3C{-}C{-}CH_3$
    $\quad\;\overset{\|}{O}$ $\qquad\qquad\overset{\|}{O}$

9.2 a. 3-methylbutanal   b. 2-butenal
         (no number is necessary   (the number locates the double
         for the aldehyde function)   bond between C-2 and C-3)
     b. cyclobutanone   d. 4-methyl-2-pentanone or isobutyl
                                                  methyl ketone

9.3 a. 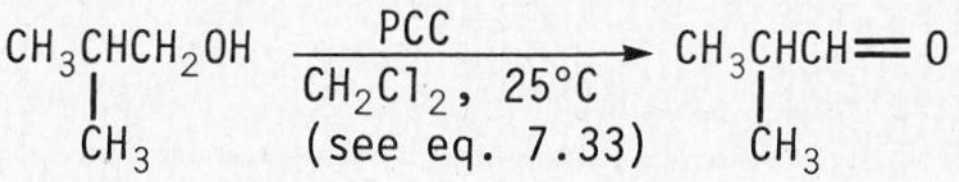 CH_3CHCH_2OH $\xrightarrow[\substack{CH_2Cl_2,\ 25°C \\ (see\ eq.\ 7.33)}]{PCC}$ $CH_3CHCH{=}0$
         $\;\;\overset{|}{CH_3}$ $\;\;\;\overset{|}{CH_3}$

b. 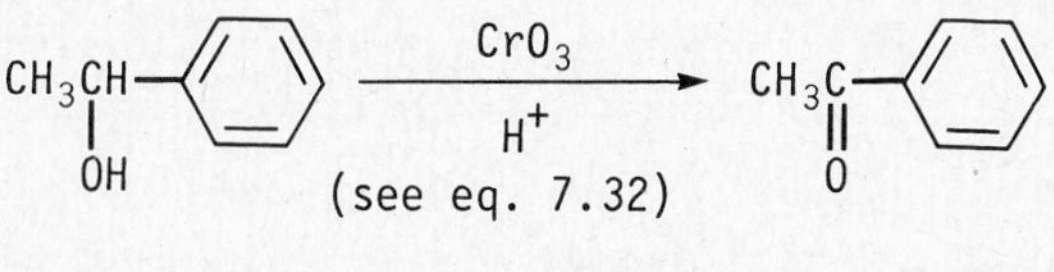 $CH_3CH{-}\langle\bigcirc\rangle$ $\xrightarrow[\substack{H^+ \\ (see\ eq.\ 7.32)}]{CrO_3}$ $CH_3C{-}\langle\bigcirc\rangle$
    $\;\;\;\overset{|}{OH}$ $\;\;\;\overset{\|}{O}$

9.4 $\xrightarrow[\substack{2.\ \textbf{Zn, H}^+ \\ (see\ eq.\ 3.46)}]{1.\ \textbf{O}_3}$

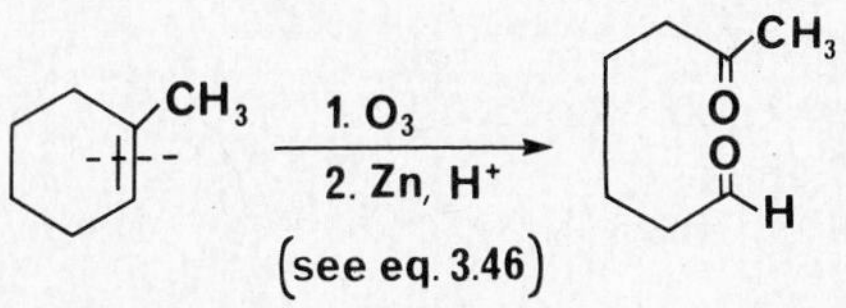

9.5 $\quad HC\equiv CCH_2CH_2CH_2CH_3 \xrightarrow[\substack{Hg^{2+} \\ \text{(see eq. 3.52)}}]{H_3O^+} CH_3CCH_2CH_2CH_2CH_3$
$\quad\quad\quad\quad\quad\quad\quad\quad\quad\quad\quad\quad\quad\quad\quad\quad\quad\quad\quad\overset{\parallel}{O}$

9.6   a. CH_3—⬡—CH_3 < ⬡—$CH=O$ < ⬡—CH_2OH < HO—⬡—OH

         p-xylene   benzaldehyde   benzyl alcohol   hydroquinone
         bp 138°C   bp 179°C   bp 205°C   bp 286°C

        The carbonyl compound is more polar than the hydrocarbon, but cannot form hydrogen bonds with itself. The compound with two hydroxyl groups hydrogen-bonds more effectively than the compound with just one hydroxyl group.

      b.   Water solubility increases in the same order, because of increasing possibilities for hydrogen bonding with water.

9.7 $\quad CH_3CH_2OH + CH_3CH=O \underset{H^+}{\rightleftharpoons} CH_3CH\overset{OH}{\underset{OCH_2CH_3}{<}}$

                                 hemiacetal

Mechanism:

$\quad CH_3CH=O + H^+ \rightleftharpoons [CH_3CH=\overset{+}{O}H \longleftrightarrow CH_3\overset{+}{C}H-OH]$

$\quad CH_3CH_2\overset{..}{\underset{..}{O}}H + CH_3CH=\overset{+}{\underset{..}{O}}H \rightleftharpoons CH_3CH\overset{OH}{\underset{\underset{H}{\overset{|}{:}\overset{+}{O}-CH_2CH_3}}{<}}$

$\quad\quad\quad\quad\quad\quad\quad\quad\quad\quad\quad\quad\quad\quad\quad\quad\quad\quad \updownarrow -H^+$

$\quad\quad\quad\quad\quad\quad\quad\quad\quad\quad\quad\quad CH_3CH\overset{OH}{\underset{OCH_2CH_3}{<}} + H^+$

9.8
$$\underset{\overset{|}{OH}}{CH_3CHOCH_2CH_3} \;+\; CH_3CH_2OH \;\xrightarrow{\;H^+\;}\; CH_3CH\overset{\displaystyle OCH_2CH_3}{\underset{\displaystyle OCH_2CH_3}{<}} \;+\; H_2O$$

Mechanism:

$$\underset{\overset{|}{OH}}{CH_3CHOCH_2CH_3} \;+\; H^+ \;\rightleftharpoons\; CH_3CHOCH_2CH_3$$

with :Ö(H)(+)(H) on the carbon bearing, $\downarrow$ $-H_2O$

$$-CH_3CH\!-\!\ddot{O}CH_2CH_3 \quad \overset{+}{}$$

$$\updownarrow$$

$$-CH_3CH\!=\!\overset{+}{\ddot{O}}CH_2CH_3$$

$$\underset{\overset{|}{OCH_2CH_3}}{CH_3CH}\overset{H}{\underset{}{:}}\overset{+}{O}\!-\!CH_2CH_3 \quad \xleftarrow{\;CH_3CH_2\ddot{O}H\;}$$

$$\downarrow\; -H^+$$

$$CH_3CH\overset{\displaystyle OCH_2CH_3}{\underset{\displaystyle OCH_2CH_3}{<}}$$

9.9   The proton first adds to one of the oxygens:

$$Ph\!-\!CH\overset{\displaystyle \ddot{O}CH_3}{\underset{\displaystyle \ddot{O}CH_3}{<}} \;+\; H^+ \;\rightleftharpoons\; Ph\!-\!CH\overset{\displaystyle \overset{H}{\underset{|}{\overset{+}{O}CH_3}}}{\underset{\displaystyle OCH_3}{<}}$$

Loss of methanol gives a resonance-stabilized carbocation:

$$Ph\!-\!CH\overset{\displaystyle \overset{H}{\underset{|}{\overset{+}{O}CH_3}}}{\underset{\displaystyle OCH_3}{<}} \;\longrightarrow\; CH_3OH \;+\; Ph\!-\!\overset{+}{C}H\!-\!\ddot{O}CH_3$$

$$\updownarrow$$

$$Ph\!-\!CH\!=\!\overset{+}{\ddot{O}}CH_3$$

The carbocation reacts with water, which is a nucleophile and is present in large excess:

hemiacetal

The sequence is then repeated, beginning with protonation of the methoxyl oxygen of the hemiacetal:

The whole process is driven forward because water is present in excess.

9.10   In the first step, one bromine is replaced by a hydroxyl group:

Loss of HBr gives acetone:

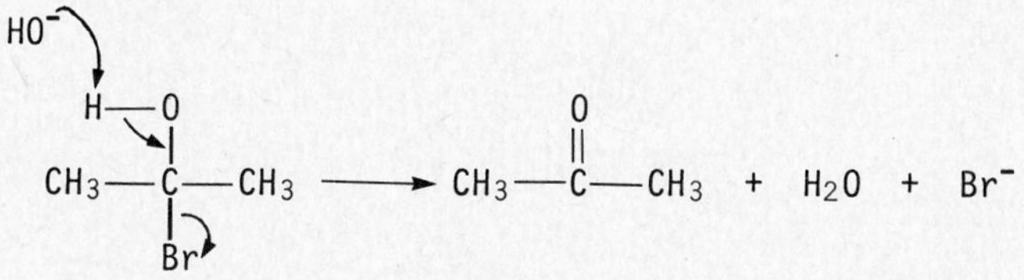

$$CH_3-\overset{\overset{\displaystyle H-O}{|}}{\underset{\underset{\displaystyle Br}{|}}{C}}-CH_3 \longrightarrow CH_3-\overset{\overset{\displaystyle O}{||}}{C}-CH_3 \;+\; H_2O \;+\; Br^-$$

(HO$^-$)

Even if both bromines were replaced by hydroxyl groups, the resulting diol (acetone hydrate) would lose water, since acetone does not form a stable hydrate.

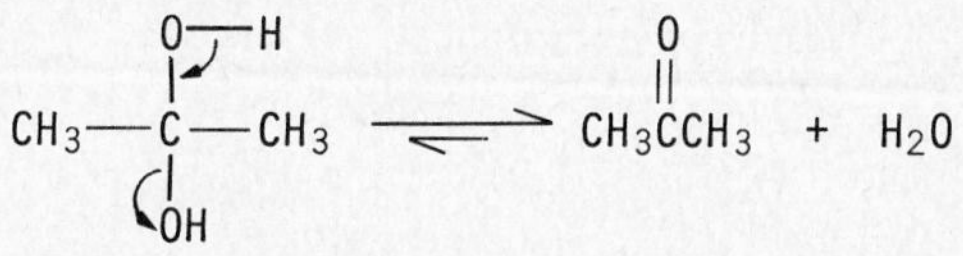

$$CH_3-\overset{\overset{\displaystyle O-H}{|}}{\underset{\underset{\displaystyle OH}{|}}{C}}-CH_3 \;\rightleftharpoons\; CH_3\overset{\overset{\displaystyle O}{||}}{C}CH_3 \;+\; H_2O$$

9.11   a.   The alcohol is primary, so formaldehyde must be used as the carbonyl component:

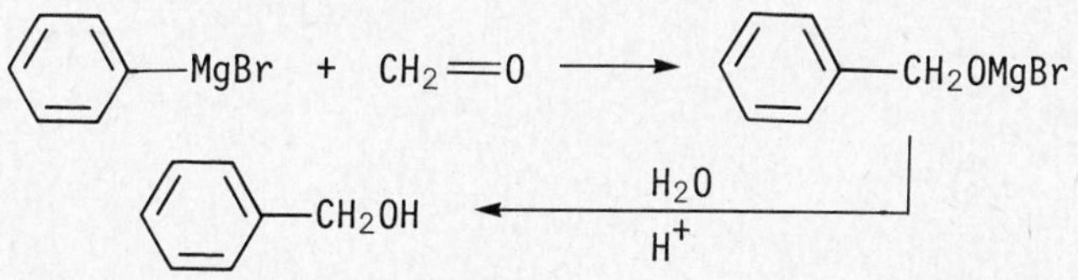

$$\langle\text{Ph}\rangle-MgBr \;+\; CH_2{=}0 \longrightarrow \langle\text{Ph}\rangle-CH_2OMgBr$$

$$\langle\text{Ph}\rangle-CH_2OH \xleftarrow[\;H^+\;]{\;H_2O\;}$$

b.

$$\langle\text{Ph}\rangle-\overset{\overset{\displaystyle CH_3}{|}}{\underset{\underset{\displaystyle CH_3}{|}}{C}}-OH$$

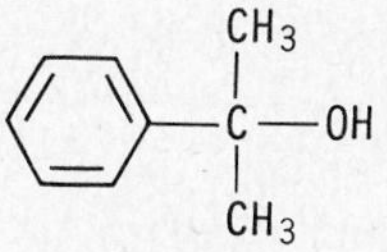

Only one R group attached to $-\overset{|}{\underset{|}{C}}-OH$ comes from the Grignard reagent. The alcohol is tertiary, so the carbonyl component must be a ketone.

187

Two possibilities:

9.12   a. $CH_3CH{=}O$ + HCN $\longrightarrow$ $CH_3CH\overset{\displaystyle OH}{\underset{\displaystyle CN}{<}}$

b.

9.13

9.14   a. $CH_3CH_2CH{=}O$ + $H_2N{-}OH$ $\longrightarrow$ $CH_3CH_2CH{=}NOH$ + H_2O

b.

9.15   a.   A metal hydride will reduce the $\overset{\diagdown}{\diagup}C{=}O$ bond but not the $\overset{\diagdown}{\diagup}C{=}C\overset{\diagup}{\diagdown}$ bond:

b. Catalytic hydrogenation reduces both double bonds:

9.16 Follow eq. 9.33 as a guide, replacing R with H.

$$CH_2{=}O + 2\ Ag(NH_3)_2^+ + 3\ OH^- \longrightarrow HCO_2^- + 2\ Ag + 4\ NH_3 + 2\ H_2O$$

9.17 a.

b.

Remove an α-hydrogen and place it on the oxygen; make the carbon-oxygen bond single, and make the bond between the α-carbon and what was the carbonyl carbon double.

9.18 a.

b.

9.19 In each case, only the α-hydrogens can be readily exchanged.

a.

There are three exchangeable hydrogens, indicated by the arrows.

b.

$$CH_3-\underset{\underset{CH_3}{|}}{\overset{\overset{CH_3}{|}}{C}}-\overset{\overset{O}{||}}{C}-CH_3$$

Only the three methyl hydrogens indicated by the arrow can be readily exchanged. The remaining methyl hydrogens are β, not α, with respect to the carbonyl group.

9.20 Follow eqs. 9.43-9.45 as a guide.

$$CH_3CH_2CH{=}O \; + \; OH^- \; \rightleftharpoons \; CH_3\overset{-}{C}HCH{=}O \; + \; H_2O$$

$$CH_3\overset{-}{C}HCH{=}O \; + \; CH_3CH_2CH{=}O \; \rightleftharpoons \; CH_3CH_2\underset{\underset{CH_3}{|}}{\overset{\overset{O^-}{|}}{C}HCHCH}{=}O$$

$$OH^- \; + \; CH_3CH_2\underset{\underset{CH_3}{|}}{\overset{\overset{OH}{|}}{C}HCHCH}{=}O \; \overset{H-OH}{\rightleftharpoons}$$

9.21 Only the acetaldehyde has an α-hydrogen, so it reacts with the basic catalyst to produce an enolate anion:

$$CH_3CH{=}O \; + \; OH^- \; \rightleftharpoons \; \overset{\ominus}{C}H_2CH{=}O \; + \; H_2O$$

The enolate anion then attacks the carbonyl group of benzaldehyde:

Dehydration occurs by an elimination mechanism:

9.22   Propanal has the α-hydrogens. The aldol is

and its dehydration product is

Additional Problems

9.23   a.   3-pentanone   b.   hexanal
        c.   benzophenone   d.   p-bromobenzaldehyde
             (diphenyl ketone)
        e.   cyclopentanone   f.   2,2-dimethylpropanal
        g.   dicyclobutyl ketone   h.   3-penten-2-one
        i.   bromoacetone   j.   2,3-pentanedione
             (or bromopropanone)

9.24 a. $$CH_3\overset{O}{\overset{||}{C}}CH_2CH_2CH_2CH_2CH_3$$ b. $(CH_3)_2CHCH_2CH_2CHO$

c.

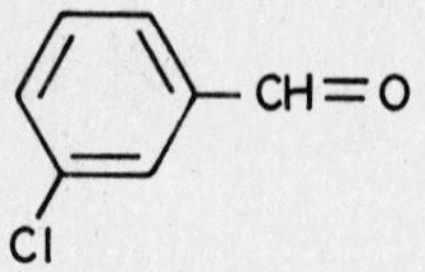

— CH=O, with Cl substituent

d.

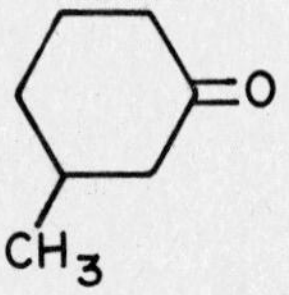

cyclohexanone with CH_3

e. $CH_3CH\!=\!CHCHO$ f.

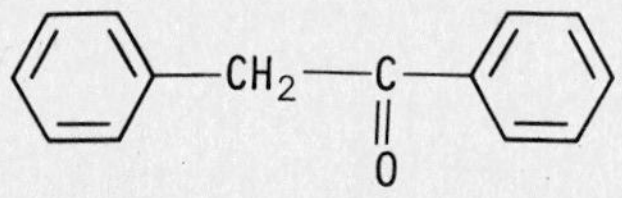

—CH_2—$\overset{}{\underset{O}{C}}$—

g. 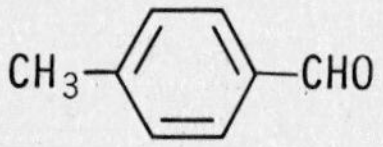 CH_3—

—CHO   h. 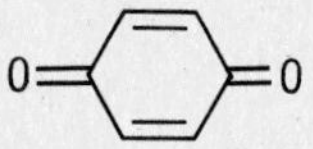 O=

=O

i. $CH_3(CH_2)_3CBr_2CH\!=\!O$ j.

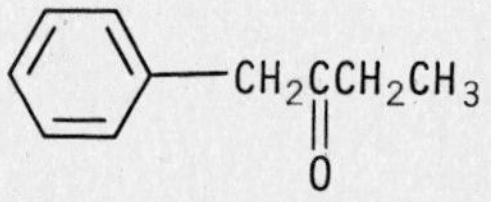

—$CH_2\overset{}{\underset{O}{C}}CH_2CH_3$

9.25 In each case, see the indicated section of the text for typical examples.

a. Sec. 9.8   b. Sec. 9.8   c. Sec. 9.8
d. Sec. 9.8   e. Sec. 9.11   f. Sec. 9.12
g. Sec. 9.12   h. Sec. 9.12   i. Sec. 9.15
j. Sec. 9.15

9.26 a. $$CH_3\underset{OH}{\overset{|}{C}H}CH_2CH_2CH_3 \xrightarrow[\substack{H^+ \\ \text{(see eq. 7.32)}}]{CrO_3} CH_3\overset{O}{\overset{||}{C}}CH_2CH_2CH_3$$

b. $$HC\!\equiv\!CCH_2CH_2CH_3 \xrightarrow[\substack{Hg^{2+} \\ \text{(see eq. 3.52)}}]{H_3O^+} CH_3\overset{O}{\overset{||}{C}}CH_2CH_2CH_3$$

9.27 a. 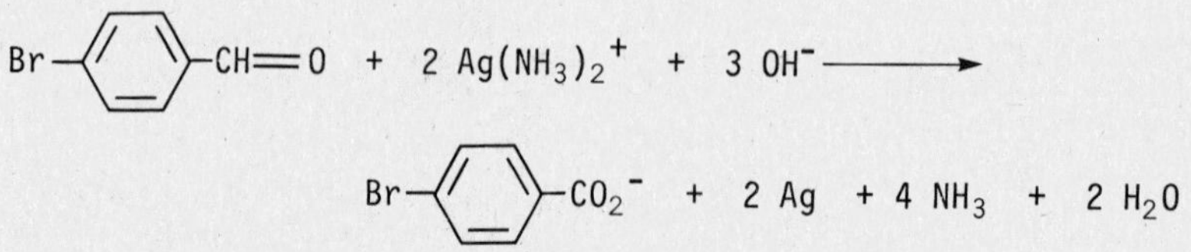 Br—

—CH=O + 2 Ag(NH$_3$)$_2$$^+$ + 3 OH$^-$ ⟶

Br—

—CO$_2$$^-$ + 2 Ag + 4 NH$_3$ + 2 H$_2$O

p-bromobenzoate ion

b. 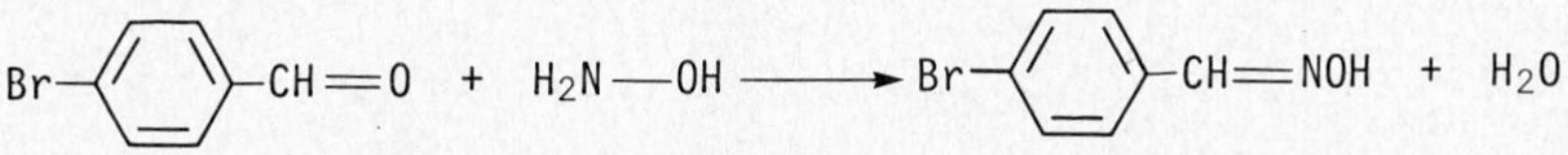 Br—⬡—$CH{=}O$ + H_2N—OH ⟶ Br—⬡—$CH{=}NOH$ + H_2O

p-bromobenzaldoxime

c. 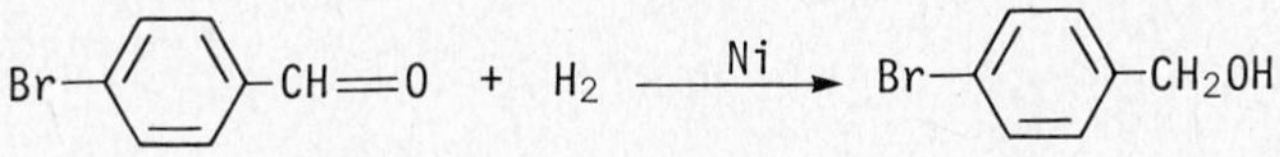 Br—⬡—$CH{=}O$ + H_2 $\xrightarrow{\text{Ni}}$ Br—⬡—CH_2OH

p-bromobenzyl alcohol

d. 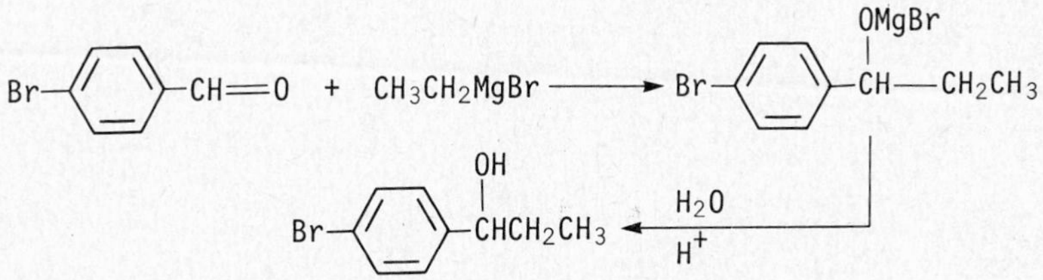 Br—⬡—$CH{=}O$ + CH_3CH_2MgBr ⟶ Br—⬡—$\overset{\displaystyle OMgBr}{CH}$—$CH_2CH_3$

Br—⬡—$\overset{\displaystyle OH}{CH}CH_2CH_3$ $\xleftarrow[\text{H}^+]{\text{H}_2\text{O}}$

1-(*p*-bromophenyl)-1-propanol

e. 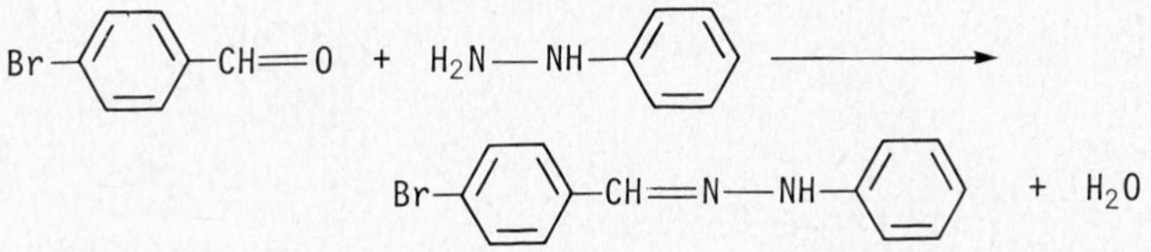 Br—⬡—$CH{=}O$ + H_2N—NH—⬡ ⟶

Br—⬡—$CH{=}N$—NH—⬡ + H_2O

p-bromobenzaldehyde phenylhydrazone

f. 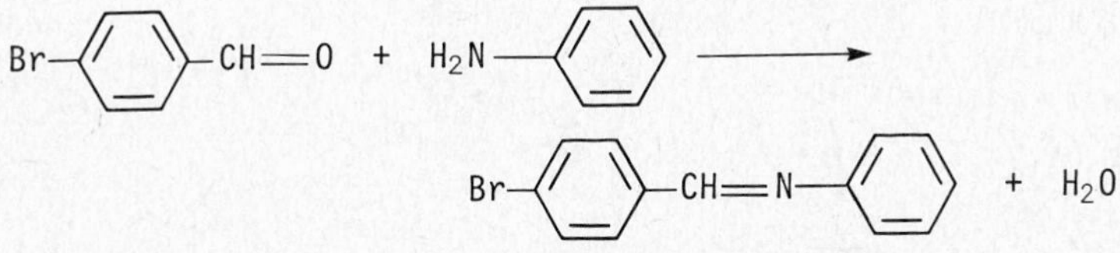 Br—⬡—$CH{=}O$ + H_2N—⬡ ⟶

Br—⬡—$CH{=}N$—⬡ + H_2O

p-bromobenzaldehyde anil

g. 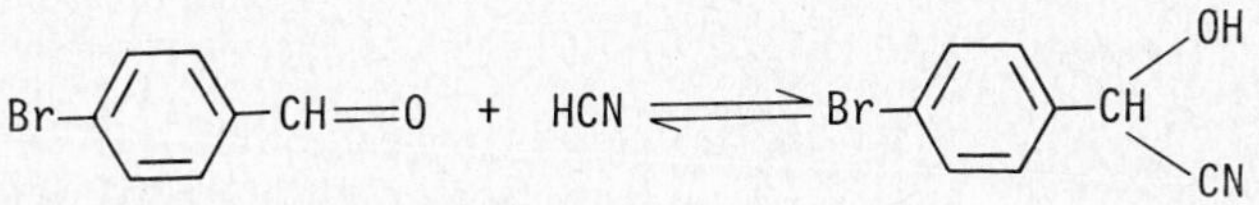 Br—⬡—$CH{=}O$ + HCN ⇌ Br—⬡—$\overset{\displaystyle OH}{\underset{\displaystyle CN}{CH}}$

p-bromobenzaldehyde cyanohydrin

h. $Br-\langle\bigcirc\rangle-CH=O$ + 2 CH_3OH $\xrightarrow{HCl}$ $Br-\langle\bigcirc\rangle-CH\begin{smallmatrix}OCH_3\\\\OCH_3\end{smallmatrix}$ + H_2O

p-bromobenzaldehyde
dimethylacetal

i. $Br-\langle\bigcirc\rangle-CH=O$ + $HOCH_2CH_2OH$ $\xrightarrow{H^+}$ $Br-\langle\bigcirc\rangle-CH\begin{smallmatrix}O-CH_2\\|\\O-CH_2\end{smallmatrix}$ + H_2O

2-p-bromophenyl-1,3-dioxolane
(p-bromobenzaldehyde ethylene
glycol acetal)

j. $Br-\langle\bigcirc\rangle-CH=O$ + $LiAlH$ $\longrightarrow$ $Br-\langle\bigcirc\rangle-CH_2OH$

p-bromobenzyl alcohol

9.28 a. Use Tollens', Fehling's, or Benedict's test. The hexanal (an
aldehyde) will react, whereas 2-hexanone (a ketone) will not.
b. Again use the aldehyde tests (Tollens', etc.). Alcohols (such
as benzyl alcohol) do not react.
c. Both compounds are ketones, but 2-cyclopentenone has a carbon-
carbon double bond and will be easily oxidized by potassium
permanganate. The saturated ketone, cyclopentanone, will not
react.

9.29 a. $\langle\bigcirc\rangle-CH=CHCH=O$ + 2 $Ag(NH_3)_2^+$ + 3 OH^- $\longrightarrow$

$\langle\bigcirc\rangle-CH=CHCO_2^-$ + 2 Ag + 4 NH_3 + 2 H_2O

b. $HO-\langle\bigcirc\rangle\!\!\begin{smallmatrix}CH_3O\\\end{smallmatrix}-CH=O$ + H_2N-OH $\longrightarrow$ $HO-\langle\bigcirc\rangle\!\!\begin{smallmatrix}CH_3O\\\end{smallmatrix}-CH=N-OH$ + H_2O

Only the aldehyde function reacts with the reagent.

c.

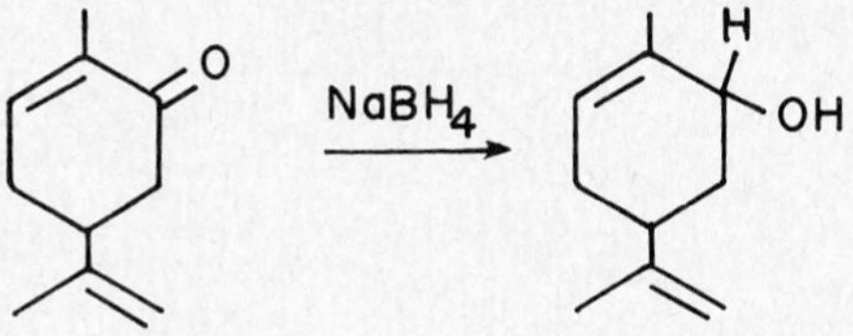

d.

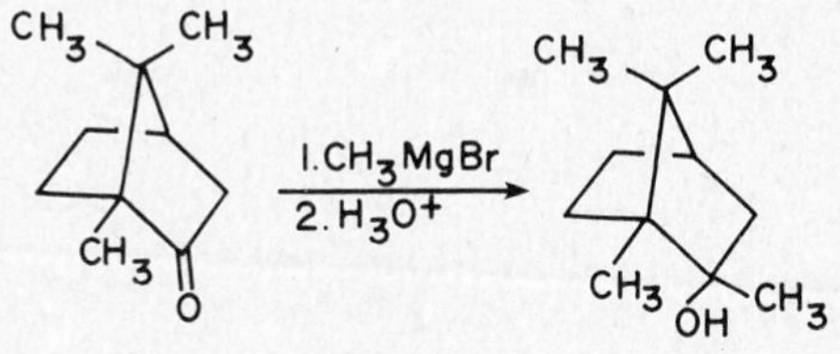

9.30 $CH_3CH_2CH_2CH_2CH_2CH_2CH=O$ $CH_3CH_2CH_2\overset{\underset{\|}{O}}{C}CH_2CH_2CH_3$ $CH_3\overset{\overset{\displaystyle CH_3}{|}}{C}H\overset{\underset{\|}{O}}{C}\overset{\overset{\displaystyle CH_3}{|}}{C}HCH_3$

             bp 155°C   bp 144°C   bp 124°C

The compounds are isomers and have identical molecular weights. Each has a carbonyl group that, because of its polarity, can associate as follows as a consequence of intermolecular attraction between opposite charges:

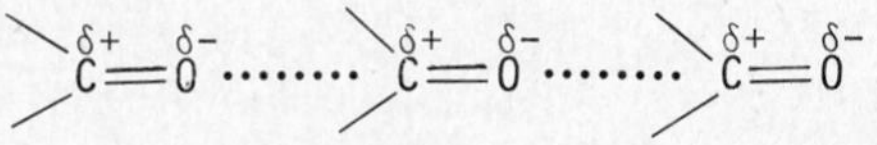

or

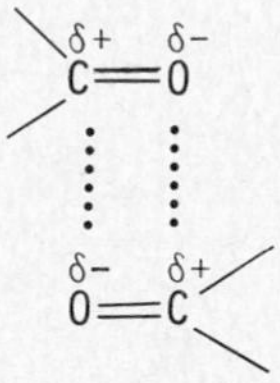

As we go from left to right in the series (as shown above), the carbonyl group is more and more hindered, or buried in the structure. Thus association is more difficult, and the boiling point decreases.

9.31 The mechanism is analogous to those discussed in Sec. 9.8, except that the last stages are *intra*molecular.

$$CH_3-\overset{\overset{\textstyle O}{\|}}{C}-CH_3 \ + \ H^+ \ \rightleftharpoons \ CH_3-\overset{\overset{\textstyle +O-H}{\|}}{C}-CH_3$$

$$\Updownarrow H\ddot{O}CH_2CH_2OH$$

$$CH_3-\overset{\overset{\textstyle :\ddot{O}H}{|}}{\underset{\underset{\textstyle CH_3}{|}}{C}}-OCH_2CH_2OH \ \overset{H^+}{\rightleftharpoons} \ CH_3-\overset{\overset{\textstyle OH}{|}}{\underset{\underset{\textstyle H_3C\ \ H}{|}}{C}}-\overset{+}{\ddot{O}}CH_2CH_2OH$$

$$H^+\Updownarrow$$

$$CH \ -\overset{\overset{\textstyle H\ddot{O}\overset{+}{H}}{|}}{\underset{\underset{\textstyle CH_3}{|}}{C}}-OCH_2CH_2OH \ \overset{-H_2O}{\rightleftharpoons} \ CH_3-\overset{+}{\underset{\underset{\textstyle CH_3}{|}}{C}}-OCH_2CH_2\ddot{O}H$$

$$\Updownarrow$$

$$\begin{array}{c} \overset{\textstyle CH_2}{} \\ O \quad CH_2 \\ CH_3-\overset{|}{C}-O \\ \underset{\textstyle CH_3}{|} \end{array} \quad \overset{-H^+}{\rightleftharpoons} \quad \begin{array}{c} H \quad CH_2 \\ \overset{+}{O} \quad CH_2 \\ CH_3-\overset{|}{C}-O \\ \underset{\textstyle CH_3}{|} \end{array}$$

9.32 All parts of this problem involve the preparation or hydrolysis of hemiacetals or acetals (or the corresponding ketone derivatives). See Sec. 9.7.

a. $CH_3CH_2CH_2CH{=}O \ + \ 2\ CH_3CH_2OH \ \overset{H^+}{\longrightarrow} \ CH_3CH_2CH_2(OCH_2CH_3)_2 \ + \ H_2O$

b. $CH_3CH(OCH_3)_2 \ \xrightarrow[H^+]{H_2O} \ CH_3CH{=}O \ + \ 2\ CH_3OH$

c. In this case, the acetal is cyclic, and the product is a hydroxy aldehyde, which may exist in its cyclic hemiacetal form:

$$\xrightarrow[H^+]{H_2O} \qquad + \ CH_3OH$$

(or $HOCH_2CH_2CH_2CH_2CH{=}O$)

d. The starting material is a "double" acetal. It is completely hydrolyzed to two equivalents of the hydroxy aldehyde.

$$+ H_2O \xrightarrow{H^+} 2\ HOCH_2CH_2CH_2CH_2CH{=}O$$

The product, as in part c, may exist in its cyclic hemiacetal form. The reaction is analogous to the hydrolysis of certain carbohydrates (for example, disaccharides to monosaccharides; see Chapter 13). For practice, write the steps in the mechanism for this reaction.

e. In this reaction, a hemiacetal is converted to an acetal.

$$+ CH_3OH \xrightarrow{H^+} \qquad + H_2O$$

This is the reverse of the reaction in part c.

9.33 The product is a cyclic thioacetal:

Compare with eq. 9.11.

9.34 For guidance, review Sec. 9.10.

a.

b.

$$\text{(phenyl)}-\overset{\overset{\displaystyle O}{\|}}{C}CH_3 \quad \xrightarrow[\text{2. } H_2O,\ H^+]{\text{1. } CH_3\,MgBr} \quad \text{(phenyl)}-\overset{\overset{\displaystyle OH}{|}}{\underset{\underset{\displaystyle CH_3}{|}}{C}}-CH_3$$

c. $CH_2=O \quad \xrightarrow[\text{2. } H_2O,\ H^+]{\text{1. } CH_3MgBr} \quad CH_3CH_2OH$

d.

$$\text{(cyclohexyl)}=O \quad \xrightarrow[\text{2. } H_2O,\ H^+]{\text{1. } CH_3\,MgBr} \quad \text{(cyclohexyl with }OH\text{ and }CH_3)$$

9.35 In each case write the structure of the alcohol:

$$R_1-\overset{\overset{\displaystyle R_2}{|}}{\underset{\underset{\displaystyle R_3}{|}}{C}}-O-H$$

One of the R groups comes from the Grignard reagent. The rest of the molecule comes from the carbonyl compound. For example, if we select R_1 as the alkyl group to come from the Grignard reagent, then the carbonyl compound is $R_2-\overset{\overset{\displaystyle }{}}{\underset{\underset{\displaystyle O}{\|}}{C}}-R_3$.

a. $CH_3CH_2CH_2CH_2\overset{\overset{\displaystyle H}{|}}{\underset{\underset{\displaystyle H}{|}}{C}}-OH$

$$CH_3CH_2CH_2CH_2MgX \;+\; \overset{\overset{\displaystyle H}{|}}{\underset{\underset{\displaystyle H}{|}}{C}}=O \;\longrightarrow\; CH_3CH_2CH_2CH_2\overset{\overset{\displaystyle H}{|}}{\underset{\underset{\displaystyle H}{|}}{C}}-O^-\overset{+}{MgX}$$

$$\Big\downarrow\; H_2O,\ H^+$$

$$CH_3CH_2CH_2CH_2-\overset{\overset{\displaystyle H}{|}}{\underset{\underset{\displaystyle H}{|}}{C}}-OH$$

For the remaining cases we will not write the equations, but simply show how the initial reactants are derived.

b. 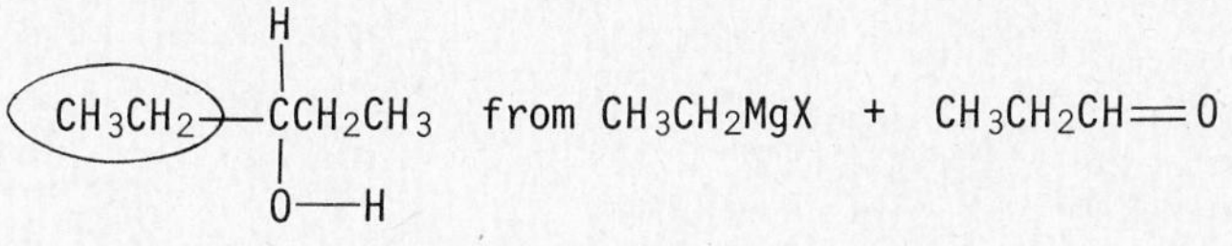

c. 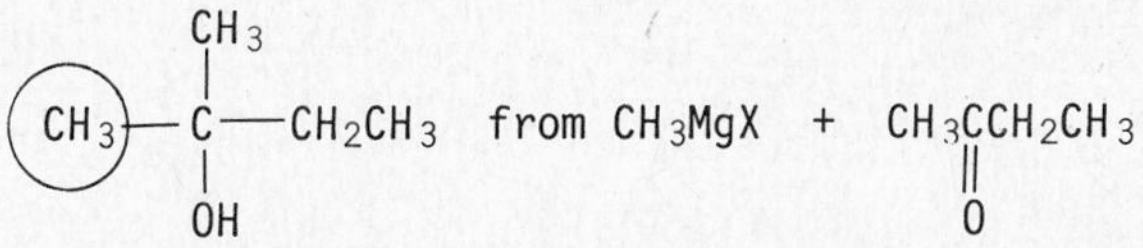

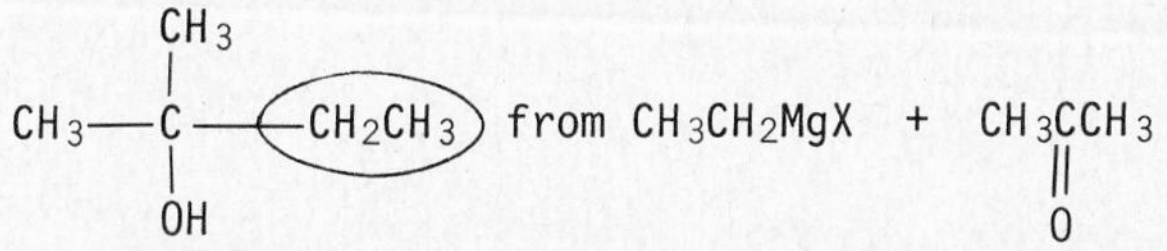

Either of these combinations of reagents will work.

d. 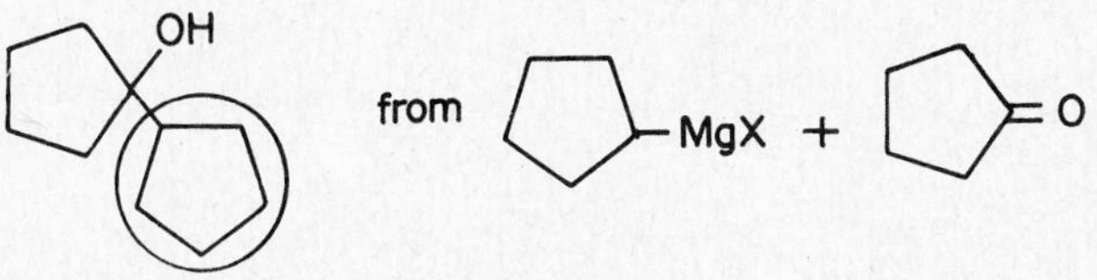

In this case, the "free-standing" R group is selected to come from the Grignard reagent.

e. 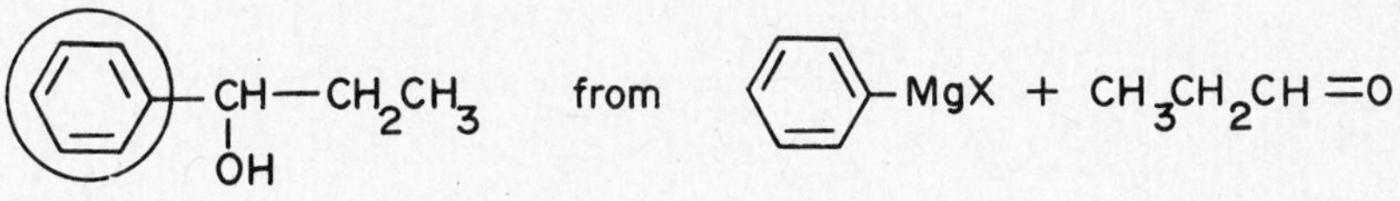

or

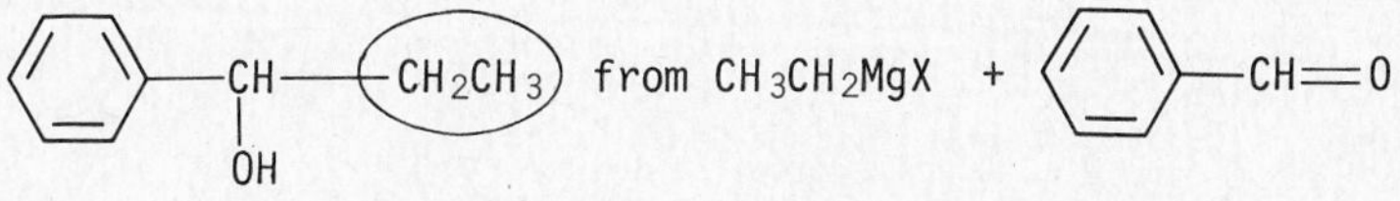

f. 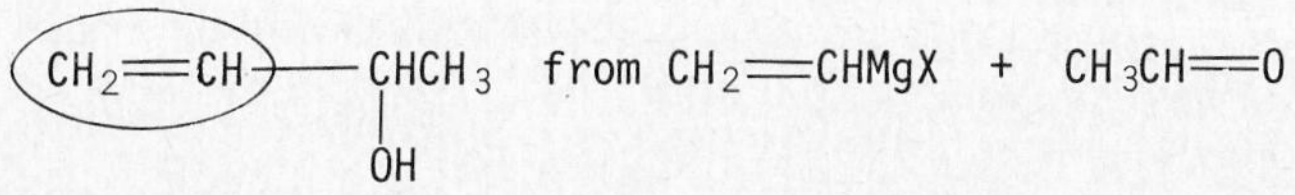

or

$CH_2{=}CH{-}\underset{\underset{OH}{|}}{CH}{-}\boxed{CH_3}$ from CH_3MgX + $CH_2{=}CH{-}CH{=}O$

Vinyl Grignard reagents are known, although they are a bit more difficult to prepare than simple alkyl Grignard reagents. Either pair of reagents will work.

9.36 a. The reaction is similar to that of a Grignard reagent with a ketone (see eq. 9.18).

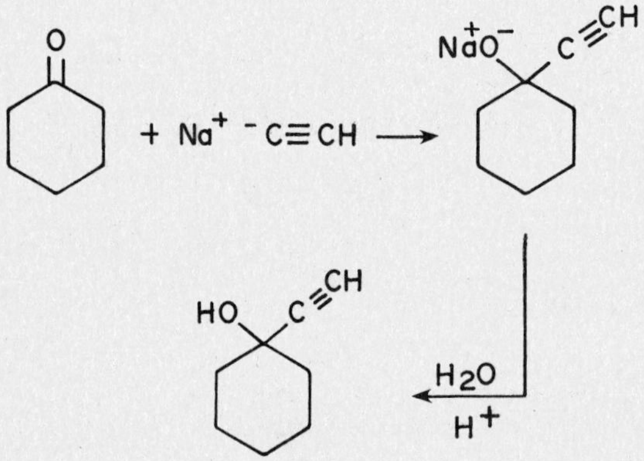

b.

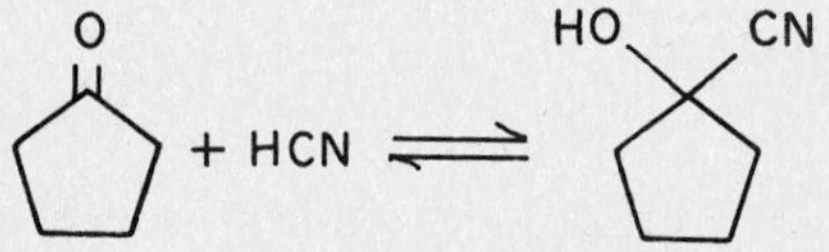

cyclopentanone
cyanohydrin

c. See Table 9.1 for guidance.

$CH_3\underset{\underset{O}{\|}}{C}CH_2CH_3$ + NH_2OH $\rightleftharpoons$ $CH_3\underset{\underset{NOH}{\|}}{C}CH_2CH_3$ + H_2O

d. See Sec. 9.12 for examples.

$$CH_3 - \langle \bigcirc \rangle - CH{=}O \; + \; H_2N{-}CH_2 - \langle \bigcirc \rangle \longrightarrow$$

$$CH_3 - \langle \bigcirc \rangle - CH{=}N{-}CH_2 - \langle \bigcirc \rangle \; + \; H_2O$$

e. See Table 9.1.

$$CH_3CH_2CH{=}O \; + \; H_2N{-}NH - \langle \bigcirc \rangle \longrightarrow$$

$$CH_3CH_2CH{=}N{-}NH - \langle \bigcirc \rangle \; + \; H_2O$$

9.37 a. The first step involves nucleophilic addition to the carbonyl group:

Proton transfer to the alkoxide part of the molecule is fast:

Dehydration gives the observed product:

b.   The mechanism is similar to that in part a.

$$(CH_3)_2C=O + H_2\ddot{N}-OH \longrightarrow \begin{array}{c} CH_3 \\ CH_3 \end{array}\!\!C\!\!\begin{array}{c} O^- \\ \overset{+}{N}H_2-OH \end{array}$$

$$H^+ + H_2O + \begin{array}{c} CH_3 \\ CH_3 \end{array}\!\!C=N-OH \longleftarrow \begin{array}{c} CH_3 \\ CH_3 \end{array}\!\!C\!\!\begin{array}{c} OH \quad H^+ \\ N-OH \\ H \end{array}$$

9.38   a.   See Sec. 9.13.

$$CH_3-\overset{\overset{O}{\|}}{C}-C_6H_5 \xrightarrow[\text{2.}H_2O, H^+]{\text{1.LiAlH}_4} CH_3-\overset{OH}{\underset{|}{CH}}-C_6H_5$$

b.   See Sec. 9.17.

$$CH_3CH_2\overset{\overset{O}{\|}}{C}-C_6H_5 \xrightarrow[\substack{CH_3OD \\ \text{(excess)}}]{CH_3O^-Na^+} CH_3CD_2\overset{\overset{O}{\|}}{C}-C_6H_5$$

Only the α-hydrogens exchange.

c.   See Sec. 9.13.

$$C_6H_5-CH=CH-CH=O \xrightarrow[\text{Ni}]{\text{excess } H_2} C_6H_5-CH_2CH_2CH_2OH$$

Usually the aromatic ring will not be reduced, although under certain reaction conditions even this is possible.

d.   The carbonyl group is reduced, but the carbon-carbon double bond and aromatic ring are not reduced.

$$CH_2=CH-C_6H_4-CH=O \xrightarrow[\text{2. } H_2O, H^+]{\text{1. NaBH}_4} CH_2=CH-C_6H_4-CH_2OH$$

9.39 Review Sec. 9.14.

a.

$$\overset{OH}{\underset{|}{CH_2\!=\!C\!-\!CH_2CH_3}} \quad \text{and} \quad \overset{OH}{\underset{|}{CH_3\!-\!C\!=\!CHCH_3}}$$

There are two types of α-hydrogens in 2-butanone, and either may enolize:

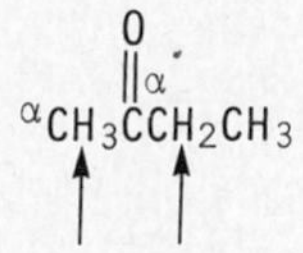

$$\overset{O}{\underset{\alpha CH_3\overset{\|}{C}CH_2CH_3}{}}$$

b. ⬡—CH₂—CH=O; the enol is ⬡—CH=CH—OH

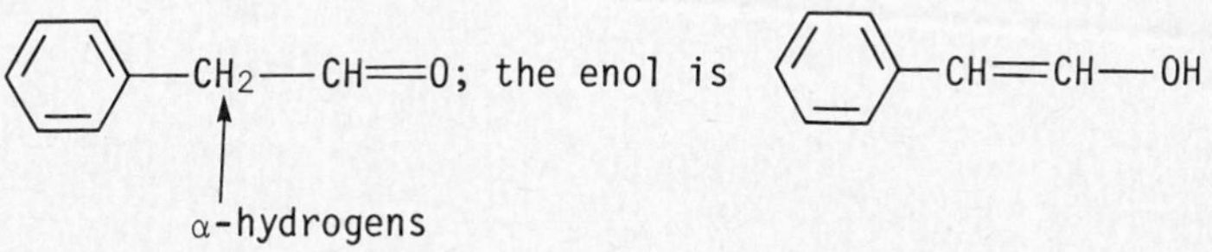

α-hydrogens

c. CH₃—C—CH₂—C—CH₃
         ‖   ‖
         O   O
      α   α   α

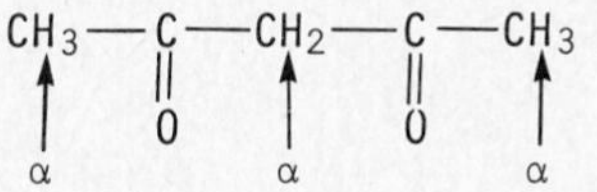

All of the hydrogens are α to a carbonyl group. The CH₂ hydrogens are α to *two* carbonyl groups and are most likely to enolize.

$$\overset{O}{\underset{CH_3\overset{\|}{C}-CH=C-CH_3}{}} \overset{OH}{\underset{|}{}} \longleftarrow \text{favored}$$

$$\overset{OH}{\underset{CH_2=C-CH_2-\overset{\|}{C}-CH_3}{|}} \overset{O}{\underset{}{}} \longleftarrow \text{not favored}$$

9.40 Review Sec. 9.17. Only hydrogens α to a carbonyl group will be replaced by deuterium.

a.

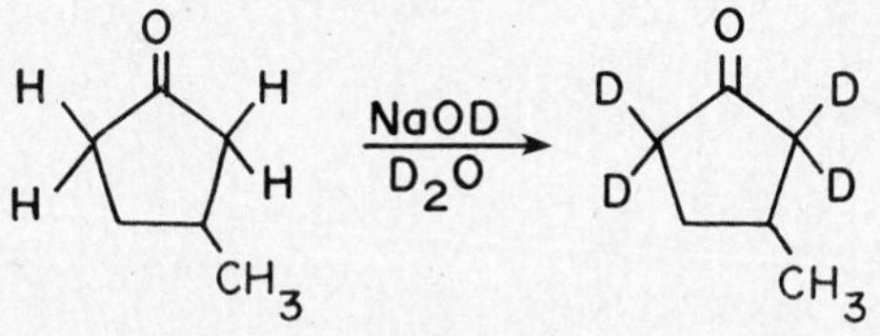

b. CH₃CH₂CHCH=O $\xrightarrow[\text{D}_2\text{O}]{\text{NaOD}}$ CH₃CH₂CDCH=O
          | |
          CH₃   CH₃

9.41   Follow eqs. 9.43-9.45. The steps in the mechanism are as follows:

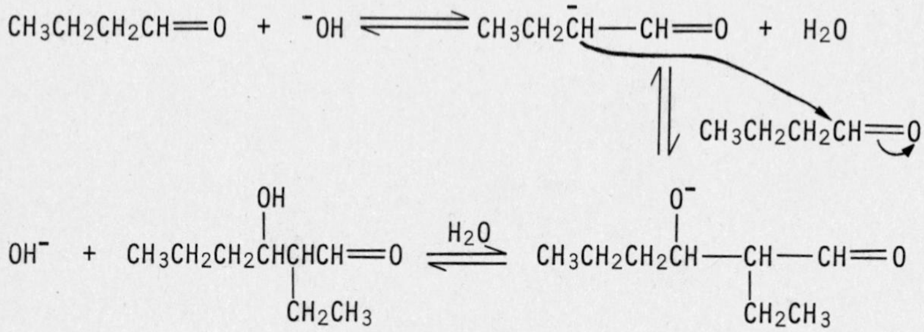

$CH_3CH_2CH_2CH{=}O$ + ^-OH $\rightleftharpoons$ $CH_3CH_2\overset{-}{C}H{-}CH{=}O$ + H_2O

$CH_3CH_2CH_2CH{=}O$

OH^- + $CH_3CH_2CH_2\overset{\displaystyle OH}{\underset{\displaystyle CH_2CH_3}{\overset{|}{\underset{|}{C}}HCHCH}}{=}O$ $\underset{H_2O}{\rightleftharpoons}$ $CH_3CH_2CH_2\overset{\displaystyle O^-}{\underset{\displaystyle CH_2CH_3}{\overset{|}{\underset{|}{C}}H}}{-}CH{-}CH{=}O$

9.42   The first step is an aldol condensation:

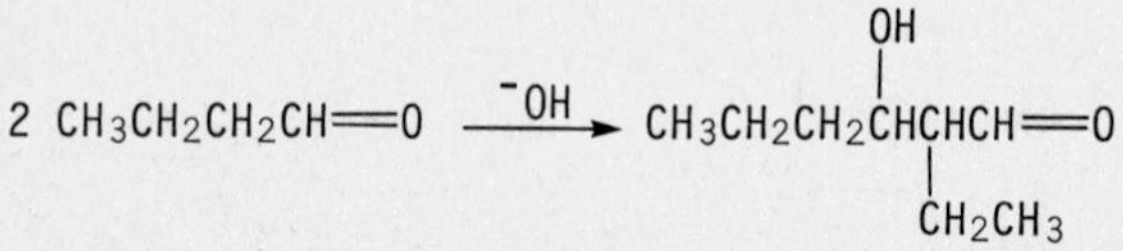

2 $CH_3CH_2CH_2CH{=}O$ $\xrightarrow{^-OH}$ $CH_3CH_2CH_2\overset{\displaystyle OH}{\underset{\displaystyle CH_2CH_3}{\overset{|}{\underset{|}{C}}HCHCH}}{=}O$

Dehydration gives the unsaturated aldehyde:

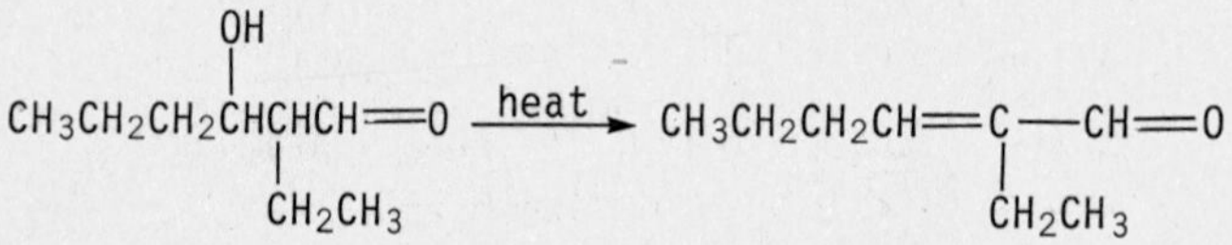

$CH_3CH_2CH_2\overset{\displaystyle OH}{\underset{\displaystyle CH_2CH_3}{\overset{|}{\underset{|}{C}}HCHCH}}{=}O$ $\xrightarrow{heat}$ $CH_3CH_2CH_2CH{=}\underset{\displaystyle CH_2CH_3}{\underset{|}{C}}{-}CH{=}O$

Catalytic hydrogenation saturates the $C{=}C$ and $C{=}O$ bonds.

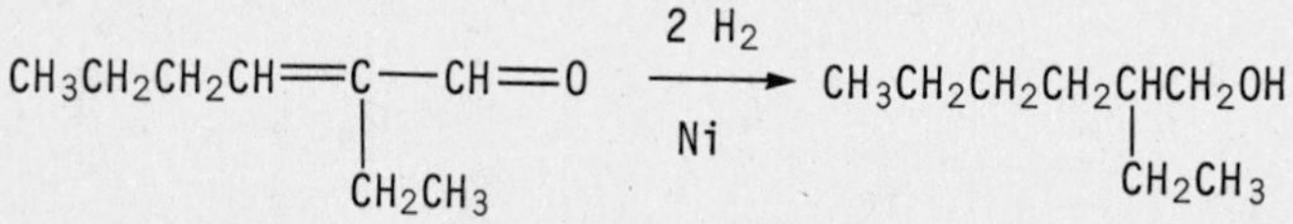

$CH_3CH_2CH_2CH{=}\underset{\displaystyle CH_2CH_3}{\underset{|}{C}}{-}CH{=}O$ $\xrightarrow[Ni]{2\ H_2}$ $CH_3CH_2CH_2CH_2\underset{\displaystyle CH_2CH_3}{\underset{|}{C}}HCH_2OH$

9.43   The reaction occurs by an *intra*molecular aldol condensation, followed by dehydration of the resulting aldol:

The starting diketone can form several different enolate anions; however, the one that can react most favorably with the second carbonyl group to form a five-membered ring is the one shown. All the equilibria are driven in the forward direction by the final dehydration step.

9.44 The product has 17 carbons, which suggests that it is formed from two benzaldehyde molecules (2 x 7 = 14 carbons) + one acetone molecule (3 carbons). The product forms by a double mixed aldol condensation:

The product is yellow because of the extended conjugated system of double bonds.

9.45 Step A: The reagents are ethylene glycol ($HOCH_2CH_2OH$) and H^+; compare with eq. 9.11.

Step B: The reagent is chromic acid; compare with eq. 7.32.

Step C: The reagent is sodium acetylide, $HC \equiv C^- Na^+$; compare with eq. 9.20.

Step D: The reagent is dilute acid, to hydrolyze the ketal; compare with eq. 9.13.

The carbonyl group in the six-membered ring must be "protected" so that in Step C there will be only one carbonyl group (the one that is introduced into the five-membered ring) available for reaction with the sodium acetylide.

Although one might expect only Enovid to be formed, its double-bond isomer Norlutin is also formed, through an acid-catalyzed enolization (acid is the reagent used to hydrolyze the ketal):

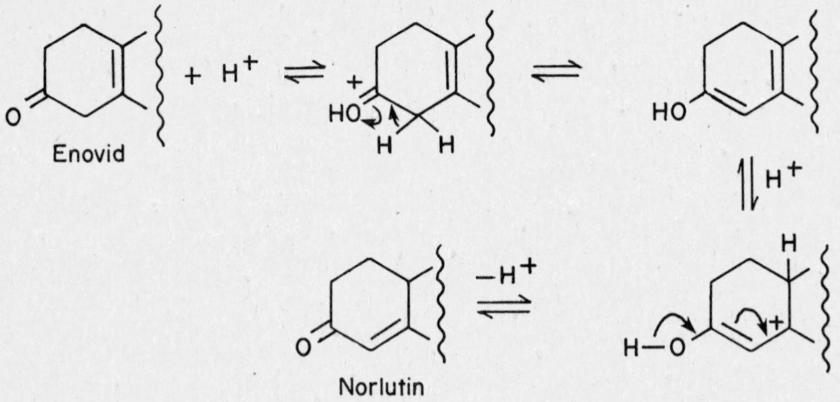

Enovid

Norlutin

CHAPTER SUMMARY

Carboxylic acids, the most important class of organic acids, contain the **carboxyl group,** $-C\overset{\displaystyle O}{\underset{\displaystyle OH}{\Big\langle}}$. The IUPAC ending for the names of these compounds is *-oic acid* but many common names (such as formic acid and acetic acid) are also used. An **acyl group,** $R-\overset{O}{\overset{\|}{C}}-$, is named by changing the *-ic* ending of the corresponding acid to *-yl* ($CH_3\overset{O}{\overset{\|}{C}}-$ is acetyl).

The carboxyl group is polar and readily forms hydrogen bonds. A carboxylic acid dissociates to a **carboxylate anion** and a proton. In the carboxylate anion, the negative charge is delocalized equally over both oxygens. The pK_a's of simple carboxylic acids are about 4-5 , but the acidity can be increased by electron-withdrawing substituents (such as chlorine) close to the carboxyl function.

Carboxylic acids react with bases to give salts. These are named by naming the cation first, then the carboxylate anion. The name of the anion is obtained by changing the *-ic* ending of the acid name to *-ate* (acet*ic* becomes acet*ate.*)

Carboxylic acids are prepared by four methods: (1) by oxidation of primary alcohols or aldehydes, (2) by oxidation of an aromatic side chain, (3) from a Grignard reagent and carbon dioxide, or (4) by hydrolysis of a nitrile, $RC\equiv N$.

Carboxylic acid derivatives are compounds in which the carboxyl $-OH$ group is replaced by other groups. Examples include **esters, acyl halides, anhydrides,** and **amides.**

Esters, RCO_2R', are named as salts are; the R' group is named first, followed by the name of the carboxylate group (for example, $CH_3CO_2CH_2CH_3$ is ethyl acetate). Esters can be prepared from an acid and an alcohol, with a mineral acid catalyst (**Fischer esterification**). The key step of the mechanism is nucleophilic attack by the alcohol on the protonated carbonyl group of the acid. Many esters are used as flavors and perfumes.

Saponification is the base-catalyzed hydrolysis of an ester, yielding its component carboxylate salt and alcohol. **Ammonolysis** of esters gives amides. Esters react with Grignard reagents to give tertiary alcohols. With lithium aluminum hydride, on the other hand, they are reduced to primary alcohols.

Acid derivatives undergo nucleophilic substitution. The mechanism is as follows: the nucleophile adds to the trigonal carbonyl carbon to form a **tetrahedral intermediate**, which, through loss of a leaving group, becomes the trigonal product. The reaction can be regarded as an **acyl transfer**, the transfer of an acyl group from one nucleophile to another. The reactivity order of acid derivatives toward nucleophiles is acyl halides > anhydrides > esters > amides.

Acyl chlorides are prepared from acids and either $SOCl_2$ or PCl_5. They react rapidly with water to give acids, with alcohols to give esters, and with ammonia to give amides. **Acid anhydrides** react similarly, but less rapidly. **Thioesters** are nature's acylating agents; they react with nucleophiles less rapidly than anhydrides do but more rapidly than ordinary esters do.

Amides can be prepared from ammonia and other acid derivatives. They can also be prepared by heating ammonium salts. They are named by replacing the -*ic* or -*oic* acid ending by -*amide*.

Because of resonance, the C—N bond in amides has considerable C=N character. Rotation about that bond is restricted, and the amide group is planar. Amides are polar, form hydrogen bonds, and have high boiling points considering their molecular weights.

Amides react slowly with nucleophiles (such as water and alcohols). They are reduced to amines by $LiAlH_4$ and can be dehydrated to nitriles. **Urea,** made from CO_2 and NH_3, is an important fertilizer.

Some reactions of acid derivatives are summarized in Table 10.4.

REACTION SUMMARY

Acids

$$RCO_2H \rightleftharpoons RCO_2^- + H^+ \quad \text{(ionization)}$$

$$RCO_2H + NaOH \longrightarrow RCO_2^-Na^+ + H_2O \quad \text{(salt formation)}$$

REACTION SUMMARY

Preparation of Acids

$$RCH_2OH \xrightarrow{Cr^{6+}} RCH{=\!=}O \xrightarrow[\text{or } Ag^+]{Cr^{6+}} RCO_2H$$

$$ArCH_3 \xrightarrow[\text{or } O_2,\ Co^{3+}]{KMnO_4} ArCO_2H$$

$$RMgX + CO_2 \longrightarrow RCO_2MgX \xrightarrow{H_3O^+} RCO_2H$$

$$RC{\equiv}N + 2H_2O \xrightarrow[^-OH]{H^+ \text{ or}} RCO_2H + NH_3$$

Fischer Esterification

$$RCO_2H + R'OH \underset{\longleftarrow}{\overset{H^+}{\rightleftharpoons}} RCO_2R' + H_2O$$

Saponification

$$RCO_2R' + NaOH \longrightarrow RCO_2^-Na^+ + R'OH$$

Ammonolysis of Esters

$$RCO_2R' + NH \longrightarrow RCONH_2 + R'OH$$

Esters and Grignard Reagents

$$RCO_2R' \xrightarrow[\text{2. } H_3O^+]{\text{1. 2 } R''MgX} R-\overset{\displaystyle R''}{\underset{\displaystyle R''}{C}}-OH \ (+ R'OH)$$

Reduction of Esters

$$RCO_2R' + LiAlH_4 \longrightarrow RCH_2OH \ (+R'OH)$$

Preparation of Acyl Chlorides

$$RCO_2H + SOCl_2 \longrightarrow R\overset{O}{\overset{\|}{C}}-Cl + HCl + SO_2$$

$$RCO_2H + PCl_5 \longrightarrow R\overset{O}{\overset{\|}{C}}-Cl + HCl + POCl_3$$

Reactions of Acyl Halides (or Anhydrides)

$$
RC\overset{O}{\|}-Cl \text{ (or } RC\overset{O}{\|}-O-C\overset{O}{\|}-R)
\begin{cases}
\xrightarrow{H_2O} RCO_2H + HCl \text{ (or } RCO_2H) \\
\xrightarrow{R'OH} RCO_2R' + HCl \text{ (or } RCO_2H) \\
\xrightarrow{NH_3} RCONH_2 + NH_4^+Cl^- \text{ (or } RCO_2H)
\end{cases}
$$

Amides from Ammonium Salts

$$RCO_2^-NH^+ \xrightarrow{heat} RCONH_2 + H_2O$$

Reactions of Amides

$$RCONH_2 + H_2O \xrightarrow[\text{or } ^-OH]{H^+} RCO_2H + NH_3$$

$$RCONH_2 \xrightarrow{LiAlH_4} RCH_2NH_2$$

$$RCONH_2 \xrightarrow{P_2O_5} RC\equiv N$$

Urea

$$CO_2 + 2\ NH_3 \xrightarrow[\text{pressure}]{heat} H_2N-\overset{O}{\overset{\|}{C}}-NH_2 + H_2O$$

MECHANISM SUMMARY

Nucleophilic Addition-Elimination

| trigonal reactant | tetrahedral intermediate | trigonal product |

LEARNING OBJECTIVES

1. Know the meaning of: carboxylic acid, carboxyl group, acyl group, carboxylate anion, acidity or ionization constant, inductive effect.

2. Know the meaning of: carboxylate salt, ester, acyl halide, acid anhydride, primary amide.

3. Know the meaning of: Fischer esterification, nucleophilic addition-elimination, tetrahedral intermediate, saponification, ammonolysis, acyl transfer.

4. Given the IUPAC name of a carboxylic acid, salt, ester, amide, acyl halide, or anhydride, write its structural formula; and given the structure, write the name.

5. Know the common names of the monocarboxylic acids listed in Table 10.1.

6. Know the systems for designating carbons in a carboxylic acid chain by numbers (IUPAC) or by Greek letters (common).

7. Know how to name acyl groups, and how to write a formula given a name that includes acyl group nomenclature.

8. Given the formula of a carboxylic acid, write an expression for its ionization constant, K_a.

9. Write the resonance structures of a carboxylate anion.

10. Given two or more carboxylic acids with closely related structures, rank them in order of increasing (or decreasing) acidities (pK_a's).

11. Tell whether a particular substituent will increase or decrease the acidity of a carboxylic acid.

12. Given a carboxylic acid and a base, write the equation for salt formation.

13. Given a carboxylic acid, tell what aldehyde or primary alcohol is needed for its preparation by oxidation.

14. Given an aromatic compound with alkyl substituents, tell what aromatic acid would be obtained from its oxidation.

15. Given a carboxylic acid, write an equation for its synthesis by hydrolysis of a nitrile (cyanide) or by the Grignard method.

16. Given an alcohol and an acid, write the equation for formation of the corresponding ester.

17. Write the steps in the mechanism for the acid-catalyzed (Fischer) esterification of a given carboxylic acid with a given alcohol.

18. Given the name or the structure of an ester, write the structure of the alcohol and acid from which it is derived.

19. Write an equation for the reaction of a given ester with aqueous base (saponification).

20. Write an equation for the reaction of a given ester with ammonia, a Grignard reagent, or lithium aluminum hydride.

21. Given a particular acid halide, write an equation for its preparation from an acid.

22. Write the equation for the reaction of a given acid halide or anhydride with a given nucleophile (especially with water, an alcohol, or ammonia).

23. Write equations for the preparation of a given amide from an acyl halide, acid anhydride, or ammonium salt.

24. Given a particular amide, write equations for its hydrolysis, reduction with lithium aluminum hydride, and dehydration to a nitrile.

25. Given a particular product that can be prepared by any of the reactions in this chapter, deduce the structures of the reactants required for its preparation, and write the equation for the reaction.

ANSWERS TO PROBLEMS

Problems Within the Chapter

10.1 a. $\overset{4}{C}H_3\overset{3}{C}H\overset{2}{C}H_2\overset{1}{C}O_2H$ b. $\overset{\gamma}{C}H_2\overset{\beta}{C}H_2\overset{\alpha}{C}H_2CO_2H$
                    | |
                    Cl   OH

10.2 a. 3-phenylpropanoic acid (IUPAC)
          β-phenylpropionic acid (common)
      b. trichloroethanoic acid (IUPAC)
          trichloroacetic acid (common)

10.3 a.

b.

10.4 a. cyclopropanecarboxylic acid
b. *p*-toluic acid (or *p*-methylbenzoic acid)
c. 2,4,6-trichlorobenzoic acid

10.5 a.

b.

c. $CH_3CH_2CH_2\overset{\overset{\displaystyle O}{\|}}{C}$—Cl

d.

10.6 K_a is 1.8×10^{-5} for acetic acid and 1.5×10^{-3} for chloroacetic acid. K_a is larger for chloroacetic acid; it is the stronger acid. The ratio is

$$\frac{1.5 \times 10^{-3}}{1.8 \times 10^{-5}} = 0.83 \times 10^2$$

In other words, chloroacetic acid is 83 times stronger than acetic acid.

10.7 K_a for benzoic acid is 6.6×10^{-5} or 0.66×10^{-4}. For *o*-, *m*-, and *p*-chlorobenzoic acids, K_a is 12.5, 1.6, and 1.0×10^{-4}, respectively. All three chloro acids are stronger than benzoic acid. However, the difference is greatest for the *ortho* isomer, since in this isomer the chloro substituent is closest to the carboxyl group and exerts the maximum electron-withdrawing inductive effect. The effect decreases as the distance between the chloro substituent and the carboxyl group increases.

10.8 $\underset{\underset{\displaystyle Br}{|}}{CH_2}CH_2CO_2H$ + K^+OH^- $\longrightarrow$ $\underset{\underset{\displaystyle Br}{|}}{CH_2}CH_2CO_2^-K^+$ + H_2O

10.9

$$\text{C}_6\text{H}_5\text{-Br} \xrightarrow[\text{ether}]{\text{Mg}} \text{C}_6\text{H}_5\text{-MgBr} \xrightarrow{\text{CO}_2}$$

$$\text{C}_6\text{H}_5-\overset{\overset{\displaystyle O}{\|}}{\text{C}}-\text{OMgBr} \xrightarrow[\text{H}^+]{\text{H}_2\text{O}} \text{C}_6\text{H}_5-\text{CO}_2\text{H}$$

10.10

$$\text{C}_6\text{H}_5-\text{CH}_2\text{Br} \xrightarrow[\text{ether}]{\text{Mg}} \text{C}_6\text{H}_5-\text{CH}_2\text{MgBr} \xrightarrow[\text{2. H}_2\text{O, H}^+]{\text{1. CO}_2} \text{C}_6\text{H}_5-\text{CH}_2\text{CO}_2\text{H}$$

$$\text{C}_6\text{H}_5-\text{CH}_2\text{Br} \xrightarrow{\text{NaCN}} \text{C}_6\text{H}_5-\text{CH}_2\text{CN} \xrightarrow[\text{heat}]{\text{H}_2\text{O, H}^+} \text{C}_6\text{H}_5-\text{CH}_2\text{CO}_2\text{H}$$

10.11 a. methyl formate (common)   b. *n*-propyl propionate (common)
         methyl methanoate (IUPAC)   1-propyl propanoate (IUPAC)

10.12 a. $\text{CH}_3\overset{\overset{\displaystyle O}{\|}}{\text{C}}\text{OCH}_2\text{CH}_2\text{CH}_2\text{CH}_2\text{CH}_3$ b. $\text{CH}_3\overset{\displaystyle \text{CH}}{\underset{\displaystyle \text{CH}_3}{|}}\overset{\overset{\displaystyle O}{\|}}{\text{C}}\text{OCH}_2\text{CH}_3$

10.13 $\text{CH}_3\overset{\overset{\displaystyle O}{\|}}{\text{C}}\text{-OH} + \text{HOCH}_2\text{CH}_2\text{CH}_2\text{CH}_3 \underset{}{\overset{\text{H}^+}{\rightleftharpoons}} \text{CH}_3\overset{\overset{\displaystyle O}{\|}}{\text{C}}\text{-O-CH}_2\text{CH}_2\text{CH}_2\text{CH}_3 + \text{H}_2\text{O}$

     acetic   1-butanol   *n*-butyl acetate
     acid

10.14

10.15

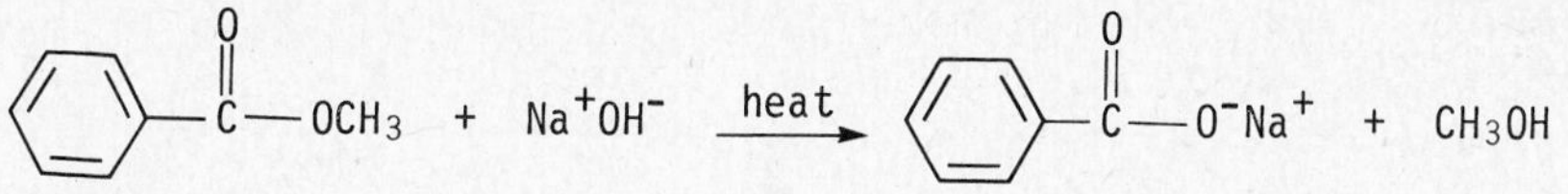

$$\text{(methyl benzoate)} \quad \phi\text{—C(=O)—OCH}_3 \;+\; Na^+OH^- \;\xrightarrow{\text{heat}}\; \phi\text{—C(=O)—O}^-Na^+ \;+\; CH_3OH$$

methyl benzoate   sodium benzoate   methanol

10.16 $CH_3\overset{O}{\overset{\|}{C}}\text{—OCH}_2CH_3 \;+\; NH_3 \;\longrightarrow\; CH_3\overset{O}{\overset{\|}{C}}\text{—NH}_2 \;+\; CH_3CH_2OH$

ethyl acetate   acetamide

10.17   The Grignard reagent provides two of the three R groups attached to the hydroxyl-bearing carbon of the tertiary alcohol; the ester provides the third R group. So, from

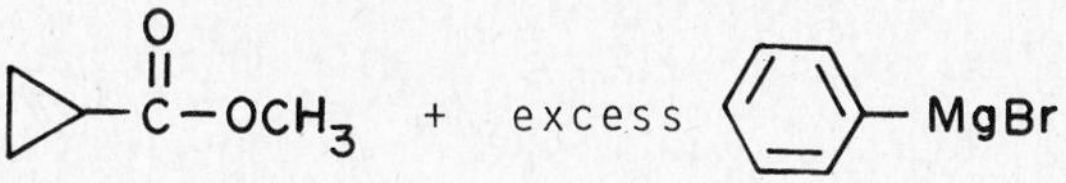

$$\triangleright\text{—C(=O)—OCH}_3 \;+\; \text{excess} \; \phi\text{—MgBr}$$

we get the tertiary alcohol

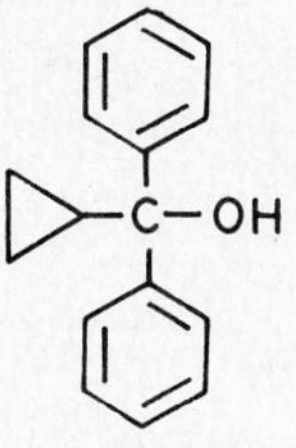

$$\triangleright\text{—C(—OH)}(\phi)(\phi)$$

10.18

$$\phi\text{—C(=O)—OH} \;+\; SOCl_2 \;\longrightarrow\; \phi\text{—C(=O)—Cl} \;+\; HCl \;+\; SO_2$$

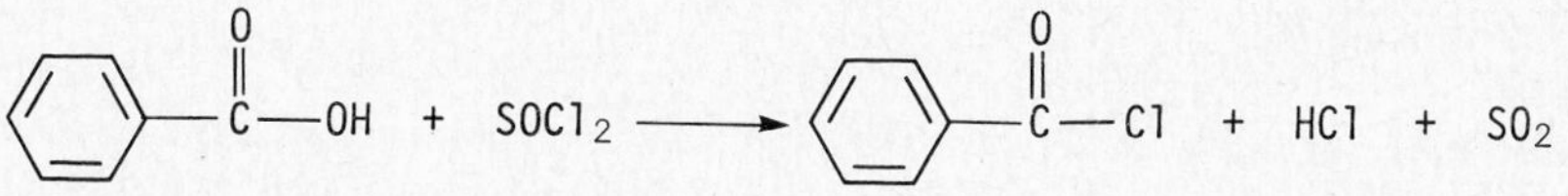

10.19   When acyl halides come in contact with the moist membranes of the nose, they hydrolyze, producing HCl, a severe irritant.

10.20   a. $CH_3CH_2CH_2\overset{O}{\overset{\|}{C}}\text{—O—}\overset{O}{\overset{\|}{C}}CH_2CH_2CH_3$

    b. 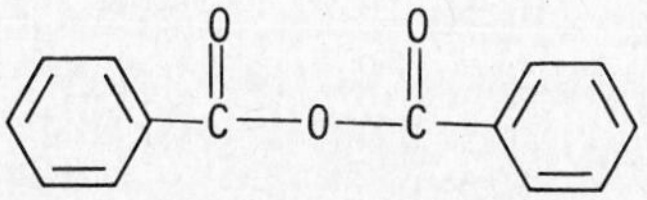 $\phi\text{—}\overset{O}{\overset{\|}{C}}\text{—O—}\overset{O}{\overset{\|}{C}}\text{—}\phi$

10.21 Use the middle part of eq. 10.29 as a guide.

$$CH_3\overset{O}{\overset{\|}{C}}-O-\overset{O}{\overset{\|}{C}}CH_3 \ + \ HOCH_2CH_2CH_2CH_3 \longrightarrow$$

$$CH_3\overset{O}{\overset{\|}{C}}-O-CH_2CH_2CH_2CH_3 \ + \ CH_3CO_2H$$

10.22 Follow eq. 10.31, with R = CH_3.

$$CH_3\overset{O}{\overset{\|}{C}}NH_2 \ + \ H_2O \ \xrightarrow[\text{or } OH^-]{H^+} \ CH_3\overset{O}{\overset{\|}{C}}OH \ + \ NH_3$$

With acid catalysis, the products are CH_3CO_2H + NH_4^+.
With base catalysis, the products are $CH_3CO_2^-$ + NH_3.

10.23 $R\overset{O}{\overset{\|}{C}}-Cl \ \xrightarrow{\text{LiAlH}_4} \ R\overset{O}{\overset{\|}{C}}-H \ \xrightarrow{\text{LiAlH}_4} R-CH_2OH$

Compare with eq. 10.22. The reaction is difficult to stop at the aldehyde stage, although some hydrides that are milder reducing agents than lithium aluminum hydride can accomplish this.

Additional Problems

10.24 a. $CH_3CH_2\overset{3\ \ 2\ \ 1}{\underset{\underset{CH_3}{|}}{CHCH_2}CO_2H}$

b. $CH_3CH_2CCl_2CO_2H$

c. $CH_3\overset{\gamma\ \ \ \beta\ \ \ \alpha}{\underset{\underset{OH}{|}}{CHCH_2}CH_2CO_2H}$

d. CH_3—⬡—CO_2H

e. ☐—CO_2H

f. benzene ring with position 1 CO_2H and position 2 $\overset{}{\underset{\underset{O}{\|}}{C}}CH_2CH_3$

g. ⬡—CH_2CO_2H

h. naphthalene ring with position 1 and position 2 CO_2H

10.25   a. 4-methylpentanoic acid   b. 3-bromo-2-methylbutanoic acid
        c. *o*-nitrobenzoic acid   d. 2-phenylpropanoic acid (or
                                            α-phenylpropionic acid)

        e. propenoic acid   f. cyclohexanecarboxylic acid
            (or acrylic acid)

        g. 2,2-difluoropropanoic   h. 3-ketocyclopentanecarboxylic
            acid   acid (or 3-carboxycyclopentanone)

        i. 4-isobutyrylbenzoic acid   j. 2-butynoic acid

10.26   a. The molecular weights are identical (74). But acids hydrogen-
            bond more effectively than alcohols do.

            $CH_3CH_2CO_2H$ > $CH_3CH_2CH_2CH_2OH$

             bp 141°C   bp 118°C

        b. Chain branching generally lowers the boiling point. Thus for
            these isomeric acids the order is

            $CH_3CH_2CH_2CH_2CO_2H$ > $(CH_3)_3CCO_2H$

               bp 187°C   bp 164°C

10.27   The factors that affect acidity of carboxylic acids are discussed
        in Sec. 10.6.

        a. CH_2ClCO_2H; both substituents, chlorine and bromine, are approxi-
            mately the same distance from the carboxyl group, but chlorine
            is more electronegative than bromine.
        b. *o*-Bromobenzoic acid; the bromine is closer to the carboxyl group,
            and is an electron-withdrawing substituent. Compare the pK_a's
            of the corresponding chloro acids, given in Table 10.3.
        c. CF_3CO_2H; fluorine is more electronegative than chlorine.
        d. Benzoic acid; the methoxy group is an electron-releasing sub-
            stituent when in the *para* position, and may destabilize the anion
            because of the presence of structures such as

        which bring two negative charges near one another.

e. $CH_3CHClCO_2H$; the chlorine, which is electron-withdrawing, is closer to the carboxyl group.

10.28 See Sec. 10.7 if you have any difficulty.

a. $ClCH_2CO_2H$ + K^+OH^- ⟶ $ClCH_2CO_2^-K^+$ + H_2O

Salt formation occurs at room temperature. If the reagents are heated for some time, an S_N2 displacement on the primary chloride may also occur, giving the salt of hydroxyacetic acid, $HOCH_2CO_2^-K^+$.

b. $2\ CH_3(CH_2)_8CO_2H$ + $Ca(OH)_2$ ⟶ $[CH_3(CH_2)_8CO_2^-]_2Ca^{2+}$ + $2\ H_2O$

10.29 a. $CH_3CH_2CH_2CH_2OH \xrightarrow{Na_2Cr_2O_7} CH_3CH_2CH_2CO_2H$

b. $CH_3CH_2CH_2OH \xrightarrow{HBr} CH_3CH_2CH_2Br$

$CH_3CH_2CH_2Br \xrightarrow{NaCN} CH_3CH_2CH_2CN \xrightarrow[H^+]{H_2O} CH_3CH_2CH_2CO_2H$

$CH_3CH_2CH_2Br \xrightarrow{Mg} CH_3CH_2CH_2MgBr \xrightarrow[\text{2. }H_2O]{\text{1. }CO_2} CH_3CH_2CH_2CO_2H$

c.

d.

chlorocyclo-
pentane

e.

f.

10.30 Catalysis by acid:

$$R-C\equiv N: \ + \ \overset{+}{H} \rightleftharpoons R-C\equiv \overset{+}{N}-H$$

Nucleophilic attack by water on the protonated carbon-nitrogen triple bond, followed by proton transfers, gives the amide:

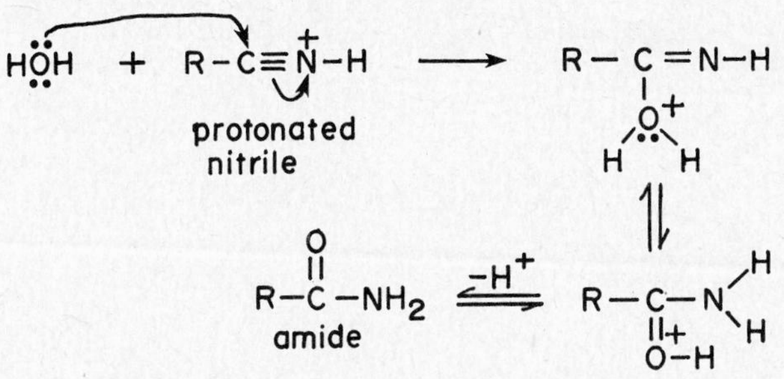

Acid once again catalyzes the hydrolysis of the amide to the carboxylic acid and ammonia. In the acidic medium, the ammonia is immediately protonated.

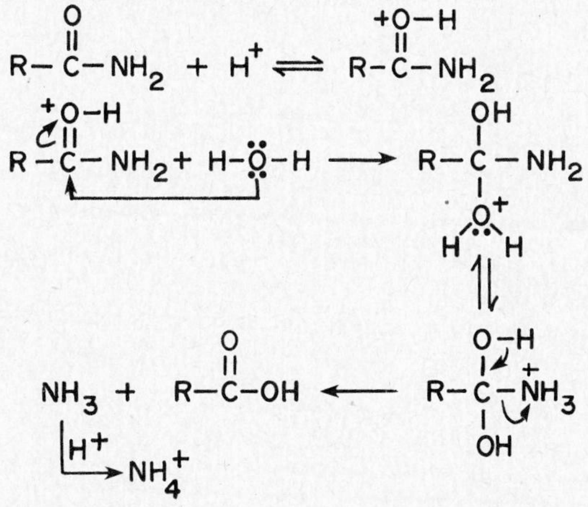

10.31 The cyanide route would involve an S_N2 displacement of bromide by cyanide, a highly unlikely step when the alkyl halide is tertiary. The Grignard route, on the other hand, works well for all alkyl halides, primary, secondary, and tertiary.

10.32 a. $CH_3CHCO_2^-Na^+$
$\quad\quad\quad\;\;|$
$\quad\quad\quad\;\;Cl$

b. $(CH_3CO_2^-)_2Ca^{2+}$

c. $CH_3\overset{O}{\overset{||}{C}}-OCH(CH_3)_2$

d. $H\overset{O}{\overset{||}{C}}-OCH_2CH_3$

e. 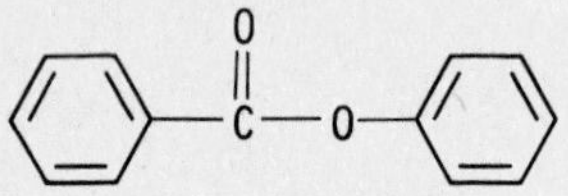 ⬡$-\overset{O}{\overset{||}{C}}-O-$⬡

f. 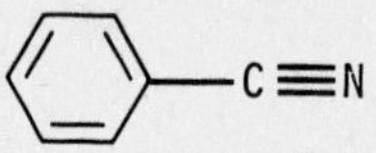 ⬡$-C\equiv N$

g. $CH_3CH_2\overset{O}{\overset{||}{C}}-O-\overset{O}{\overset{||}{C}}CH_2CH_3$

h. 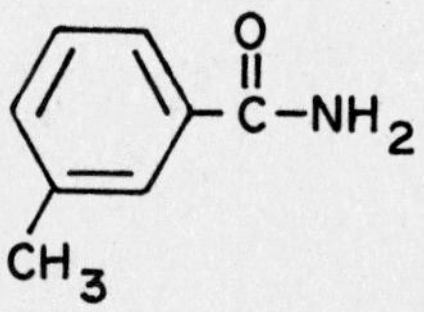 ⬡$-\overset{O}{\overset{||}{C}}-NH_2$
$\quad\;\;CH_3$

i. $CH_2CH_2CH_2\overset{O}{\overset{||}{C}}-Cl$
$\quad\;|$
$\quad\;Cl$

j. $H-\overset{O}{\overset{||}{C}}-F$

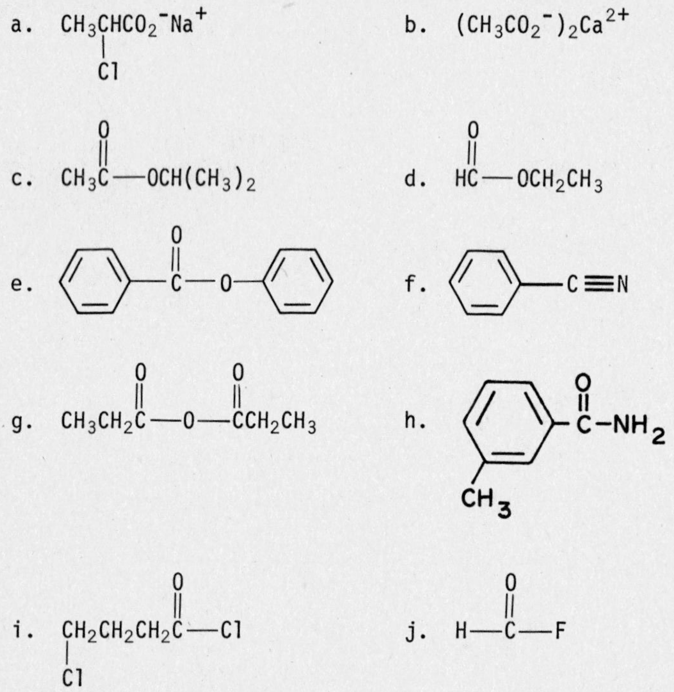

10.33 a. ammonium *p*-chlorobenzoate
     c. phenyl isobutyrate
        (or phenyl 2-methylpropanoate)
     e. thioacetic acid
     g. formamide

     b. calcium butanoate
     d. methyl trifluoroacetate

     f. methyl thioacetate
     h. cyclopropanecarboxylic
        anhydride

10.34 Compare with the answer to Problem 10.14.

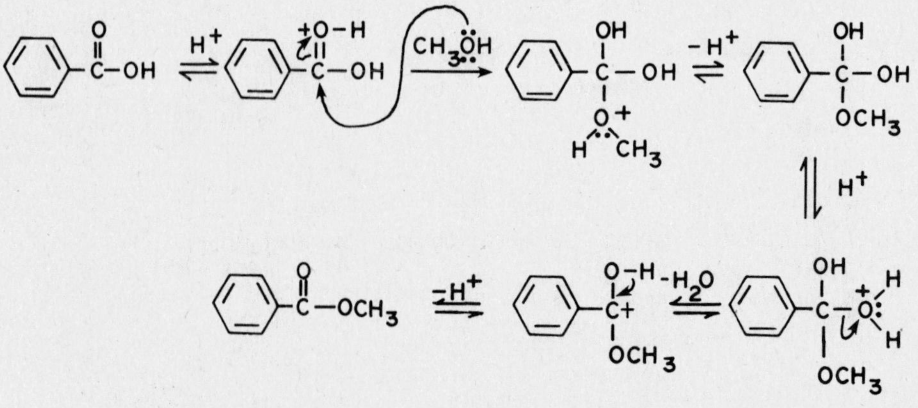

10.35 a.

See eq. 10.17, where R = phenyl and R' = ethyl.

b. Compare with eq. 10.19.

benzamide

c. See Sec. 10.15.

d.

10.36 a.

b.

10.37 a. The carbonyl group in esters is less reactive than the carbonyl group of ketones because of the possibility of resonance in esters:

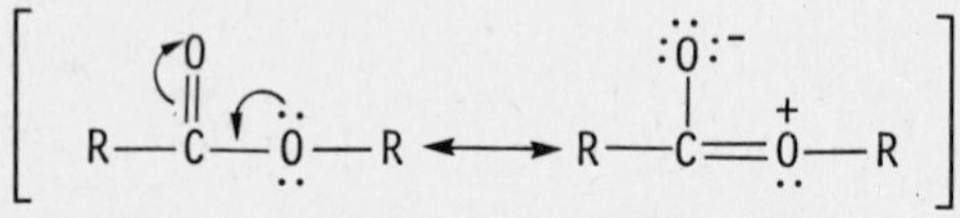

This delocalizes to the "ether" oxygen some of the positive charge usually associated with the carbonyl carbon atom:

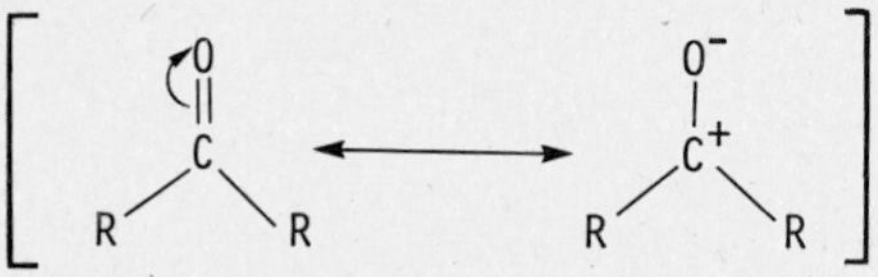

The carbonyl carbon in esters is therefore less susceptible to nucleophilic attack than is the carbonyl carbon of ketones.

b. Compare

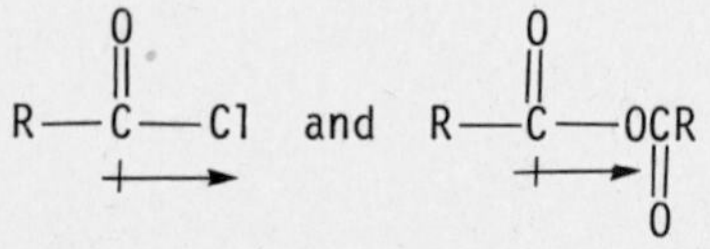

In both, a dipole in the direction shown enhances the positive charge on the carbonyl carbon, making the carbonyl groups shown more susceptible to nucleophilic attack than the carbonyl groups of esters or acids. But chlorine is more electronegative than oxygen; therefore, acyl chlorides are usually more reactive toward nucleophiles than are acid anhydrides.

c. In benzoyl chloride, the positive charge on the carbonyl carbon can be delocalized in the aromatic ring:

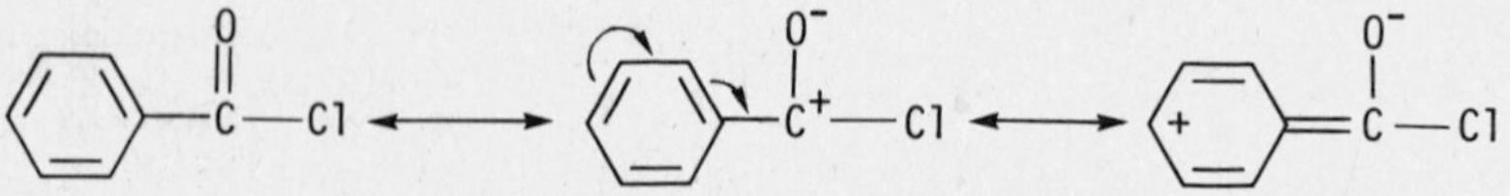

Such delocalization is not possible in cyclohexanecarbonyl chloride or any other aliphatic acid chloride. For this reason, aryl acid chlorides are usually less reactive toward nucleophiles than are aliphatic acid chlorides.

10.38

$$R-\overset{\overset{\displaystyle :\overset{..}{O}:}{\|}}{C}-OR' \quad + \quad :NH_3 \quad \rightleftharpoons \quad R-\overset{\overset{\displaystyle :\overset{..}{O}:^-}{|}}{\underset{\underset{\displaystyle +NH_3}{|}}{C}}-OR'$$

$$R-\overset{\overset{\displaystyle O}{\|}}{C}-NH_2 \quad + \quad R'OH \quad \longleftarrow \quad R-\overset{\overset{\displaystyle :\overset{..}{O}-H}{|}}{\underset{\underset{\displaystyle NH_2}{|}}{C}}-OR'$$

10.39 In each case, the two like organic groups attached to the hydroxyl-bearing carbon come from the Grignard reagent, and the third group comes from the acid part of the esters.

a. $CH_3CH_2CH_2MgBr$ + [benzene ring]$-\overset{\overset{\displaystyle O}{\|}}{C}-OR$

b. [benzene ring]$-MgBr$ + $CH_3CH_2CH_2\overset{\overset{\displaystyle O}{\|}}{C}-OR$

The identity of the R group in the ester does not affect the product structure. R is usually CH_3 or CH_3CH_2.

10.40 a. CH_3COCl + H_2O $\longrightarrow$ CH_3CO_2H + HCl

b. [benzene ring]$-\overset{\overset{\displaystyle O}{\|}}{C}-Cl$ + CH_3OH $\longrightarrow$ [benzene ring]$-\overset{\overset{\displaystyle O}{\|}}{C}-OCH_3$ + HCl

c. $CH_3\overset{\overset{\displaystyle O}{\|}}{C}-O-\overset{\overset{\displaystyle O}{\|}}{C}CH_3$ + $HOCH_2CH_2CH_2CH_2CH_3$ $\longrightarrow$

$CH_3\overset{\overset{\displaystyle O}{\|}}{C}-O-CH_2CH_2CH_2CH_2CH_3$ + CH_3CO_2H

d. $\underset{\underset{\displaystyle Br}{|}}{CH_2CH_2CH_2}\overset{\overset{\displaystyle O}{\|}}{C}-Br$ + $2\ NH_3$ $\longrightarrow$ $\underset{\underset{\displaystyle Br}{|}}{CH_2CH_2CH_2}\overset{\overset{\displaystyle O}{\|}}{C}-NH_2$ + $NH_4{}^+Br^-$

Note that the acyl bromide is much more reactive toward nucleophiles than is the alkyl bromide.

e. $CH_3CH_2CH_2CH_2\overset{\overset{O}{\|}}{C}\!\!-\!\!OH$ + CH_3CH_2OH $\xrightarrow{H^+}$

$CH_3CH_2CH_2CH_2\overset{\overset{O}{\|}}{C}\!\!-\!\!OCH_2CH_3$ + H_2O

10.41 a. $CH_3CH_2CH_2\overset{\overset{O}{\|}}{C}\!\!-\!\!OH$ + PCl_5 $\longrightarrow$ $CH_3CH_2CH_2\overset{\overset{O}{\|}}{C}\!\!-\!\!Cl$ + HCl + $POCl_3$

(see eq. 10.25)

b. $CH_3(CH_2)_8CO_2H$ + $SOCl_2$ $\longrightarrow$ $CH_3(CH_2)_8\overset{\overset{O}{\|}}{C}\!\!-\!\!Cl$ + HCl + SO_2

(see eq. 10.24)

c. 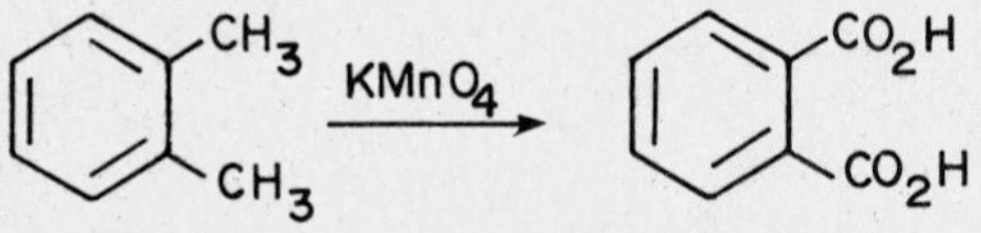 $\xrightarrow{KMnO_4}$

(see eq. 10.8)

d. 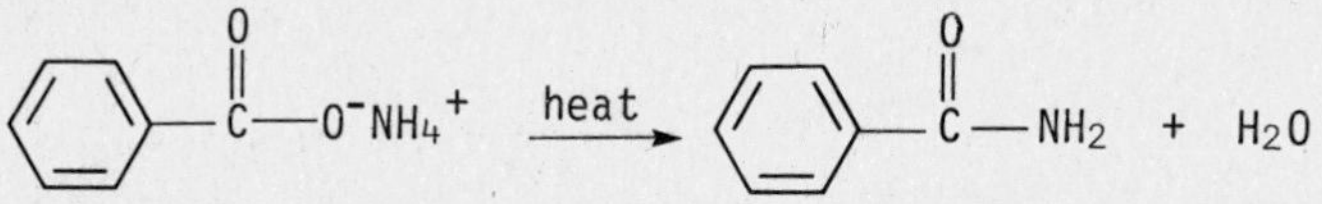 $\xrightarrow{heat}$ + H_2O

(see eq. 10.30)

e. $CH_3(CH_2)_5\overset{\overset{O}{\|}}{C}\!\!-\!\!NH_2$ $\xrightarrow{LiAlH_4}$ $CH_3(CH_2)_5CH_2NH_2$

(see eq. 10.32)

f. 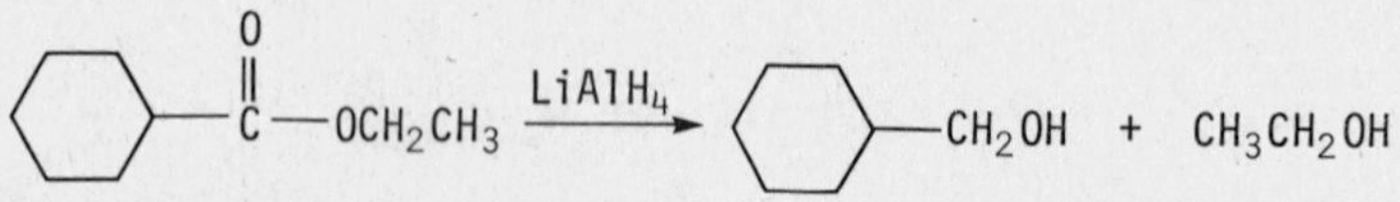 $\xrightarrow{LiAlH_4}$ $-\!\!CH_2OH$ + CH_3CH_2OH

(see eq. 10.21)

10.42 Review Sec. 10.20.

$$\left[\quad CH_3CH_2C\underset{\underset{H}{\overset{\vert}{N}}-H}{\overset{\overset{\cdot\cdot}{O}\cdot}{\diagdown}} \quad \longleftrightarrow \quad CH_3CH_2C\underset{\underset{H}{\overset{\vert}{N}}-H}{\overset{\overset{\cdot\cdot}{O}:^-}{\diagdown}} \quad \right]$$

The six atoms that lie in one plane are

$$\overset{O^-}{\underset{C}{\diagdown}}C=\overset{+}{\underset{H}{N}}\overset{H}{\diagup}$$

10.43 Consider the leaving group in each nucleophilic substitution:

$$R-\overset{\overset{O}{\parallel}}{C}-SR' \ + \ Nu: \longrightarrow R-\overset{\overset{O}{\parallel}}{C}-Nu \ + \ ^{\ominus}SR'$$

$$R-\overset{\overset{O}{\parallel}}{C}-\overset{\overset{O}{\parallel}}{O}CR' \ + \ Nu: \longrightarrow R-\overset{\overset{O}{\parallel}}{C}-Nu \ + \ ^{\ominus}O-\overset{\overset{O}{\parallel}}{C}R'$$

$$R-\overset{\overset{O}{\parallel}}{C}-Cl \ + \ Nu: \longrightarrow R-\overset{\overset{O}{\parallel}}{C}-Nu \ + \ Cl^{\ominus}$$

The K_a's of the conjugate acids increase in the order shown:

$$RSH \cong 10^{-10}; \ R\overset{\overset{O}{\parallel}}{C}OH \cong 10^{-5}; \ HCl = \text{strong acid.}$$ Therefore, the basicity of the anions decreases in the same order:

$$^{\ominus}SR' > {}^{\ominus}O\overset{\overset{O}{\parallel}}{C}R' > Cl^{\ominus}$$

The weaker the basicity of the leaving group, the more facile the nucleophilic substitution. Therefore, the reactivity order is thioesters < anhydrides < acyl chlorides.

10.44 Ketones are more reactive toward nucleophiles than esters are. Reduction therefore occurs at the ketone carbonyl group, to give

$$CH_3\overset{\overset{OH}{\vert}}{C}HCH_2CH_2CO_2CH_3$$

10.45 The method combines the formation of a cyanohydrin (Sec. 9.10) with the hydrolysis of a cyanide to an acid (Sec. 10.8d).

benzaldehyde
cyanohydrin

CHAPTER SUMMARY

Two functional groups remote from one another on a carbon framework tend to react independently, but when two functional groups are close to one another, each can influence the chemistry of the other.

Dicarboxylic acids occur in nature; their nomenclature is summarized in Table 11.1. When the two carboxyl groups in such acids are near one another, the first acidity constant K_1 is increased substantially, since one carboxyl group acts as an electron-withdrawing group with respect to the other.

Dicarboxylic acids with both carboxyl groups attached to the same carbon atom, such as **malonic acid,** are readily **decarboxylated** on heating. When there are two or three carbon atoms between the carboxyl groups, heating brings about dehydration to a **cyclic anhydride.**

Dicarboxylic acids react with diols to form **polyesters.** Important examples include **Dacron** and **Mylar,** made from terephthalic acid and ethylene glycol.

The simplest unsaturated acid is **acrylic acid** (propenoic acid). The corresponding nitrile, **acrylonitrile,** gives **Orlon** when polymerized. **Methyl methacrylate** is the monomer used to produce **Plexiglas. Maleic** and **fumaric acids** are the two geometric isomers of 2-butenedioic acid.

Important natural α-**hydroxy acids** include **glycolic, lactic, malic, tartaric,** and **citric acids. Lactones** are cyclic esters. Five- and six-membered lactone rings form when γ- or δ-hydroxy acids are heated. In contrast, β-hydroxy acids dehydrate to unsaturated acids, and α-hydroxy acids form cyclic diesters called **lactides.**

Salicylic acid (o-hydroxybenzoic acid) is manufactured from phenol, carbon dioxide, and base. Its methyl ester is **oil of wintergreen,** and its acetyl derivative is **aspirin.**

Pyruvic acid is an α-**keto acid** that is important in biological processes. β-**Keto acids,** such as **acetoacetic acid,** are important intermediates in fat metabolism. They readily decarboxylate on heating, to give ketones. They can be synthesized in the laboratory by the **Claisen condensation,** a reaction analogous to the aldol condensation but involving **ester enolates** as the reactive intermediates.

Fats and oils are triesters formed by reaction of the triol glycerol with long-chain saturated or unsaturated acids. The common acids in fats and oils have an even number of carbons, and if unsaturated, they have the Z configuration (Table 11.2 lists them, with their common names). Oils, which have a high percentage of unsaturated acids, can be hardened to solid fats by hydrogenation.

Saponification of fats and oils by boiling with strong base gives glycerol and the sodium salts of fatty acids. The latter are soaps. Soaps contain a long carbon chain that is lipophilic and a terminal polar group that is hydrophilic. Soap molecules aggregate in water to form micelles, which help emulsify droplets of oil or grease.

Two disadvantages of ordinary soaps are their alkalinity and their tendency to form insoluble salts with the Ca^{2+} or Mg^{2+} ions present in "hard" water. Synthetic detergents or syndets overcome these disadvantages. The most widely used syndets at present are straight-chain alkylbenzenesulfonates, obtained by successive alkylation and sulfonation of benzene. The straight chains are necessary for biodegradability.

Fat metabolism involves initial hydrolysis of the fat or oil to glycerol and a mixture of fatty acids. These are then degraded, two carbons at a time, to give acetate and eventually carbon dioxide and water. A key step in this process is a reverse Claisen condensation. A similar sequence, but in reverse, is involved in the biosynthesis of fats.

Phospholipids are triesters of glycerol in which one ester group is derived from an amino phosphonic acid. They are an important structural unit in cell membranes. Waxes are monoesters of long-chain acids and alcohols. Steroids are lipids with a unique four-ring structure. Important examples of steroids include cholesterol, the bile acids, and the sex hormones. Prostaglandins are twenty-carbon cyclopentane derivatives of arachidonic acid that have profound biological effects even in minute quantities.

REACTION SUMMARY

Ionization of Dicarboxylic Acids

$$HO_2C-(CH_2)_{\overline{n}}CO_2H \overset{K_1}{\rightleftharpoons} {}^-O_2C-(CH_2)_{\overline{n}}CO_2H \overset{K_2}{\rightleftharpoons} {}^-O_2C-(CH_2)_{\overline{n}}CO_2{}^-$$

Decarboxylation of Malonic-Type Acids

REACTION SUMMARY

Dehydration of a Dicarboxylic Acid

$$\underset{\substack{\text{CO}_2\text{H} \\ (\text{CH}_2)_n \\ \text{CO}_2\text{H}}}{} \xrightarrow{\text{heat}} (\text{CH}_2)_n \overset{\displaystyle C=O}{\underset{\displaystyle C=O}{\bigg\langle}} O \qquad (n = 2, 3)$$

Lactonization of a Hydroxy Acid

$$\underset{\substack{\text{CO}_2\text{H} \\ (\text{CH}_2)_n \\ \text{OH}}}{} \xrightarrow{\text{H}^+} (\text{CH}_2)_n \overset{\displaystyle C=O}{\underset{\displaystyle O}{\bigg|}} \qquad (n = 3, 4)$$

Polyester Formation

$$\text{HO} \sim\!\sim \text{OH} \;+\; \text{HO} - \overset{O}{\overset{\|}{C}} \sim\!\sim \overset{O}{\overset{\|}{C}} - \text{OH} \longrightarrow \text{H} - \text{O} \!\left(\!\overset{O}{\overset{\|}{C}} \sim\!\sim \overset{O}{\overset{\|}{C}} - \text{O} \sim\!\sim \text{O}\!\right)\!\!\text{H}$$

Synthesis of Aspirin

$$\underset{}{\text{OH}}\!\!\bigcirc \xrightarrow[\substack{\text{NaOH} \\ \text{heat, pressure}}]{\text{CO}_2} \xrightarrow{\text{H}^+} \underset{\text{salicylic acid}}{\overset{\text{OH}}{\bigcirc}\!\text{CO}_2\text{H}} \xrightarrow{(\text{CH}_3\overset{O}{\overset{\|}{C}})_2\text{O}} \underset{\text{aspirin}}{\overset{\text{O}-\overset{O}{\overset{\|}{C}}\text{CH}_3}{\bigcirc}\!\text{CO}_2\text{H}}$$

Decarboxylation of β-Keto Acids

$$R - \overset{O}{\overset{\|}{C}} - \overset{|}{\underset{|}{C}} - \text{CO}_2\text{H} \xrightarrow{\text{heat}} R - \overset{O}{\overset{\|}{C}} - \underset{\underset{\text{ketone}}{|}}{\overset{|}{C}} - H \;+\; \text{CO}_2$$

Claisen Condensation of Esters

$$2\; R\text{CH}_2\overset{O}{\overset{\|}{C}}\!\!-\!\!\text{OR}' \xrightarrow{\text{NaOR}'} R\text{CH}_2\overset{O}{\overset{\|}{C}}\!\!-\!\!\underset{\underset{R}{|}}{\overset{|}{C}}\text{H}\!-\!\text{CO}_2\text{R}' \;+\; \text{R}'\text{OH}$$

β-keto ester

Saponification of a Triglyceride

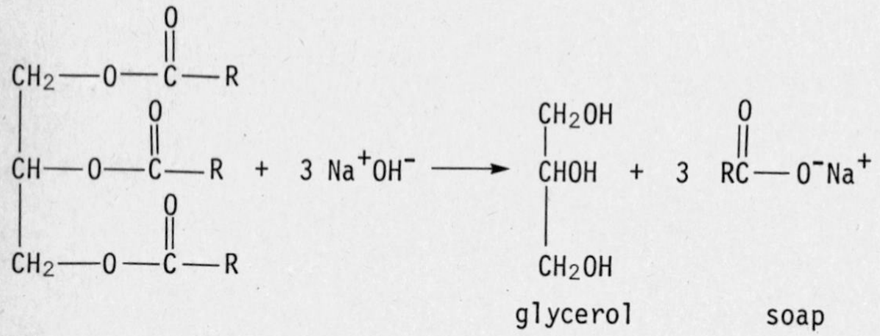

$$CH_2-O-\underset{\underset{O}{\|}}{C}-R$$
$$CH-O-\underset{\underset{O}{\|}}{C}-R \;+\; 3\;Na^+OH^- \longrightarrow$$
$$CH_2-O-\underset{\underset{O}{\|}}{C}-R$$

$$CH_2OH$$
$$CHOH \;+\; 3\; RC\underset{\underset{O}{\|}}{}-O^-Na^+$$
$$CH_2OH$$

glycerol   soap

MECHANISM SUMMARY

Decarboxylation of Malonic-Type Acids

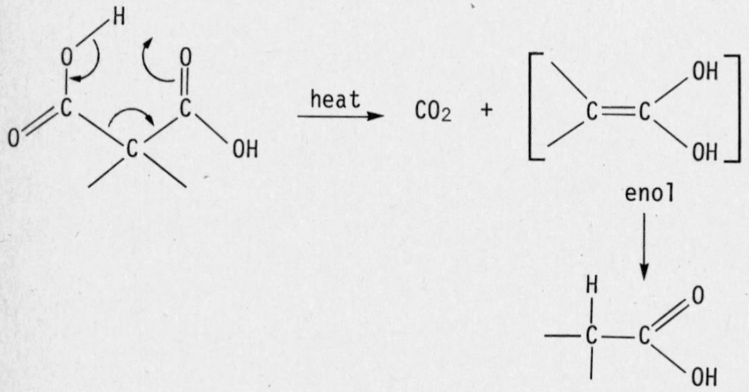

$$\xrightarrow{\text{heat}} CO_2 \;+\; \left[\quad C=C \begin{matrix} OH \\ OH \end{matrix} \right]$$

enol

Decarboxylation of β-Keto Acids

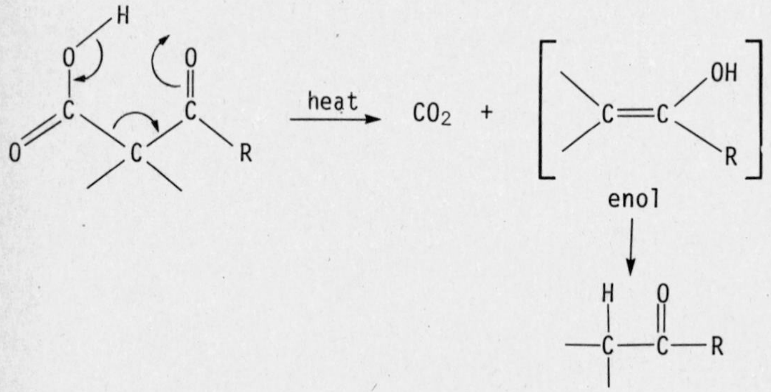

$$\xrightarrow{\text{heat}} CO_2 \;+\; \left[\quad C=C \begin{matrix} OH \\ R \end{matrix} \right]$$

enol

LEARNING OBJECTIVES

1. Be able to write the structures of the following dicarboxylic acids: oxalic, malonic, succinic, glutaric, adipic, phthalic, isophthalic, terephthalic.

2. Given a dicarboxylic acid, write equations for its first and second ionization. Explain the difference in the values of the ionization constants, K_1 and K_2.

3. Write an equation for the decarboxylation (loss of CO_2) of a compound with two carboxyl groups on the same carbon atom. Write a mechanism for this reaction.

4. Predict the product when a given dicarboxylic acid is heated.

5. Draw the structure of the polyester formed from a given dicarboxylic acid and diol. Draw the structure of Dacron.

6. Know the meaning of: acrylonitrile, methyl methacrylate, Orlon, Plexiglas, maleic acid, fumaric acid.

7. Be able to write the structures of the following hydroxy acids: glycolic, lactic, malic, tartaric, citric, salicylic.

8. Predict the product when a given hydroxy acid is heated.

9. Write equations for the preparation of: a lactone, salicylic acid, aspirin.

10. Write an equation for the effect of heat on a given β-keto acid.

11. Given a particular ester, write the structure of the β-keto ester obtained from its self-condensation (Claisen condensation).

12. Write the steps in the mechanism of a particular Claisen condensation.

13. Know the meaning of: triglyceride, fatty acid, fat, oil, hardening of a vegetable oil, soap, saponification.

14. Know the structures and common names of the acids listed in Table 11.2.

15. Given the name of a glyceride, write its structure.

16. Given the name or structure of a carboxylic acid, write the formula for the corresponding glyceride.

17. Given the structure of a glyceride, write the equation for its saponification.

18. Given the structure of an unsaturated glyceride, write equations (including catalysts) for its hydrogenation and hydrogenolysis.

19. Explain the difference between the structure of a fat and that of a vegetable oil.

20. Describe the structural features essential for a good soap or detergent.

21. Explain, with the aid of a diagram, how a soap emulsifies fats and oils.

22. Explain, with the aid of equations, what happens when an ordinary soap is used in hard water and how synthetic detergents overcome this difficulty.

23. Know the meaning of: lipophilic, hydrophilic, sodium alkyl sulfate, alkylbenzenesulfonate.

24. Know the meaning of: lipid, phospholipid, wax, steroid, sex hormone, estrogen, androgen, prostaglandin.

ANSWERS TO PROBLEMS

Problems Within the Chapter

11.1   The K_1/K_2 ratio for phthalic acid is 333, whereas the same ratio for isophthalic acid and for terephthalic acid is only 11 and 21, respectively. The main reason for the difference is that the carboxyl groups are much closer to one another in phthalic acid than in its isomers. Therefore, the effect of one carbonyl group on the other is greatest in this isomer. In each case, K_1 is larger and K_2 is smaller than the ionization constant of benzoic acid (6.6×10^{-5}; Table 10.3), but the differences are greatest in each case for phthalic acid, resulting in the largest K_1/K_2 ratio.

11.2   In each case, the product structure can be deduced by replacing one of the carboxyl groups with a hydrogen atom (eq. 11.4).

a.
⬡—CH$_2$COOH

b.
◇ H
COOH

11.3

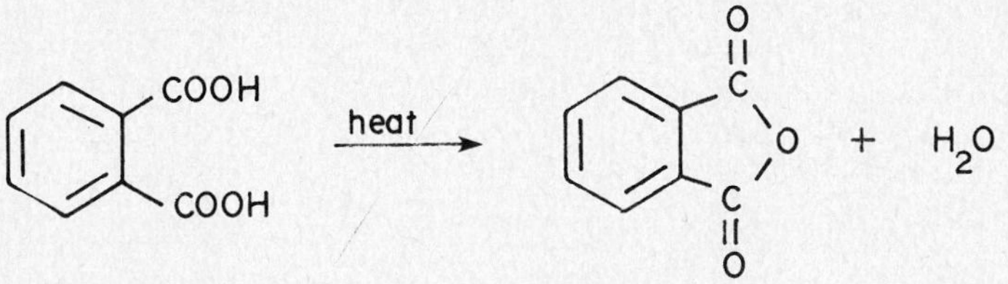

11.4 The monomers are HOC—⟨benzene⟩—COH and HOCH$_2$—⟨cyclohexane⟩—CH$_2$OH.

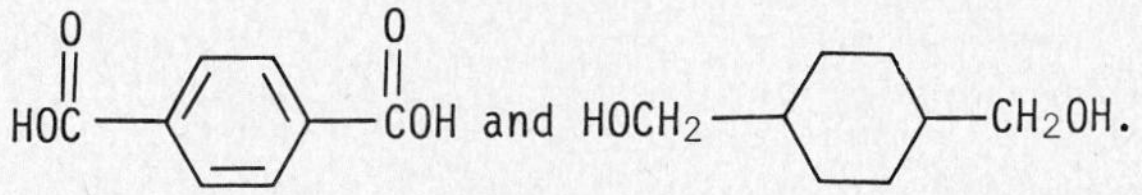

Note that the diol used for Kodel can be made by complete reduction of the dicarboxylic acid (terephthalic acid).

11.5

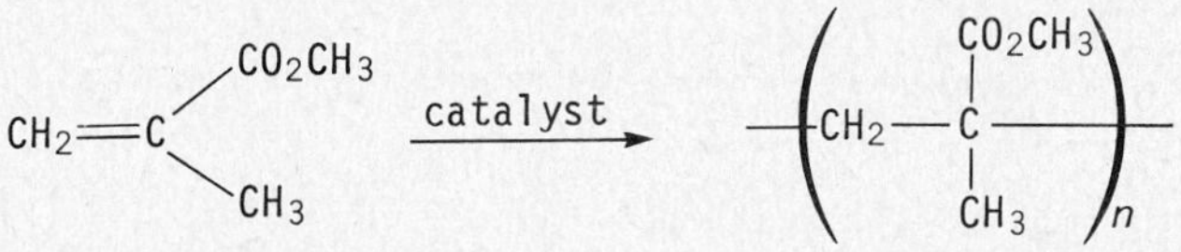

methyl methacrylate   Plexiglas (also called Lucite)

11.6 The product structure depends on the distance between the two functional groups.

a. 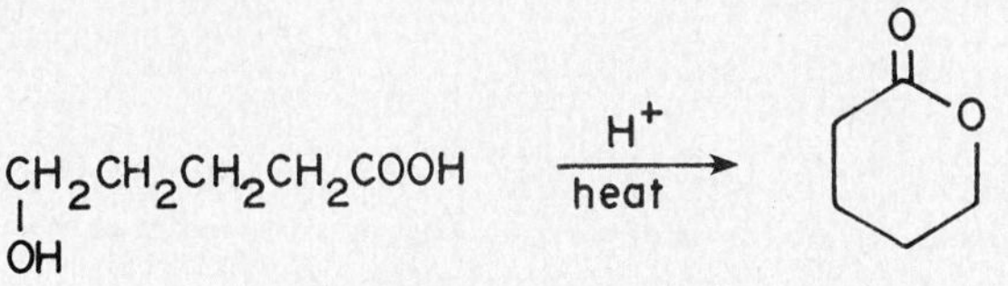 $CH_2CH_2CH_2CH_2COOH$ (with OH on first carbon) $\xrightarrow[\text{heat}]{H^+}$

b. 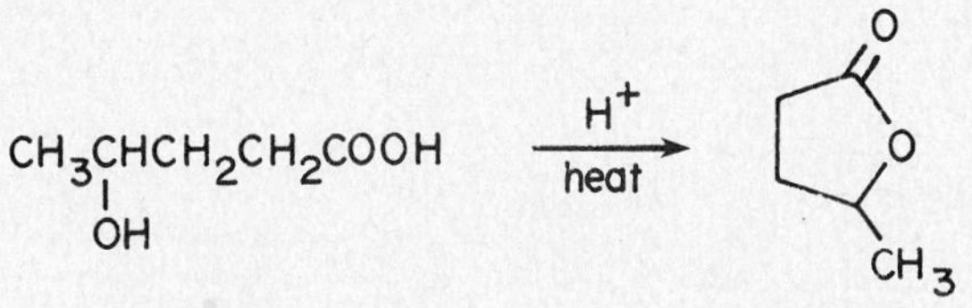 $CH_3CHCH_2CH_2COOH$ (with OH on first carbon) $\xrightarrow[\text{heat}]{H^+}$

c. $CH_3CH_2CHCH_2COOH$ (with OH) $\xrightarrow[\text{heat}]{H^+}$ $CH_3CH_2CH{=}CHCOOH$

d. 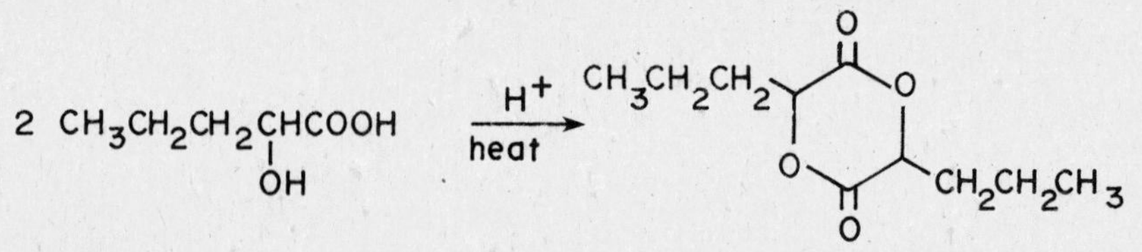   2 $CH_3CH_2CH_2\overset{\displaystyle |}{\underset{\displaystyle OH}{C}}HCOOH$ $\xrightarrow[\text{heat}]{H^+}$

11.7 Fischer esterification gives the product:

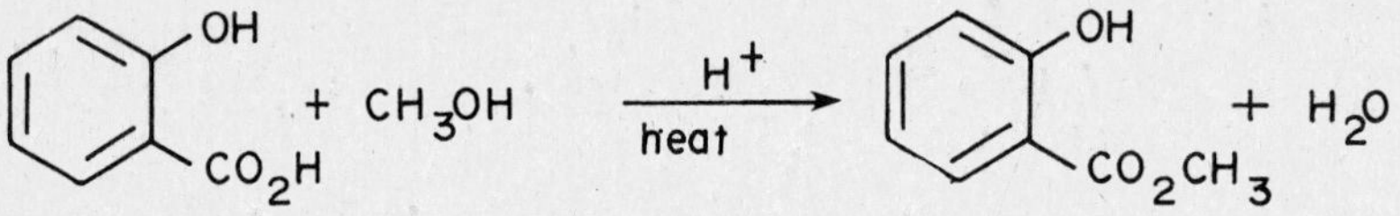

$$\text{(structure: benzene ring with OH and CO}_2\text{H)} + CH_3OH \xrightarrow[\text{heat}]{H^+} \text{(structure: benzene ring with OH and CO}_2CH_3\text{)} + H_2O$$

11.8 The product structure is deduced by replacing the carboxyl group with a hydrogen atom (eq. 11.16).

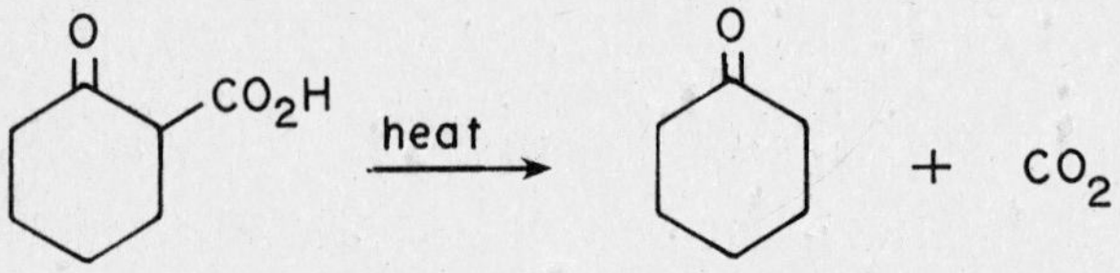

$$\text{(cyclohexanone with CO}_2\text{H)} \xrightarrow{\text{heat}} \text{(cyclohexanone)} + CO_2$$

The reaction mechanism is

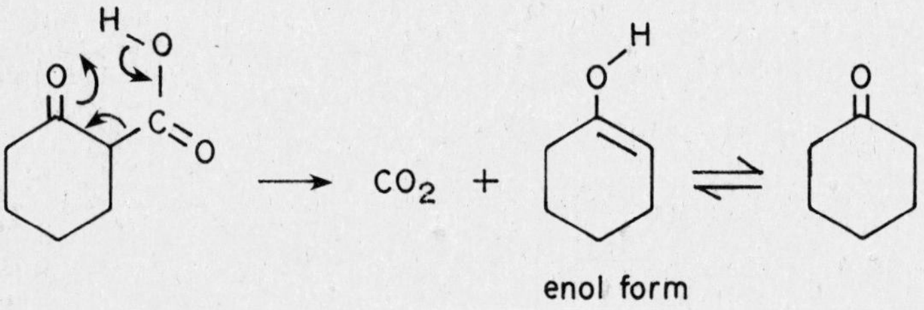

$$\longrightarrow \; CO_2 \; + \; \text{(enol form)} \; \rightleftharpoons \; \text{(cyclohexanone)}$$

enol form

11.9 <u>Step 1:</u>

$$CH_3CH_2\overset{\displaystyle O}{\overset{\displaystyle \|}{C}}\!-\!OCH_2CH_3 \; + \; Na^{+-}OCH_2CH_3 \; \rightleftharpoons \; CH_3\overset{\displaystyle -}{C}H\overset{\displaystyle O}{\overset{\displaystyle \|}{C}}\!-\!OCH_2CH_3 \; + \; HOCH_2CH_3$$
$$Na^+$$

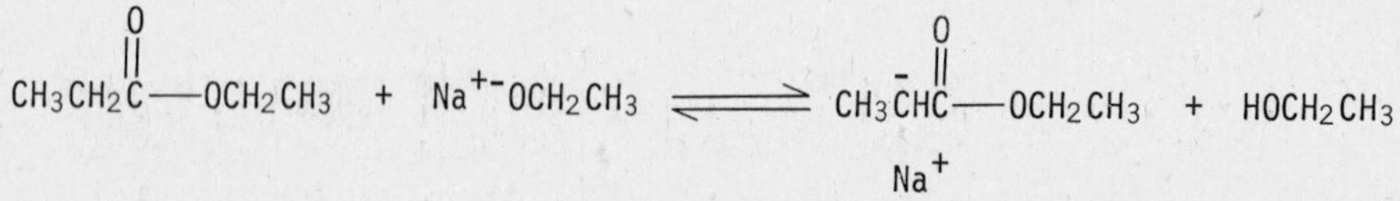

Step 2:

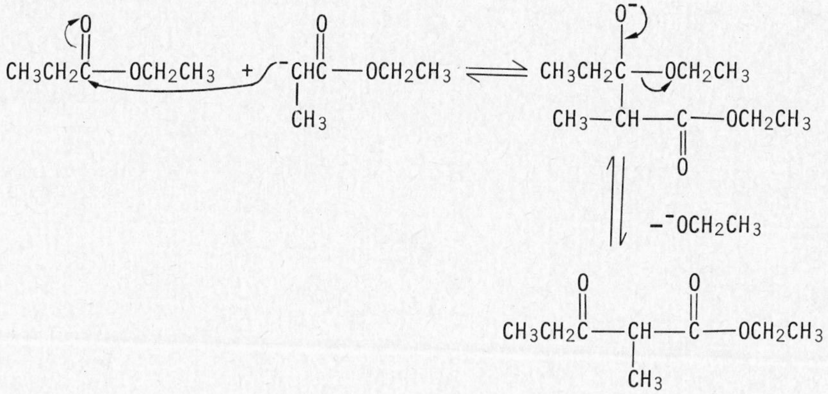

$$CH_3CH_2C-OCH_2CH_3 + \ ^-CHC-OCH_2CH_3 \ \rightleftharpoons \ CH_3CH_2C-OCH_2CH_3$$

with CH_3 and reaction structures leading to

$$CH_3-CH-C-OCH_2CH_3$$

$$\Big\Downarrow \quad -^-OCH_2CH_3$$

$$CH_3CH_2C-CH-C-OCH_2CH_3$$
$$\underset{CH_3}{}$$

Step 3:

$$CH_3CH_2C-CH-C-OCH_2CH_3 \ + \ ^-OCH_2CH_3 \ \rightleftharpoons$$
$$\underset{CH_3}{}$$

$$CH_3CH_2C \overset{\ominus}{-\!-\!-\!C\!-\!-\!-} C-OCH_2CH_3 \ + \ HOCH_2CH_3$$
$$\underset{CH_3}{}$$

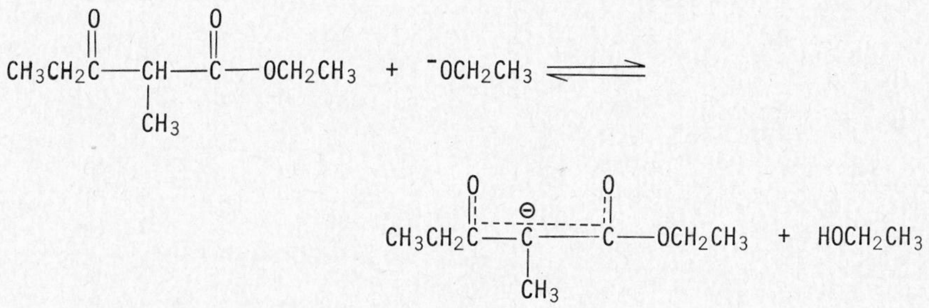

11.10

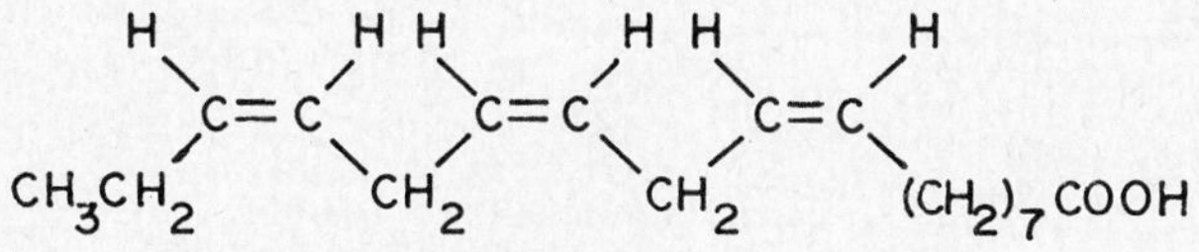

linolenic acid

11.11 a.

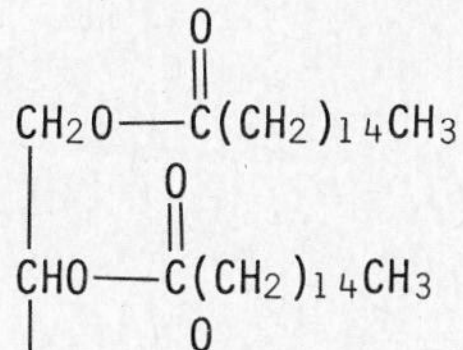

$$CH_2O-C(CH_2)_{14}CH_3$$
$$CHO-C(CH_2)_{14}CH_3$$
$$CH_2O-C(CH_2)_{14}CH_3$$

b.

$$CH_2O-C(CH_2)_{14}CH_3$$
$$CHO-C(CH_2)_7CH=CH(CH_2)_7CH_3$$
$$CH_2O-C(CH_2)_{16}CH_3$$

235

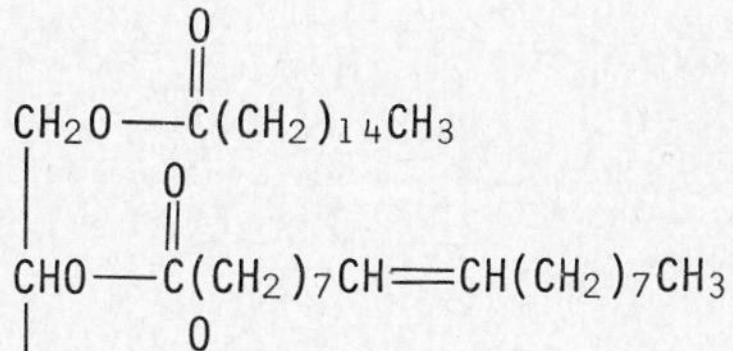

11.12 a. glycerol and sodium palmitate, $CH_3(CH_2)_{14}CO_2^-Na^+$

b. glycerol and equimolar amounts of the sodium salts of the three carboxylic acids:

$CH_3(CH_2)_{14}CO_2^-Na^+$
sodium palmitate

$CH_3(CH_2)_7CH\!=\!\!=\!\!CH(CH_2)_7CO_2^-Na^+$
sodium oleate

$CH_3(CH_2)_{16}CO_2^-Na^+$
sodium stearate

The triglyceride shown in Example 11.6 would give the same saponification products.

Additional Problems

11.13 a. $CH_3O\!-\!\overset{\overset{\displaystyle O}{\|}}{C}\!-\!\overset{\overset{\displaystyle O}{\|}}{C}\!-\!OCH_3$

b. $Cl\!-\!\overset{\overset{\displaystyle O}{\|}}{C}\!-\!(CH_2)_4\!-\!\overset{\overset{\displaystyle O}{\|}}{C}\!-\!Cl$

c.

d. $\begin{array}{c} CH_3CH_2 \qquad CO_2CH_3 \\ \diagdown C \diagup \\ CH_3CH_2 \qquad CO_2CH_3 \end{array}$

e. $HOOC\!-\!\underset{\underset{\displaystyle OH}{|}}{CH}\!-\!CH_2COOH$

f. $HOOC\!-\!\underset{\underset{\displaystyle CH_3}{|}}{CH}\!-\!\underset{\underset{\displaystyle CH_3}{|}}{CH}\!-\!COOH$

11.14 See the answer to Problem 11.1. The carboxyl groups in malonic acid are much closer to one another than are the ones in adipic acid. The electron-withdrawing effect of the carboxyl group on K_1 and the repulsion between two negative charges in the dicarboxylate dianion with its ensuing effect on K_2 are therefore both much greater in malonic acid than in adipic acid.

11.15 Compare the dissociation of acetic acid with the second dissociation of malonic acid:

$$H\!-\!CH_2CO_2H \rightleftharpoons H\!-\!CH_2CO_2^- + H^+$$

$$^-O_2C\!-\!CH_2CO_2H \rightleftharpoons {}^-O_2C\!-\!CH_2CO_2^- + H^+$$

In malonic acid, the $—CO_2^-$ substituent, being negatively charged, is an electron-releasing substituent (much more so than $—H$). Since electron-releasing substituents decrease acidity, K_2 for malonic acid is less than K_a for acetic acid. The same is true for K_2 of the larger dicarboxylic acids.

11.16

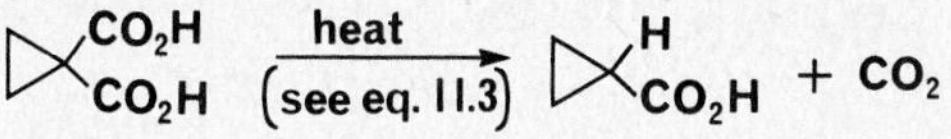

cis -1,2-Cyclopropanedicarboxylic acid forms a cyclic anhydride:

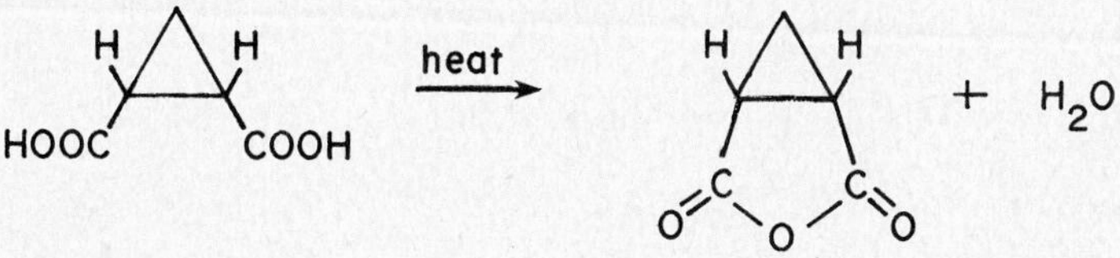

The *trans* isomer cannot form a cyclic anhydride because the carboxyl groups are too far apart. It may form a polymeric anhydride.

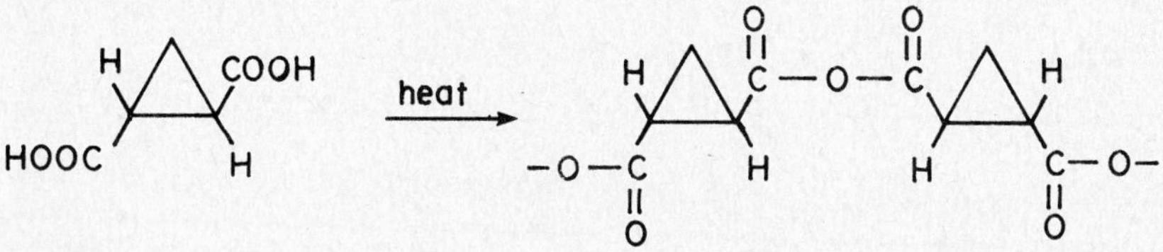

11.17 a.

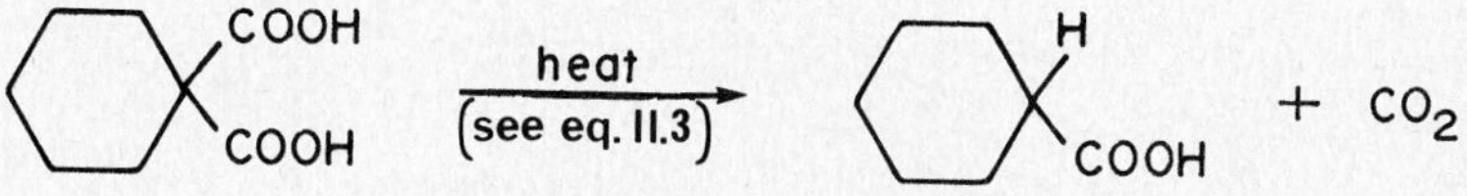

b.

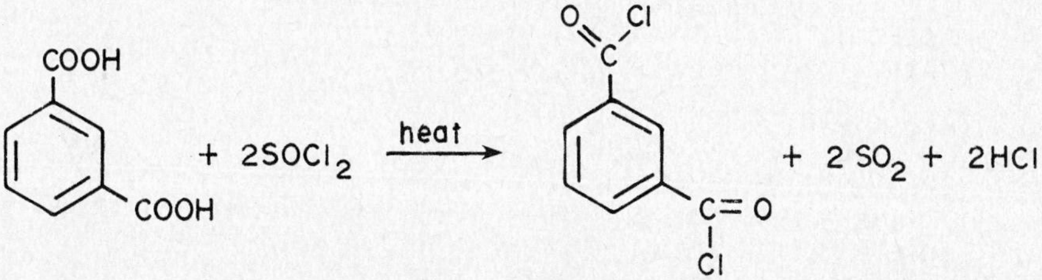

237

c.

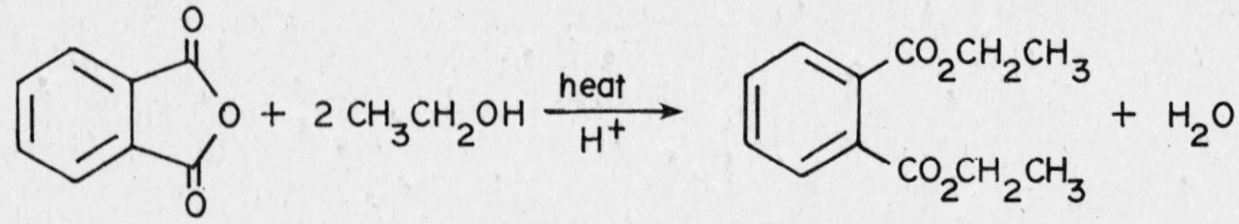

$+ \; 2 \; CH_3CH_2OH \; \xrightarrow[H^+]{heat}$ [benzene ring with $CO_2CH_2CH_3$ and $CO_2CH_2CH_3$] $+ \; H_2O$

d. $CH_2{=}CHCOOH \; + \; HBr \; \longrightarrow \; CH_2CH_2COOH$
with Br on carbon-3

The bromine ends up on carbon-3 as a consequence of the following mechanism:

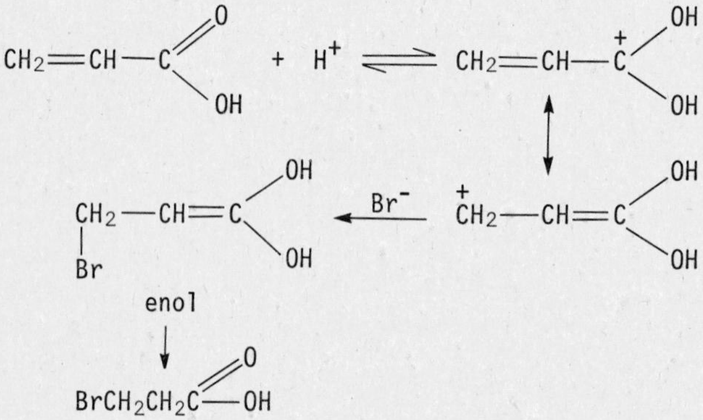

$CH_2{=}CH{-}C\overset{O}{\diagdown_{OH}} \; + \; H^+ \; \rightleftharpoons \; CH_2{=}CH{-}\overset{+}{C}\overset{OH}{\diagup_{OH}}$

$CH_2{-}CH{=}C\overset{OH}{\diagup_{OH}}$ (with Br) $\;\xleftarrow{Br^-}\; \overset{+}{C}H_2{-}CH{=}C\overset{OH}{\diagup_{OH}}$
Br ... OH ... OH

enol

$\downarrow$

$BrCH_2CH_2C\overset{O}{\diagup}{-}OH$

e. $Cl{-}\overset{O}{\overset{\|}{C}}{-}\overset{O}{\overset{\|}{C}}{-}Cl \; + \; 4 \; NH_3 \; \longrightarrow \; H_2N{-}\overset{O}{\overset{\|}{C}}{-}\overset{O}{\overset{\|}{C}}{-}NH_2 \; + \; 2 \; NH_4^{+}Cl^{-}$

11.18 $CH_3O_2C{-}$[benzene ring]${-}CO_2CH_3 \; + \; HOCH_2CH_2OH \; \xrightarrow{H^+}$

$CH_3O_2C{-}$[benzene ring]${-}\overset{O}{\overset{\|}{C}}{-}OCH_2CH_2OH \; + \; CH_3OH$

$\downarrow$ $CH_3O_2C{-}$[benzene ring]${-}CO_2CH_3, \; H^+$

$CH_3O_2C{-}$[benzene ring]${-}\overset{O}{\overset{\|}{C}}{-}OCH_2CH_2O{-}\overset{O}{\overset{\|}{C}}{-}$[benzene ring]${-}CO_2CH_3 \; + \; CH_3OH$

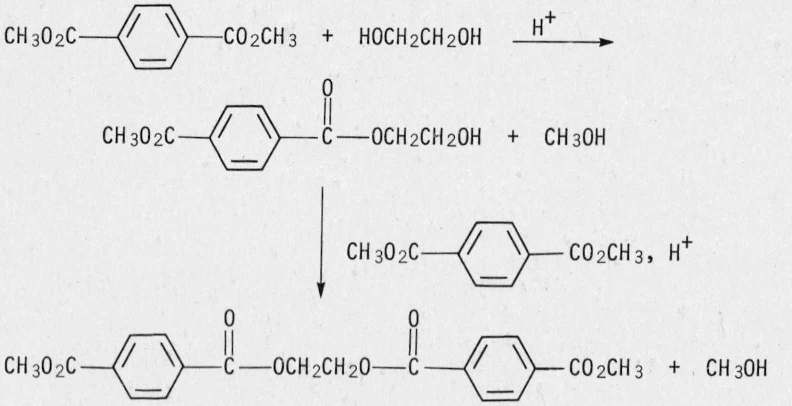

and so on

11.19 If the ester contains 2% of the *meta* isomer, then in 100 units there are, on the average, 2 *meta* units. Each *meta* unit sends the chain off at an angle of 60°. This angle arises from the geometry of the benzene ring.

para (linear)   meta (bent)

It is clear that the resulting polymer would have a very different shape from that of Dacron made with pure dimethyl terephthalate. High-purity starting materials are critical for controlling the properties of the polymers produced in polymerization reactions.

11.20 a.

b.

11.21 Because glycerol has three hydroxyl groups and phthalic acid has two carboxyl groups, the polymer will not be "linear" like Dacron. Instead it will be a complex, cross-linked, three-dimensional network. There may also be some cyclic components to the structure. There is no simple repeating unit that can be drawn, but some idea of the complexity of such a polymer is given by the following partial structure:

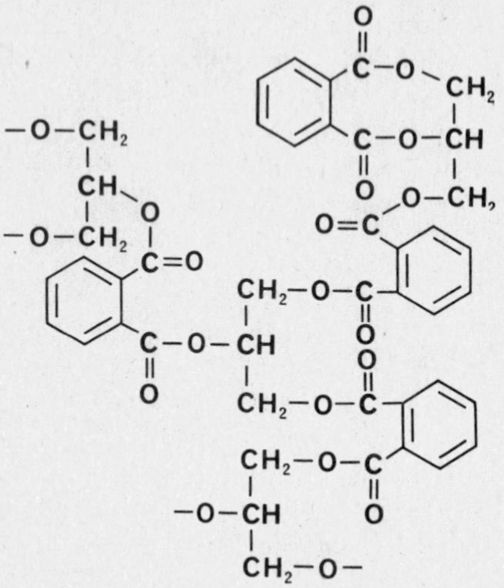

11.22

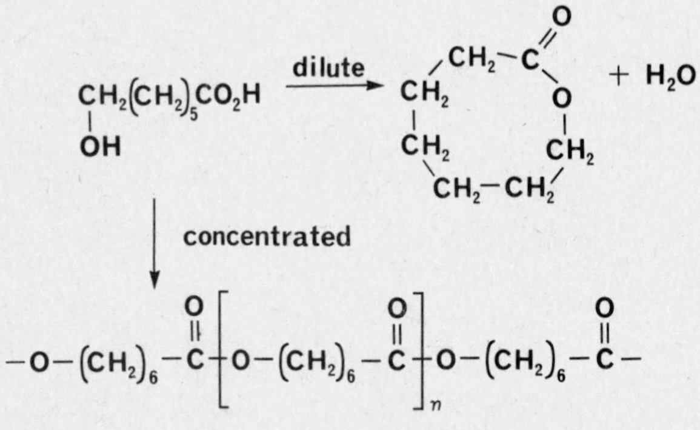

$$CH_2(CH_2)_5CO_2H \xrightarrow{\text{dilute}} \quad + H_2O$$
$$\overset{|}{OH}$$

concentrated

Polyester formation involves an *inter*molecular reaction (reaction between separate molecules), whereas lactone formation is an *intra*molecular reaction (reaction between two groups in a single molecule). The former is favored at high concentrations, when the chance for intermolecular collisions is great; the latter is favored in dilute solutions, where the likelihood of intermolecular collisions is much lower.

11.23 a. CH_3CH_2OC—CH—CH—$COCH_2CH_3$ b.

with two O double bonds on the first and last carbonyls, and OH groups below the two central CH carbons

c. K^+ ^-OC—CH—CH—C—OH d.

with two O double bonds, and OH groups below the two central CH carbons

e. $\overset{4}{CH_2}$=$\overset{3}{CH}\overset{2}{CH_2}\overset{1}{COOH}$ f.

CH_3\ α ... β γ structure

g. $CH_3\overset{O}{C}CH_2\overset{O}{C}OCH_2CH_3$ h. $CH_3\overset{O}{C}COOH$

11.24 a.   The compound is made of two isoprene units (heavy lines) linked together as shown by the dashed lines.

b.   Nepatalactone has three chiral centers:

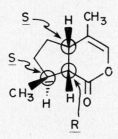

$\underline{S}$ H CH_3
$\underline{S}$
CH_3 H   H O
$\underline{R}$

11.25   Since salicylic acid

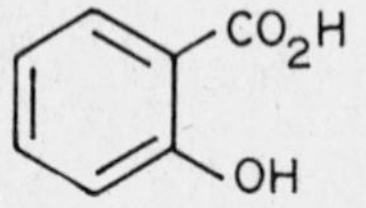

is an oxidation product, the starting formula should be

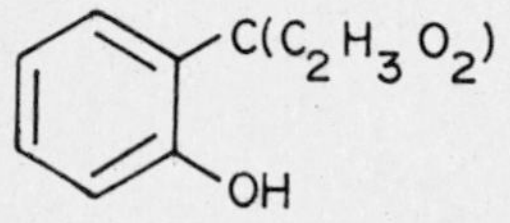

Since the side chain includes a carboxyl group (the compound is said to be a phenolic *acid*), this structure can be further refined:

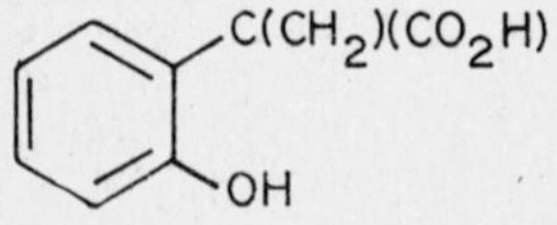

The two possible isomers are

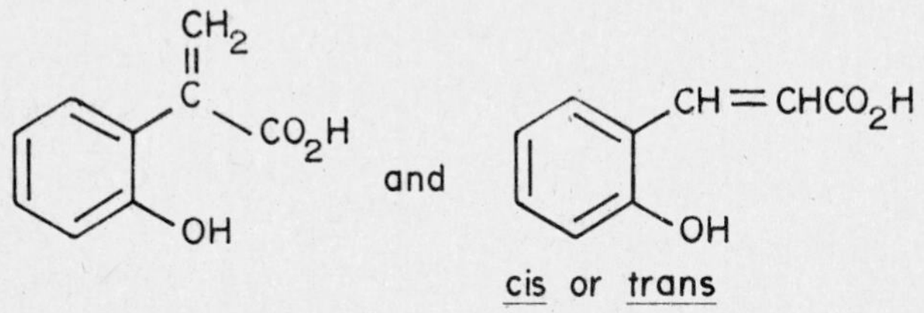

and

cis or trans

Only the second of these gives oxalic acid on oxidation:

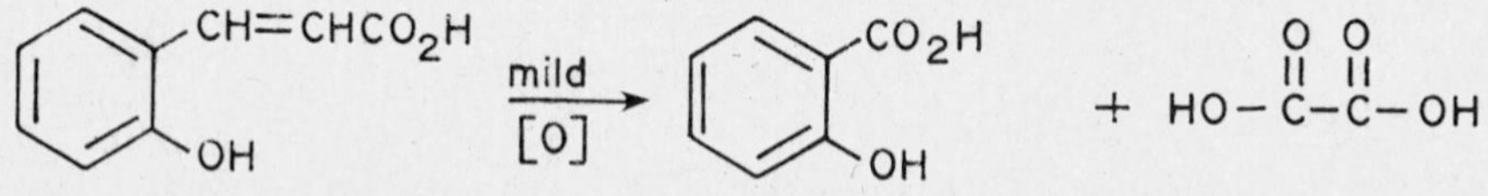

Therefore, the two isomeric forms are *cis-* and *trans-o-*hydroxycinnamic acid. Of these, the *cis* isomer easily loses water to form a lactone; the *trans* isomer cannot, because the carboxyl and hydroxyl groups are too far apart.

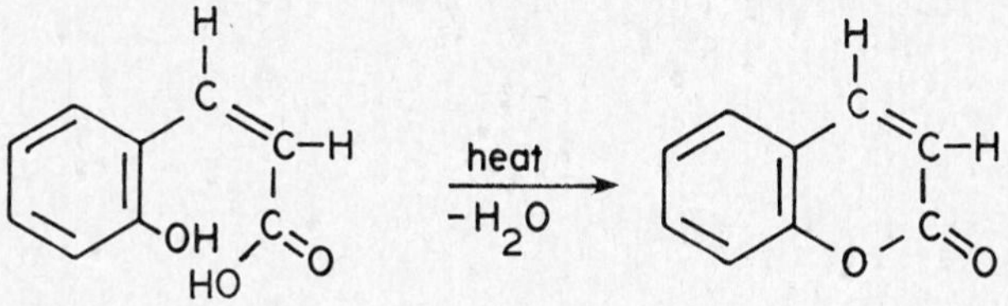

cis

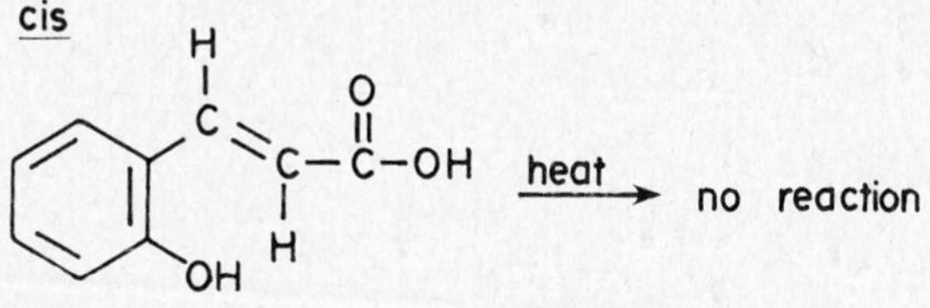

trans

11.26 Protonation of the carboxyl oxygen gives the allylic cation shown:

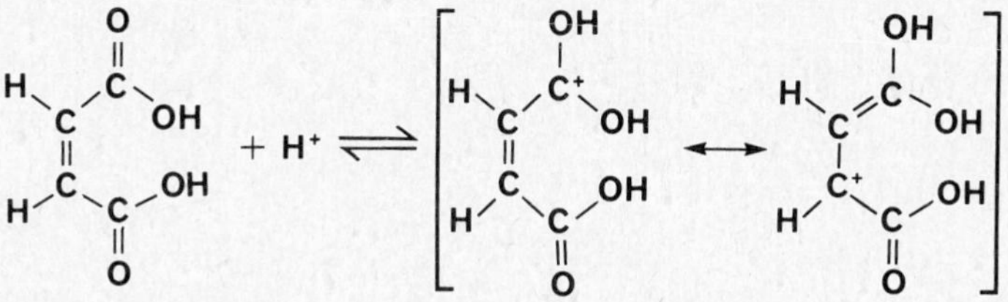

One of the resonance contributors to the cation has a single bond where the carbon-carbon double bond was; if rotation occurs around that single bond before proton loss, the product will be fumaric acid:

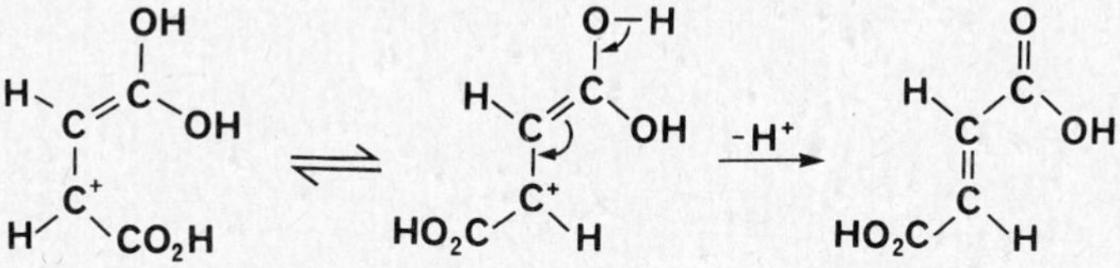

11.27 The two reactions are dehydration of the alcohol and alcoholysis of the nitrile:

11.28 Compare with the answer to Problem 10.28a.

$$ClCH_2CO_2H + NaOH \rightleftharpoons ClCH_2CO_2^-Na^+ + H_2O$$

11.29 The reaction involves nucleophilic addition to the $C\!=\!\!O$ bond of carbon dioxide:

a keto form of the phenol

11.30 Use eqs. 11.20-11.23 as a guide. The overall equation is

$$2 \quad \text{C}_6\text{H}_5\text{–CH}_2\overset{\overset{\displaystyle O}{\|}}{\text{C}}\text{–OCH}_2\text{CH}_3 \xrightarrow{\text{NaOCH}_2\text{CH}_3}$$

$$\text{C}_6\text{H}_5\text{–CH}_2\overset{\overset{\displaystyle O}{\|}}{\text{C}}\text{–}\overset{\overset{\displaystyle }{}}{\text{C}}\text{H}\overset{\overset{\displaystyle O}{\|}}{\text{C}}\text{–OCH}_2\text{CH}_3 \;+\; \text{CH}_3\text{CH}_2\text{OH}$$

The steps are as follows.

Step 1:

$$\text{C}_6\text{H}_5\text{–CH}_2\overset{\overset{\displaystyle O}{\|}}{\text{C}}\text{OCH}_2\text{CH}_3 + {}^-\text{OCH}_2\text{CH}_3 \rightleftharpoons \text{C}_6\text{H}_5\text{–}\bar{\text{C}}\text{H}\overset{\overset{\displaystyle O}{\|}}{\text{C}}\text{OCH}_2\text{CH}_3 + \text{CH}_3\text{CH}_2\text{OH}$$

Step 2:

$$\text{C}_6\text{H}_5\text{–CH}_2\overset{\overset{\displaystyle O}{\|}}{\text{C}}\text{–OCH}_2\text{CH}_3 + {}^-\bar{\text{C}}\text{H}\overset{\overset{\displaystyle O}{\|}}{\text{C}}\text{OCH}_2\text{CH}_3 \rightleftharpoons \text{C}_6\text{H}_5\text{–CH}_2\overset{\overset{\displaystyle O^-}{|}}{\text{C}}\text{–OCH}_2\text{CH}_3$$

$$\text{C}_6\text{H}_5\text{–CH}_2\overset{\overset{\displaystyle O}{\|}}{\text{C}}\text{–CH–}\overset{\overset{\displaystyle O}{\|}}{\text{C}}\text{–OCH}_2\text{CH}_3$$

Step 3:

$$\text{Ph}-CH_2\overset{O}{\underset{\|}{C}}-CH-\overset{O}{\underset{\|}{C}}-OCH_2CH_3 \;(\text{with Ph on CH}) \;+\; {}^-OCH_2CH_3 \longrightarrow$$

$$\text{Ph}-CH_2\overset{O}{\underset{\|}{C}}=\overset{\ominus}{CH}=\overset{O}{\underset{\|}{C}}-OCH_2CH_3 \;(\text{with Ph on CH}) \;+\; CH_3CH_2OH$$

11.31 The overall equation is

$$\begin{array}{c}
CH_2-\overset{O}{\underset{\|}{C}}-OCH_2CH_3 \\
CH_2 \\
CH_2-CH_2-\overset{O}{\underset{\|}{C}}-OCH_2CH_3
\end{array}
\xrightarrow{\;NaOCH_2CH_3\;}$$

$$\begin{array}{c}
CH_2-\overset{O}{\underset{\|}{C}} \\
CH_2 \\
CH_2-CH-\overset{O}{\underset{\|}{C}}-OCH_2CH_3
\end{array}
\;+\; CH_3CH_2OH$$

Sodium ethoxide removes an α-hydrogen, and the resulting enolate anion attacks the carbonyl group at the other end of the molecule. The key steps are as follows:

$$\begin{array}{c}
CH_2-\overset{O}{\underset{\|}{C}}-OCH_2CH_3 \\
CH_2 \\
CH_2-\overset{\ominus}{CH}-\overset{O}{\underset{\|}{C}}-OCH_2CH_3
\end{array}
\longrightarrow
\begin{array}{c}
CH_2-\overset{O^-}{\underset{|}{C}}-OCH_2CH_3 \\
CH_2 \\
CH_2-CH-\overset{O}{\underset{\|}{C}}-OCH_2CH_3
\end{array}
\longrightarrow \text{product}$$

11.32 Only the ethyl acetate has α-hydrogens and can form an enolate anion. The overall equation is

The key steps in the mechanism are as follows:

11.33 a. $CH_3(CH_2)_{14}CO_2{}^-K^+$

b. $[CH_3(CH_2)_7CH{=\!=}CH(CH_2)_7CO_2{}^-]_2Mg^{2+}$

c.

d.

e. $CH_3(CH_2)_4CH{=\!=}CHCH_2CH{=\!=}CH(CH_2)_7\overset{\displaystyle O}{\overset{\|}{C}}{-}O{-}(CH_2)_{13}CH_3$

$$\text{f.} \quad CH_3(CH_2)_{18}\overset{\overset{\displaystyle O}{\|}}{C}\text{——}OCH_2CH_3$$

11.34 <u>Saponification:</u>

$$
\begin{array}{l}
CH_2\text{—}O\text{—}\overset{\overset{\displaystyle O}{\|}}{C}(CH_2)_7CH\!=\!CHCH_2CH\!=\!CHCH_2CH\!=\!CHCH_2CH_3 \\[1em]
CH\text{—}O\text{—}\overset{\overset{\displaystyle O}{\|}}{C}(CH_2)_7CH\!=\!CHCH_2CH\!=\!CHCH_2CH\!=\!CHCH_2CH_3 \quad + \quad 3\ NaOH \longrightarrow \\[1em]
CH_2\text{—}O\text{—}\overset{\overset{\displaystyle O}{\|}}{C}(CH_2)_7CH\!=\!CHCH_2CH\!=\!CHCH_2CH\!=\!CHCH_2CH_3
\end{array}
$$

$$
\begin{array}{l}
CH_2OH \\
CHOH \quad + \quad 3\ CH_3CH_2CH\!=\!CHCH_2CH\!=\!CHCH_2CH\!=\!CH(CH_2)_7CO_2^{-}Na^{+} \\
CH_2OH
\end{array}
$$

<u>Hydrogenation:</u>

$$
\begin{array}{l}
CH_2\text{—}O\text{-}\overset{\overset{\displaystyle O}{\|}}{C}(CH_2)_7CH\!=\!CHCH_2CH\!=\!CHCH_2CH\!=\!CHCH_2CH_3 \\
CH\text{—}O\text{-}\overset{\overset{\displaystyle O}{\|}}{C}(CH_2)_7CH\!=\!CHCH_2CH\!=\!CHCH_2CH\!=\!CHCH_2CH_3 \quad +\ 9\ H_2 \xrightarrow[\text{heat}]{\text{Ni}} \\
CH_2\text{—}O\text{-}\overset{\overset{\displaystyle O}{\|}}{C}(CH_2)_7CH\!=\!CHCH_2CH\!=\!CHCH_2CH\!=\!CHCH_2CH_3
\end{array}
$$

$$
\begin{array}{l}
CH_2\text{—}O\text{-}\overset{\overset{\displaystyle O}{\|}}{C}(CH_2)_{16}CH_3 \\
CH\text{—}O\text{-}\overset{\overset{\displaystyle O}{\|}}{C}(CH_2)_{16}CH_3 \\
CH_2\text{-}O\text{-}\overset{\overset{\displaystyle O}{\|}}{C}(CH_2)_{16}CH_3
\end{array}
$$

<u>Hydrogenolysis:</u>

$$
\begin{array}{l}
CH_2\text{—}O\text{—}\overset{\overset{\displaystyle O}{\|}}{C}(CH_2)_7CH\!=\!CHCH_2CH\!=\!CHCH_2CH\!=\!CHCH_2CH_3 \\[1em]
CH\text{—}O\text{—}\overset{\overset{\displaystyle O}{\|}}{C}(CH_2)_7CH\!=\!CHCH_2CH\!=\!CHCH_2CH\!=\!CHCH_2CH_3 \quad +\ 6\ H_2 \xrightarrow{\substack{\text{zinc}\\\text{chromite}}} \\[1em]
CH_2\text{—}O\text{—}\overset{\overset{\displaystyle O}{\|}}{C}(CH_2)_7CH\!=\!CHCH_2CH\!=\!CHCH_2CH\!=\!CHCH_2CH_3
\end{array}
$$

$$
\begin{array}{l}
CH_2OH \\
CHOH \quad + \quad 3\ CH_3CH_2CH\!=\!CHCH_2CH\!=\!CHCH_2CH\!=\!CH(CH_2)_7CH_2OH \\
CH_2OH
\end{array}
$$

11.35 a. Compare with eq. 11.27.

$$C_{15}H_{31}\overset{\overset{O}{\parallel}}{C}\text{---}O^-Na^+ \ + \ HCl \ \longrightarrow \ C_{15}H_{31}COOH \ + \ Na^+Cl^-$$

b. Compare with eq. 11.28.

$$2 \ C_{15}H_{31}\overset{\overset{O}{\parallel}}{C}\text{---}O^-Na^+ \ + \ Mg^{2+} \ \longrightarrow \ (C_{15}H_{31}\overset{\overset{O}{\parallel}}{C}\text{---}O^-)_2 \ Mg^{2+} \ + \ 2 \ Na^+$$

11.36 $CH_3(CH_2)_7CH{=}CH_2$ + ⬡ $\xrightarrow{H^+}$ $CH_3(CH_2)_7CHCH_3$

$\downarrow H_2SO_4$

$CH_3(CH_2)_7CHCH_3$ $\xleftarrow{NaOH}$ $CH_3(CH_2)_7CHCH_3$

$SO_3^-Na^+$ SO_3H

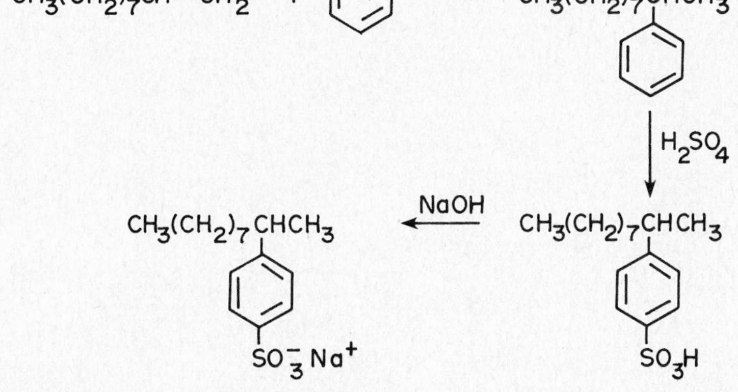

Note that in the first step Markovnikov's rule is followed, and
alkylation occurs via a secondary carbocation.

11.37 The first steps are analogous to eq. 8.15, and the last steps are
analogous to eq. 11.30.

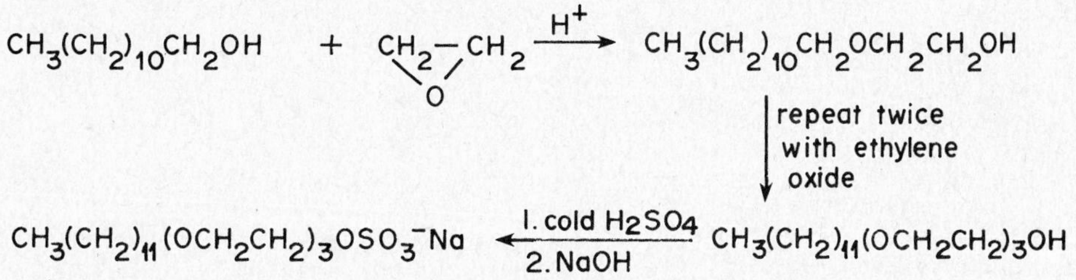

$$CH_3(CH_2)_{10}CH_2OH \ + \ \underset{\underset{O}{\diagdown}}{CH_2}{-}CH_2 \ \xrightarrow{H^+} \ CH_3(CH_2)_{10}CH_2OCH_2CH_2OH$$

$\downarrow$ repeat twice
with ethylene
oxide

$$CH_3(CH_2)_{11}(OCH_2CH_2)_3OSO_3^-Na \ \xleftarrow[2.\,NaOH]{1.\,cold \ H_2SO_4} \ CH_3(CH_2)_{11}(OCH_2CH_2)_3OH$$

11.38 This early experiment very clearly supported the idea that fatty acids are metabolized in two-carbon units. If n is an even number, then the ultimate oxidation product obtained by oxidizing the chain two carbons at a time from the carboxyl end has to be phenylacetic acid, as the following example shows:

not possible

Further oxidation is not possible, since it requires breaking a bond to the benzene ring. If n is odd, the product of a similar oxidation procedure is benzoic acid.

These results are consistent with present knowledge of the mechanisms of fatty-acid metabolism.

11.39 In the first step, the acid is esterified with coenzyme A:

$$CH_3CH_2CH_2CO_2H + CoA\text{---}SH \underset{ATP}{\overset{enzyme}{\rightleftharpoons}} CH_3CH_2CH_2\overset{O}{\overset{\|}{C}}\text{---}S\text{---}CoA + H_2O$$

Then the steps outlined in Figure 11.7 are followed:

11.40 Many answers are possible. Definitions or examples are to be found in the indicated sections of the text.

a. Sec. 11.10 b. Secs. 11.10 and 11.11
c. Sec. 11.16 d. Sec. 11.12
e. Sec. 11.13 f. Sec. 11.17
g. Sec. 11.15 h. Sec. 11.8
i. Sec. 11.6

11.41 The priority order of groups attached to the chiral center is

$$\underset{\substack{\| \\ O}}{R'C}O- \;>\; -CH_2OP \;>\; \underset{\substack{\| \\ O}}{-CH_2O}C- \;>\; H-$$

The structure of the lecithin is therefore

$$\underset{\substack{| \\ O^-}}{\underset{\substack{\| \\ O}}{R''OP}OCH_2}\overset{\substack{CH_2OCR \\ \| \\ O}}{\underset{\substack{OCR \\ \| \\ O}}{\overset{|}{C}\cdots H}}$$

11.42 $CH_3(CH_2)_{13}CH_2\underset{\substack{\| \\ O}}{C}O(CH_2)_{29}CH_3$ + Na^+OH^- $\xrightarrow{\text{heat}}$

$$CH_3(CH_2)_{13}CH_2\underset{\substack{\| \\ O}}{C}-O^-Na^+ \;+\; HO(CH_2)_{29}CH_3$$

11.43 a. The two hydroxyl groups will be esterified.

$$+ \ 2 \ CH_3\overset{O}{\underset{\|}{C}}-O-\overset{O}{\underset{\|}{C}}CH_3 \longrightarrow$$

$$+ \ 2 \ CH_3COOH$$

b. The two ketone functions will be reduced, but the carbon-carbon double bond will not be reduced. Several stereoisomers are possible.

$$\xrightarrow{\text{LiAlH}_4}$$

c. The alkene will be converted to an epoxide. Two stereoisomers are possible.

$$\xrightarrow{\text{CH}_3\text{CO}_3\text{H}}$$

d. The secondary alcohol will be oxidized to a ketone.

11.44 a. Use the steroid numbering system shown on p. 312 of the text.

Note that the "angular" methyl groups are numbered 18 and 19 and that numbering then continues with the side chain attached to C-17.

b. Chiral centers are present at C-8, C-9, C-13, C-14, and C-17. To assign configuration, you must first assign the priority order at each chiral center.

C-8: C-9 > C-14 > C-7 > H therefore *S*
C-9: C-11 > C-10 > C-8 > H therefore *S*
C-10: C-5 > C-9 > C-1 > C-19 therefore *R*
C-13: C-17 > C-14 > C-12 > C-18 therefore *S*
C-14: C-13 > C-8 > C-15 > H therefore *S*
C-17: OH > C-20 > C-13 > C-16 therefore *R*

CHAPTER SUMMARY

Amines are organic derivatives of ammonia. They may be **primary, secondary,** or **tertiary,** depending on whether one, two, or three organic groups are attached to the nitrogen. The nitrogen is sp^3-hybridized and pyramidal, nearly tetrahedral.

Amines are named by adding the suffix *-amine* to the names of the alkyl groups attached to the nitrogen. The **amino group** is ——NH_2. Aromatic amines are named as derivatives of aniline or of the aromatic ring system.

Amines form intermolecular N——H$\cdots$N bonds. Their boiling points are higher than those of alkanes but lower than those of alcohols with comparable molecular weights. Lower members of the series are water-soluble because of N$\cdots$H——O bonding.

Amines can be prepared by S_N2 alkylation of ammonia or less-substituted amines. Aromatic amines are made by reduction of the corresponding nitro compounds. Amides and nitriles can also be reduced to amines.

Amines are weak bases. Alkylamines have K_b's comparable to that of ammonia (10^{-5}), but aromatic amines are much weaker ($K_b = 10^{-10}$) as a result of delocalization of the unshared electron pair on nitrogen to the *ortho* and *para* carbons of the aromatic ring.

Amides are much weaker bases than amines because of delocalization of the unshared electron pair on nitrogen to the adjacent carbonyl oxygen. Amides are stronger Brønsted acids than amines because of the partial positive charge on the amide nitrogen and because of resonance in the **amidate anion.**

Amines react with strong acids to form **amine salts.** The fact that these salts are usually water soluble can be taken advantage of in separating amines from neutral or acidic contaminants. Chiral amines can be used to resolve enantiomeric acids, through the formation of diastereomeric salts.

Primary and secondary amines react with acid derivatives to form amides. Amides made commercially this way include **acetanilide** and

N,N-diethyl-*m*-toluamide (the insect repellent "Off"). **Methyl isocyanate** (MIC), used to prepare **carbamate pesticides**, is made similarly from methylamine and phosgene.

Tertiary amines react with alkyl halides to form **quarternary ammonium salts**. An example of this type of salt that has important biological properties is **choline** (2-hydroxyethyltrimethylammonium ion).

Depending on their class, amines react differently with nitrous acid. Tertiary aromatic amines give aromatic nitroso compounds, secondary amines give **nitrosamines**, and primary amines give **diazonium ions**, $R—\overset{+}{N}\equiv N:$. If R is alkyl, these diazonium ions lose nitrogen to give carbocations and products derived from them.

Primary aromatic amines give **aryldiazonium ions**, ArN_2^+. These are useful intermediates in synthesis of aromatic compounds. The process by which they are formed is called **diazotization**. The nitrogen in these ions can readily be replaced by various nucleophiles (OH, Cl, Br, I, CN). Diazonium ions couple with reactive aromatics, such as amines or phenols, to form **azo compounds**, which are useful as dyes.

Nylon is a **polyamide** prepared from 1,6-diaminohexane and adipic acid. Other commercial polyamides are nylon-6 (from **caprolactam**; lactams are cyclic amides), and the **aramid Kevlar** (made from *p*-phenylenediamine and terephthaloyl chloride).

Important five-membered heterocyclic amines include **pyrrole, pyrrolidine**, and **indole**. Pyrrole is a building block for **porphyrins** such as *heme* (a constituent of hemoglobin) and **chlorophyll**. The indole derivatives **tryptamine** and **serotonin** are important in brain chemistry.

Important six-membered heterocyclic amines include **pyridine, piperidine, quinoline**, and **isoquinoline**. These ring systems occur in many natural products.

Three important heterocyclic amines with more than one nitrogen are **imidazole, pyrimidine**, and **purine**. **Barbiturates** are pyrimidine derivatives, made from urea and malonic acids. **Caffeine** is a common purine. **Morphine** is the prototype of many nitrogen-containing heterocyclic drugs.

REACTION SUMMARY

Alkylation of Ammonia and Amines

$$R{-}X \ + \ 2 \ NH_3 \longrightarrow R{-}NH_2 \ + \ NH_4{}^+X^-$$

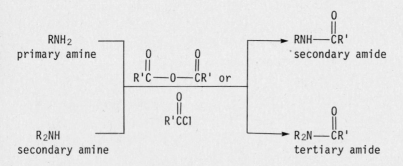

Reduction Routes to Amines

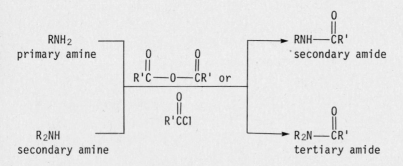

Amine Basicity

$$RNH_2 \ + \ HOH \ \rightleftharpoons \ \overset{+}{RNH_3} \ + \ {}^-OH$$

$$RNH_2 \ + \ HCl \longrightarrow \overset{+}{RNH_3} \ + \ Cl^-$$

Acylation of Primary and Secondary Amines

REACTION SUMMARY

Isocyanates and Carbamates

$$RNH_2 \ + \ Cl-\overset{\overset{\textstyle O}{\|}}{C}-Cl \ \xrightarrow{\text{base}} \ R-N\!=\!C\!=\!O \ (+ \ 2 \ HCl)$$
alkyl isocyanate

$$R-N\!=\!C\!=\!O \ + \ R'OH \ \longrightarrow \ R-NH-\overset{\overset{\textstyle O}{\|}}{C}-OR'$$
carbamate

Quaternary Ammonium Salts

$$R_3N \ + \ R'X \ \longrightarrow \ R_3\overset{+}{N}-R' \ \ X^-$$

Reactions with Nitrous Acid

$$R_2N-\!\!\left\langle\!\!\bigcirc\!\!\right\rangle \ + \ HONO \ \longrightarrow \ R_2N-\!\!\left\langle\!\!\bigcirc\!\!\right\rangle\!-\ddot{N}\!=\!\ddot{O}\!: \ + \ H_2O$$
aryl nitroso compound

$$R_2NH \ + \ HONO \ \longrightarrow \ R_2N-N\!=\!O \ + \ H_2O$$
N-nitroso compound

$$RNH_2 \ + \ HONO \ \xrightarrow{H^+} \ RN_2^+ \ + \ 2 \ H_2O$$
alkyldiazonium ion

$$\longrightarrow \ N_2 \ + \ R^+ \ \longrightarrow \ \text{products}$$

Aryldiazonium Salts

$$ArNH_2 \ + \ HONO \ \xrightarrow{HX} \ ArN_2^+X^-$$
aryldiazonium ion

$$ArN_2^+ \ + \ H_2O \ \xrightarrow{\text{heat}} \ ArOH \ + \ N_2 \ + \ H^+$$
phenols

$$+ \ HX \ \xrightarrow{Cu_2X_2} \ ArX \ \ (X = Cl, \ Br)$$

$$+ \ KI \ \longrightarrow \ ArI$$

$$+ \ KCN \ \xrightarrow{Cu_2(CN)_2} \ ArCN$$

Diazo Coupling

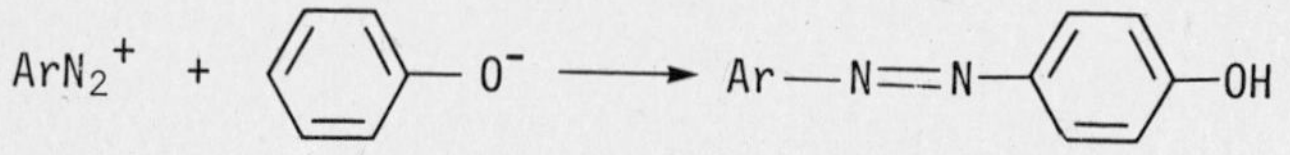

$$ArN_2^+ \; + \; \text{⟨phenol⟩—O}^- \longrightarrow Ar-N{=}N-\text{⟨ring⟩—OH}$$

azo compound

Polyamide

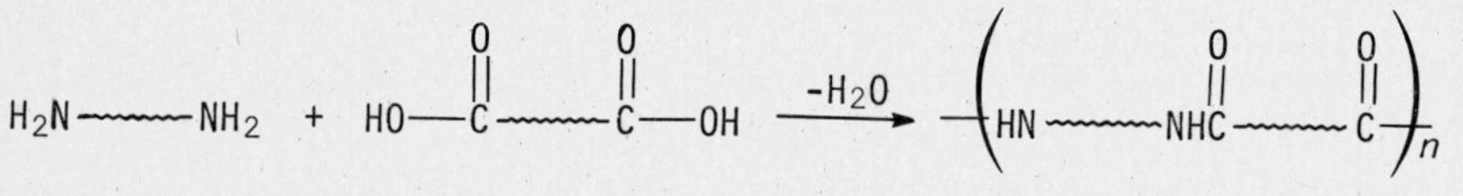

$$H_2N\text{——}NH_2 \; + \; HO-\overset{\overset{O}{\|}}{C}\text{——}\overset{\overset{O}{\|}}{C}-OH \xrightarrow{-H_2O} \left(HN\text{——}NH\overset{\overset{O}{\|}}{C}\text{——}\overset{\overset{O}{\|}}{C}\right)_n$$

Pyrrole Substitution

$$\text{⟨pyrrole⟩} \; + \; E^+ \xrightarrow{\text{facile}} \text{⟨pyrrole⟩}-E \; + \; H^+$$

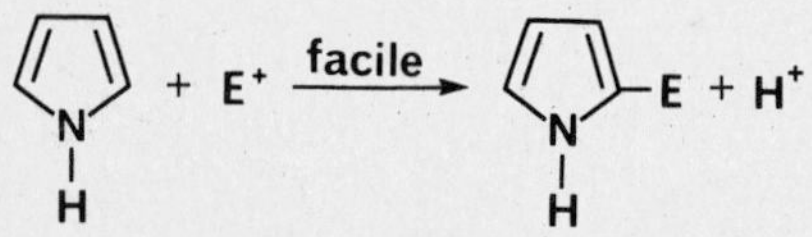

Pyridine Substitution

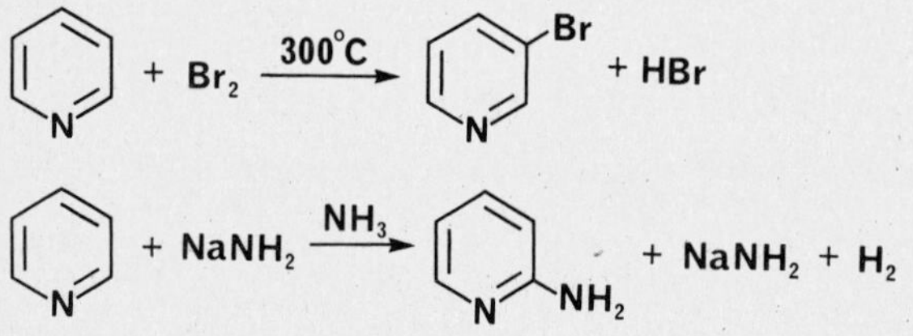

$$\text{⟨pyridine⟩} \; + \; Br_2 \xrightarrow{300^\circ C} \text{⟨pyridine-Br⟩} \; + \; HBr$$

$$\text{⟨pyridine⟩} \; + \; NaNH_2 \xrightarrow{NH_3} \text{⟨pyridine-NH}_2\text{⟩} \; + \; NaNH_2 \; + \; H_2$$

Synthesis of Barbiturates

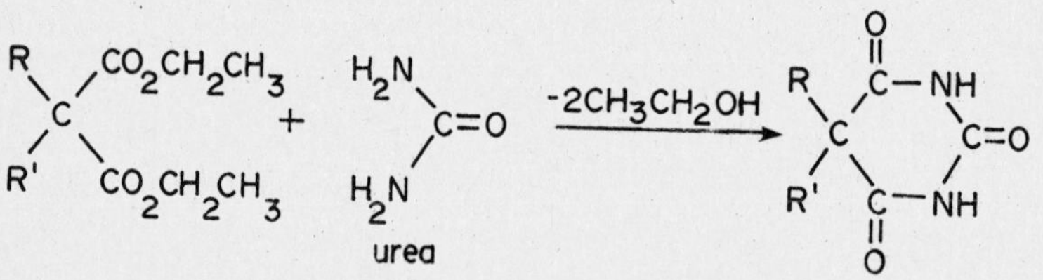

$$\underset{R'}{\overset{R}{C}}\!\!\left\langle\begin{array}{l}CO_2CH_2CH_3\\CO_2CH_2CH_3\end{array}\right. \; + \; \underset{H_2N}{\overset{H_2N}{}}\!\!C{=}O \xrightarrow{-2CH_3CH_2OH} \text{⟨barbiturate ring⟩}$$

urea

MECHANISM SUMMARY

Nitrosation

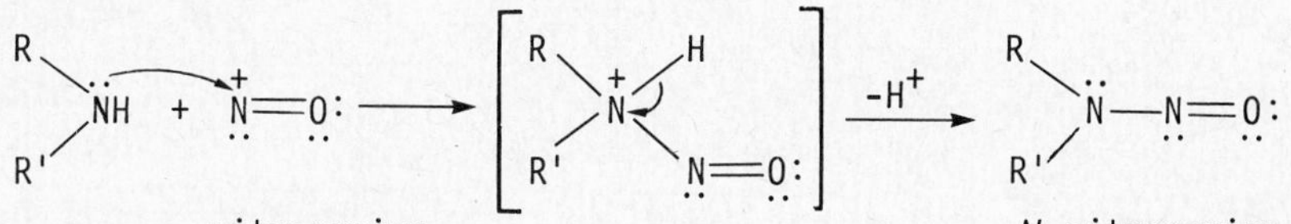

nitrosonium
ion

N-nitrosamine

Diazotization

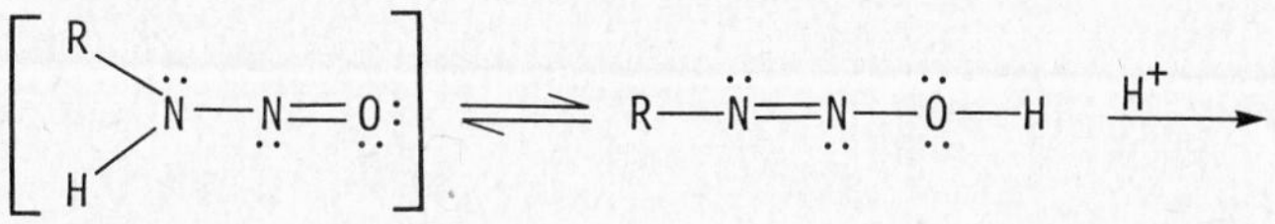

primary nitrosamine

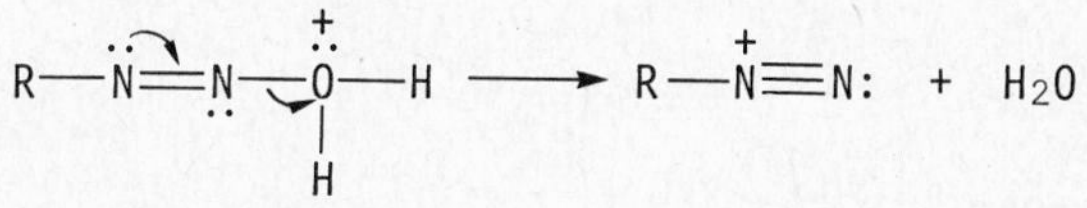

LEARNING OBJECTIVES

1. Know the meaning of: primary, secondary, and tertiary amine; amino group; aniline; dissociation constant K_b; amine salt; quaternary ammonium salt.

2. Know the meaning of: nitrous acid, nitrosonium ion, N-nitrosamine, diazonium ion, diazotization, azo coupling.

3. Know the meaning of: polyamide, nylon-6,6, nylon-6.

4. Know the structures of: pyrrole, pyrrolidine, indole, pyridine, piperidine, quinoline, isoquinoline, imidazole, pyrimidine, purine, barbiturates, and given the name, be able to write structures for derivatives of these parent heterocycles.

5. Given the structure of an amine, identify it as primary, secondary, or tertiary.

6. Given the structure of an amine, name it. Also, given the name of an amine, write its structural formula.

7. Explain the effect of hydrogen bonding on the boiling points of amines and on their solubility in water.

8. Write an equation for the reaction between ammonia or an amine of any class and an alkyl halide.

9. Write an equation for the preparation of a given aromatic amine from the corresponding nitro compound.

10. Write an equation for the preparation of a given amine of any class by reduction of the appropriate amide.

11. Write an equation for the preparation of a primary amine of the type RCH_2NH_2 or $ArCH_2NH_2$ by reduction of the appropriate nitrile.

12. Write an equation for the dissociation of an amine in water.

13. Write an expression for K_b of any amine.

14. Draw the important contributors to the resonance hybrid for an aromatic amine.

15. Given the structures of several amines, rank them in order of relative basicity.

16. Account for the difference in basicity between an aliphatic and an aromatic amine.

17. Write an equation for the reaction of a given amine of any class with a strong acid. Also, write an equation for the reaction of an amine salt with a strong base.

18. Account for the basicity and acidity difference between amines and amides.

19. Explain, with the aid of equations, how you can separate an amine from a mixture containing neutral and/or acidic compounds.

20. Explain how chiral amines can be used to resolve a mixture of enantiomeric acids.

21. Write an equation for the reaction of a given primary or secondary amine with an acid anhydride or acid halide.

22. Write the steps in the mechanism for acylation of a primary or secondary amine.

23. Write equations for the reaction of a given primary, secondary, or tertiary amine with nitrous acid.

24. Write an equation for the diazotization of a given primary aromatic amine.

25. Write the equations for the reaction of an aromatic diazonium salt with: aqueous base; $HX + Cu_2X_2$ (X = Cl, Br); and $KCN + Cu_2(CN)_2$.

26. Write an equation for the coupling of an aromatic diazonium salt with a phenol or aromatic amine.

27. Write an equation for the formation of a polyamide from a diamine and a dicarboxylic acid or acid derivative.

28. Write an equation for the formation of nylon-6,6 and nylon-6.

29. Tell whether a nitrogen atom in a particular heterocyclic amine is likely to be strongly or weakly basic.

30. Write an equation for the reaction of an isocyanate with an alcohol or amine.

ANSWERS TO PROBLEMS

Problems Within the Chapter

12.1 a. primary   b. secondary
      c. primary   d. tertiary

12.2 a. *t*-butylamine
      b. 2-aminoethanol
      c. *p*-nitroaniline

12.3 a. $(CH_3CH_2CH_2)_2NH$ b. $CH_3CH_2CHCH_2CH_2CH_3$
                                              |
                                            NH_2

c.

d.

12.4 Trimethylamine has no hydrogens on the nitrogen: $(CH_3)_3N$. Thus intermolecular hydrogen bonding is not possible. In contrast, intermolecular hydrogen bonding is possible for $CH_3CH_2CH_2NH_2$, and this raises its boiling point considerably above that of its tertiary isomer.

12.5 a. $CH_3CH_2CH_2Br + 2\ \overset{..}{N}H_3 \longrightarrow CH_3CH_2CH_2NH_2 + NH_4{}^+Br^-$

b. $CH_3CH_2I + 2\ (CH_3CH_2)_2\overset{..}{N}H \longrightarrow (CH_3CH_2)_3N + (CH_3CH_2)_2\overset{+}{N}H_2\ I^-$

c. $(CH_3)_3N: + CH_3I \longrightarrow (CH_3)_4N^+\ I^-$

d. $CH_3CH_2CH_2NH_2 + $ $—CH_2Br \longrightarrow CH_3CH_2CH_2\overset{+}{N}H_2CH_2—$ $+ Br^-$

$CH_3CH_2CH_2NHCH_2—$ $\xleftarrow{\quad CH_3CH_2CH_2NH_2 \quad}$

$+ CH_3CH_2CH_2\overset{+}{N}H_3\ \ Br^-$

12.6 a. 2 $—NH_2 + CH_3CH_2Br \longrightarrow$

$—NHCH_2CH_3 + $ $—\overset{+}{N}H_3\ \ Br^-$

b. $CH_3CH_2CH_2CH_2Br + 2\ NH_3 \longrightarrow CH_3CH_2CH_2CH_2NH_2 + NH_4{}^+\ \ Br^-$

12.7 a. $—CH_3 \xrightarrow[H_2SO_4]{HONO_2} O_2N—$ $—CH_3 \xrightarrow[\text{(see eq. 12.10)}]{SnCl_2,HCl} H_2N—$ $—CH_3$
$\underset{NO_2}{} \qquad \underset{NH_2}{}$

Nitration of toluene twice gives mainly the 2,4-dinitro product.

b. $CH_3\overset{\overset{O}{\|}}{C}N(CH_3)_2 \xrightarrow[\text{(see eq. 12.11)}]{LiAlH_4} CH_3CH_2N(CH_3)_2$

c. $—CH_2Br \xrightarrow{Na^+CN^-}$ $—CH_2CN \xrightarrow[\text{(see eq. 12.12)}]{LiAlH_4}$ $—CH_2CH_2NH_2$

12.8   Secondary amine:

$$R_2NH \; + \; H\!\!-\!\!OH \; \rightleftharpoons \; R_2\overset{+}{N}H_2 \; + \; OH^-$$

$$K_b \; = \; \frac{[R_2\overset{+}{N}H_2][OH^-]}{[R_2NH]}$$

Tertiary amine:

$$R_3N \; + \; H\!\!-\!\!OH \; \rightleftharpoons \; R_3\overset{+}{N}H \; + \; OH^-$$

$$K_b \; = \; \frac{[R_3\overset{+}{N}H][OH^-]}{[R_3N]}$$

12.9 $ClCH_2CH_2NH_2$ is a weaker base than $CH_3CH_2NH_2$. The chlorine substituent is electron-withdrawing compared to hydrogen and will
destabilize the protonated base because of the repulsion between
the positive charge on nitrogen and the partial positive charge on
C-2 due to the C——Cl bond moment:

$$\overset{\delta-}{Cl}\!\!-\!\!\overset{\delta+}{CH_2}CH_2\overset{+}{N}H_3 \quad \text{compared to} \quad CH_3CH_2\overset{+}{N}H_3$$

Ethylamine is therefore easier to protonate (more basic) than
2-chloroethylamine.

12.10 The order of the substituents by increasing electron-donating ability
is ——NO_2 < ——H < ——CH_3. Therefore, the basicities will increase
in that order:

$$O_2N\!\!-\!\!\langle\bigcirc\rangle\!\!-\!\!NH_2 \; < \; \langle\bigcirc\rangle\!\!-\!\!NH_2 \; < \; CH_3\!\!-\!\!\langle\bigcirc\rangle\!\!-\!\!NH_2$$

12.11 Amides are less basic than amines, and aromatic amines are less
basic than aliphatic amines. The order is therefore

acetanilide < aniline < cyclohexylamine

The acidity increases in the reverse direction.

12.12

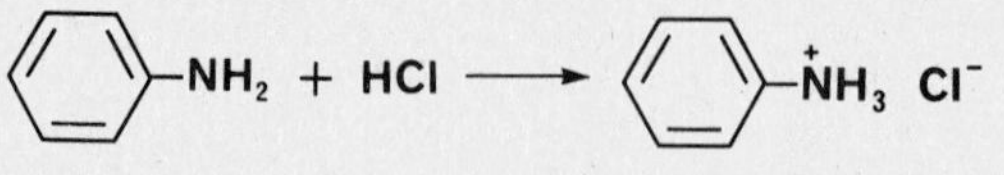

anilinium chloride

12.13 The morphine can be extracted with hydrochloric acid:

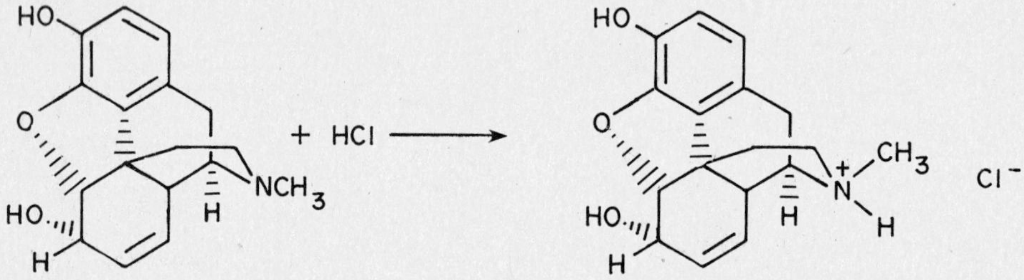

Once separated, the salt can be reconverted to the free base by
treatment with strong base:

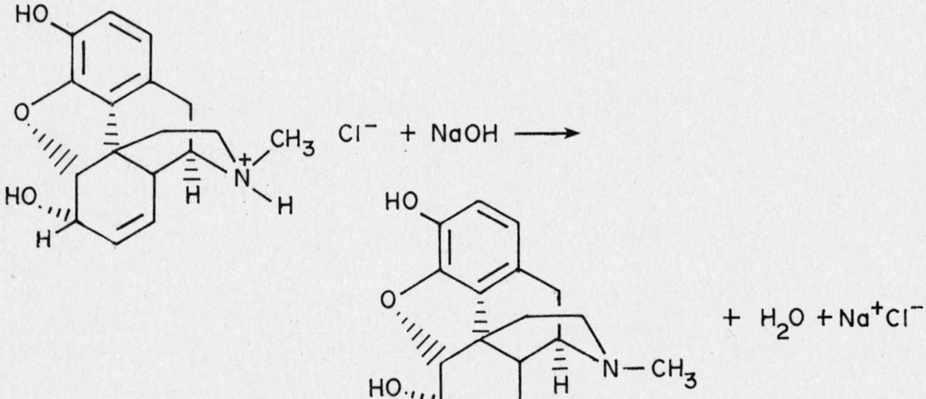

12.14

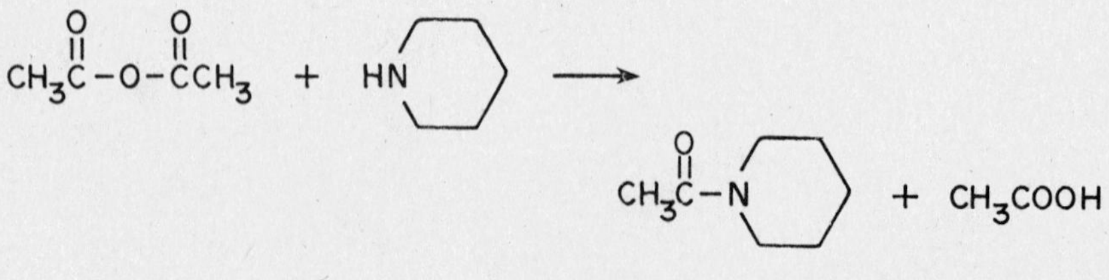

12.15

The base combines with the HCl to form sodium chloride and water. Otherwise, the HCl would protonate the diethylamine and prevent it from functioning as a nucleophile.

12.16 Cyclopentylamine is a primary amine. It will form a diazonium ion, which then loses nitrogen to give a cyclopentyl cation. This cation will either react with water (the main nucleophile present) to form cyclopentanol or will lose a proton to give cyclopentene.

12.17

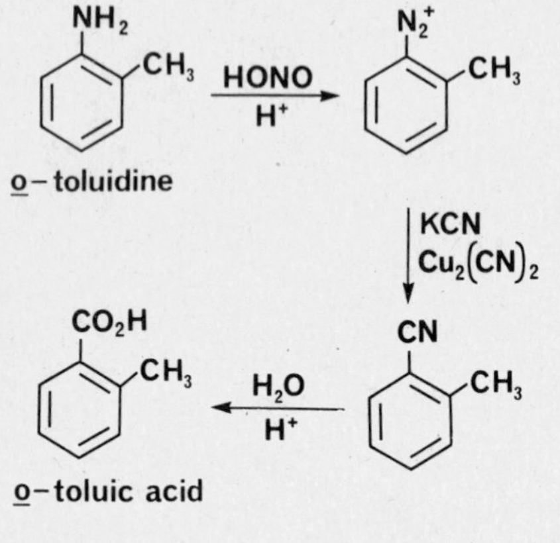

o-toluidine

o-toluic acid

12.18

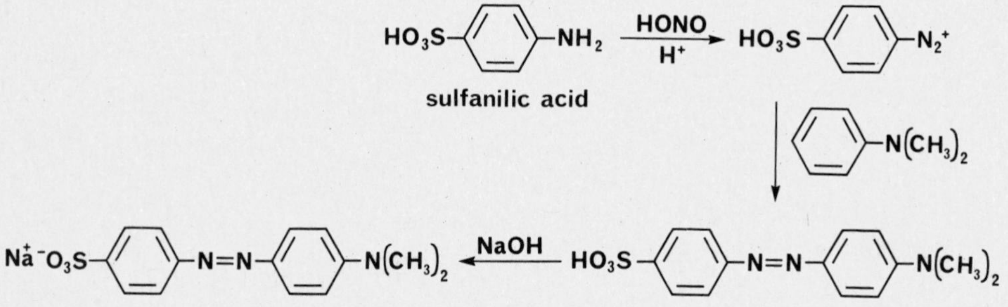

sulfanilic acid

12.19

intermediate for nitration at C-2

intermediate for nitration at C-3

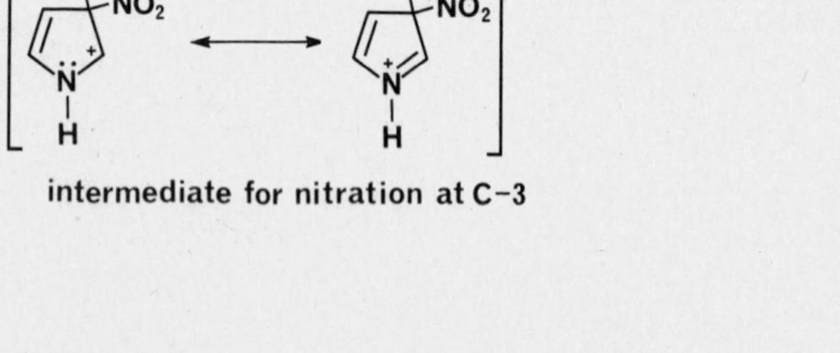

12.20

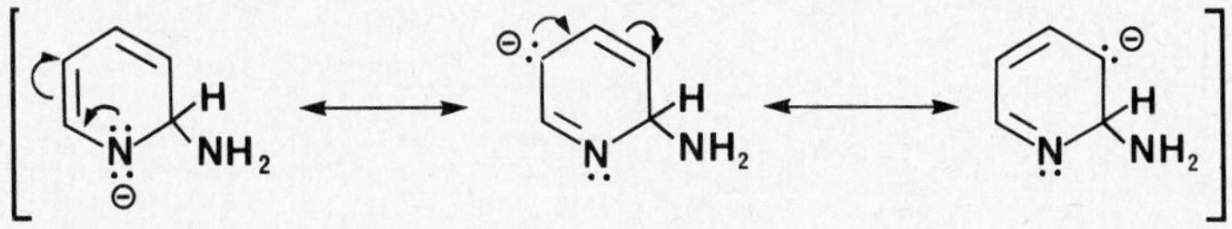

Additional Problems

12.21 Many correct answers are possible, but only one example is given in each case.

a. CH_3NH_2

   methylamine

b.

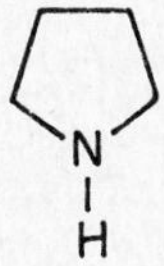

   pyrrolidine

c.

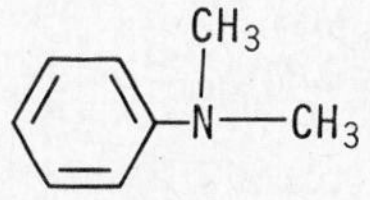

   N,N-dimethylaniline

d.

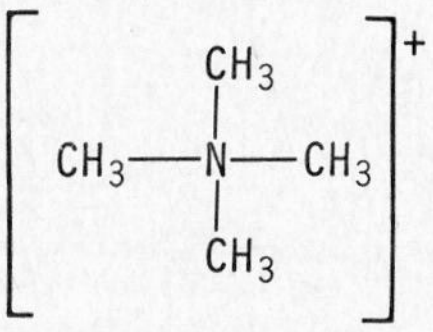

   tetramethylammonium chloride

e.

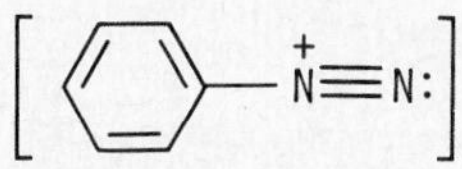

   benzenediazonium chloride

f.

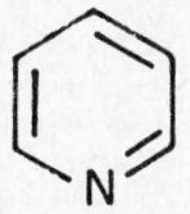

   pyridine

g.

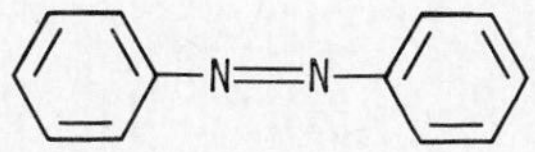

   azobenzene

h.

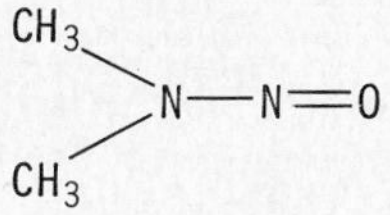

   N-nitrosodimethylamine

i. 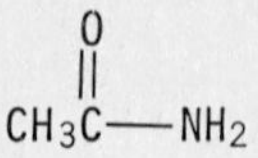
$$\begin{array}{c} O \\ \parallel \\ CH_3C\!-\!NH_2 \end{array}$$

acetamide

j.

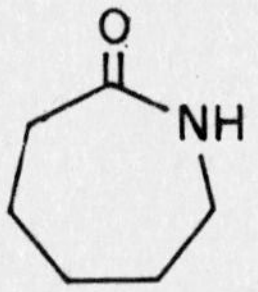

caprolactam

12.22 a.

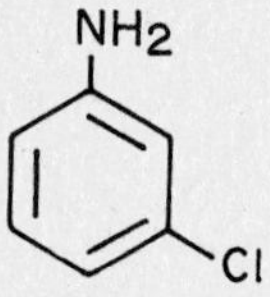

b. 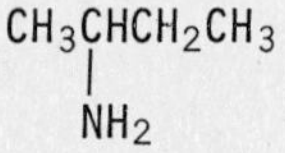 $CH_3CHCH_2CH_3$
       |
      NH_2

c. 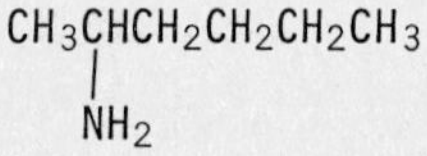 $CH_3CHCH_2CH_2CH_2CH_3$
     |
   NH_2

d.  $CH_3CH_2CH_2NCH_3$
           |
         CH_3

e.

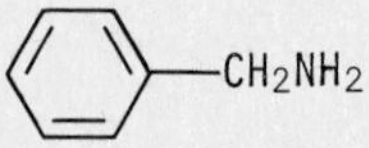

$-CH_2NH_2$

f. 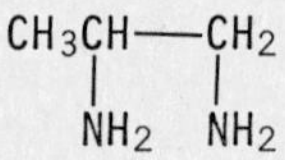 $CH_3CH\!-\!CH_2$
      | |
    NH_2 NH_2

g.

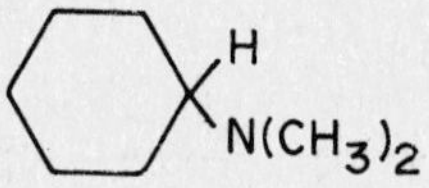

h. $(CH_3CH_2)_4N^+Br^-$

i.

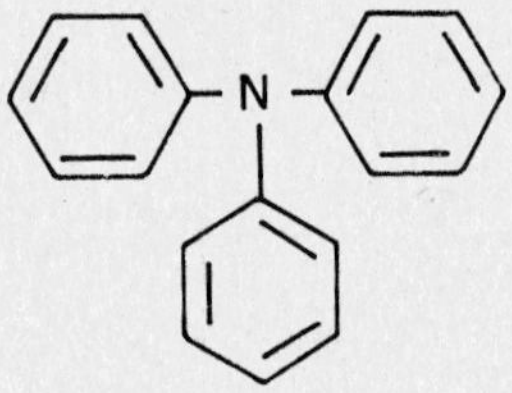

j.

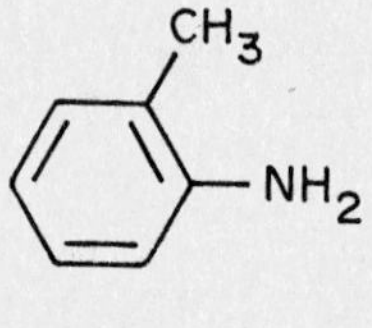

12.23 a. *p*-bromoaniline
     c. diethylmethylamine
     e. 4-amino-2-butanol
     g. *p*-bromobenzenediazonium
         chloride
     i. aminocyclopentane
        (or cyclopentylamine)

b. methylpropylamine
d. tetramethylammonium chloride
f. 4-aminocyclohexanone
h. *N*-methyl-*p*-toluidine
j. 1,6-diaminohexane
    (or hexamethylenediamine)

12.24 $CH_3CH_2CH_2CH_2NH_2$ *n*-butylamine (primary)

$CH_3CH_2CHCH_3$
|
NH_2

         sec-butylamine (primary)

$CH_3CHCH_2NH_2$
|
CH_3

         isobutylamine (primary)

$(CH_3)_3CNH_2$ *t*-butylamine (primary)

$CH_3CH_2CH_2NHCH_3$ methyl-*n*-propylamine (secondary)

$CH_3CHNHCH_3$
|
CH_3

         methylisopropylamine (secondary)

$CH_3CH_2NHCH_2CH_3$ diethylamine (secondary)

CH_3—N—CH_2CH_3
|
CH_3

         ethyldimethylamine (tertiary)

12.25 See the answer to Problem 12.4. Since intermolecular hydrogen
bonding is not possible for trimethylamine, the difference between
its boiling point and that of the structurally similar isobutane is
relatively small. On the other hand, the possibilities for intermolecular hydrogen bonding in *n*-propylamine enhance the difference
between its boiling point and that of the structurally similar
n-butane.

    The other factor is the polarity of the $\overset{\delta+}{C}$——$\overset{\delta-}{N}$ bond compared
with the C——C bond. Trimethylamine is more polar than isobutane
and tends to associate:

This association accounts for the modest (~13°C) difference in the
boiling points of these two compounds. Undoubtedly, such association contributes *part* of the boiling point difference between
n-propylamine and *n*-butane, the rest of the difference being due
to hydrogen bonding.

12.26 Ethylenediamine, with two amino groups, has more possibilities for
intermolecular hydrogen bonding than does *n*-propylamine, for example.

—N···H—N—CH_2CH_2—N—H···N—
           | |
           H   H

12.27 The boiling point order is

pentane < methyl *n*-propyl ether < 1-aminobutane < 1-butanol

O—H···O bonds are more effective than N—H···N bonds, which explains the order of the last two compounds. No hydrogen bonding is possible in the first two compounds, but C—O bonds are polar, giving the ether a higher boiling point than the alkane. The actual boiling points are pentane, 36°C; methyl *n*-propyl ether, 39°C; 1-aminobutane, 78°C; 1-butanol, 118°C.

12.28 a.

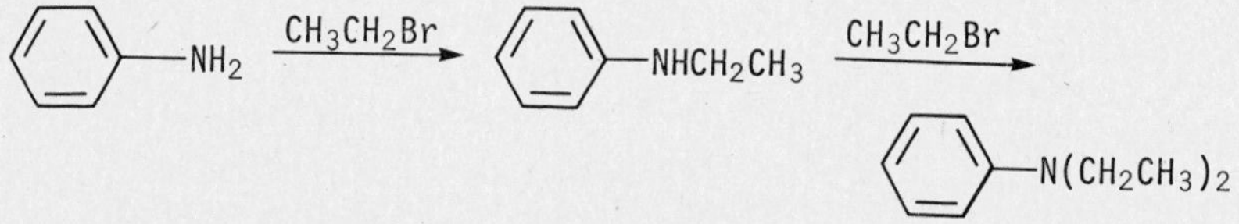

b.

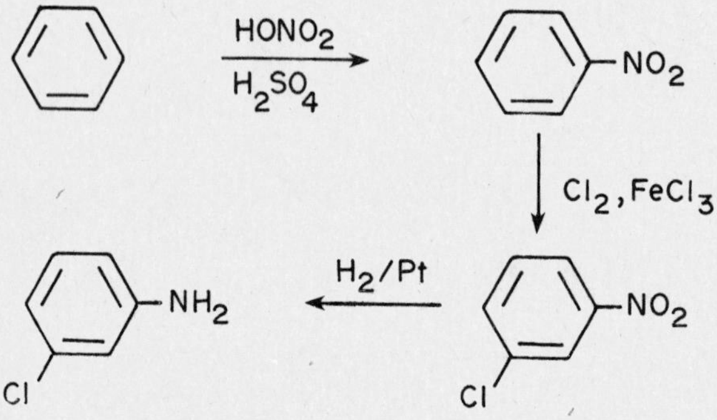

First nitrate, then chlorinate, to obtain the *meta* orientation.

c.

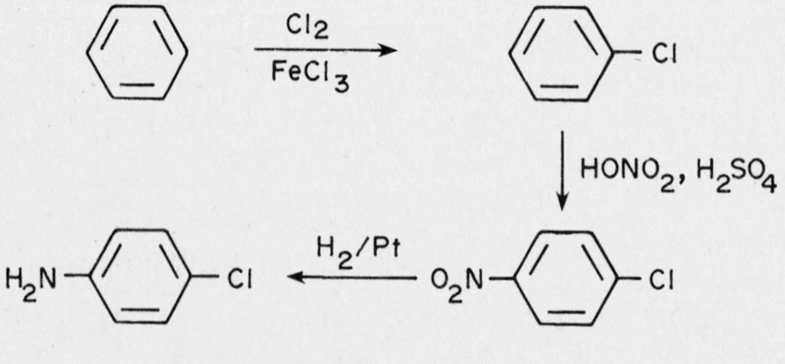

The reverse of the sequence in part b gives mainly *para* orientation.

d. $CH_3CH_2CH_2CH_2Br \xrightarrow{\text{NaCN}} CH_3CH_2CH_2CH_2C\equiv N$

$\xrightarrow{\text{LiAlH}_4} CH_3CH_2CH_2CH_2CH_2NH_2$

12.29 a. 2 $-NH_2 \; + \; CH_2{=}CHCH_2Br \xrightarrow[S_N2]{\text{heat}}$

$-NHCH_2CH{=}CH_2 \; + \;$ $-\overset{+}{N}H_3Br^-$

b. $CH_3\overset{O}{\overset{\|}{C}}{-}Cl \; + \; H_2NCH_2CH(CH_3)_2 \longrightarrow CH_3\overset{O}{\overset{\|}{C}}{-}NHCH_2CH(CH_3)_2 \xrightarrow{\text{LiAlH}_4}$

$CH_3CH_2NHCH_2CH(CH_3)_2$

c. $CH_3O\overset{O}{\overset{\|}{C}}$ $\xrightarrow[H^+]{\text{HONO}_2}$ $CH_3O\overset{O}{\overset{\|}{C}}$$NO_2$

$CH_3OH \; + \; HOCH_2$$NH_2$ $\xleftarrow{\text{LiAlH}_4}$

In the first step, the ester group is *meta*-directing. In the
second step, both the nitro group and the ester group are reduced
when excess $LiAlH_4$ is used.

d. $-CH_2Br \xrightarrow{\text{NaCN}}$ $-CH_2CN$

$\xrightarrow{\text{LiAlH}_4}$

$-CH_2CH_2\overset{O}{\overset{\|}{N}HCCH_3}$ $\xleftarrow{(CH_3CO)_2O}$ $-CH_2CH_2NH_2$

12.30

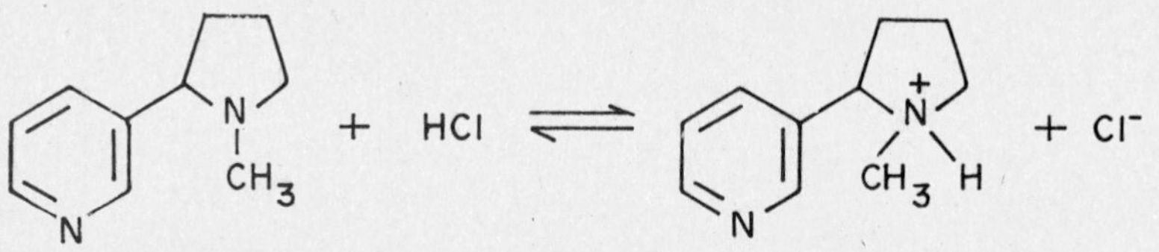

Of the two nitrogens, the aliphatic amine is the more basic. Compare, for example, the pK_b's of trimethylamine (a tertiary aliphatic amine) and pyridine (see Table 12.1).

12.31 a. Aniline is the stronger base. The p-cyano group is electron-withdrawing and therefore decreases the basicity of aniline. Note that the possibilities for delocalization of the unshared electron pair are greater in p-cyanoaniline than in aniline.

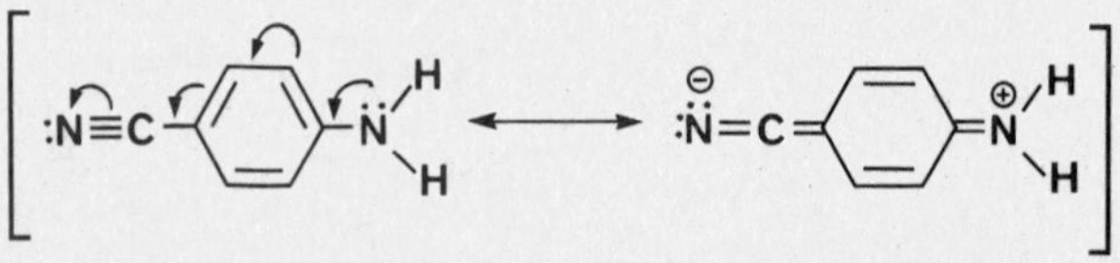

Resonance stabilizes the free base relative to its protonated form, and the effect is greater with p-cyanoaniline than with aniline.

b. The possibilities for delocalization of an electron pair are greater in diphenylamine than in aniline (two phenyl groups vs. one phenyl group). Thus aniline is the stronger base.

12.32 The mixture is first dissolved in an inert, low-boiling solvent such as ether. The following scheme describes a separation procedure:

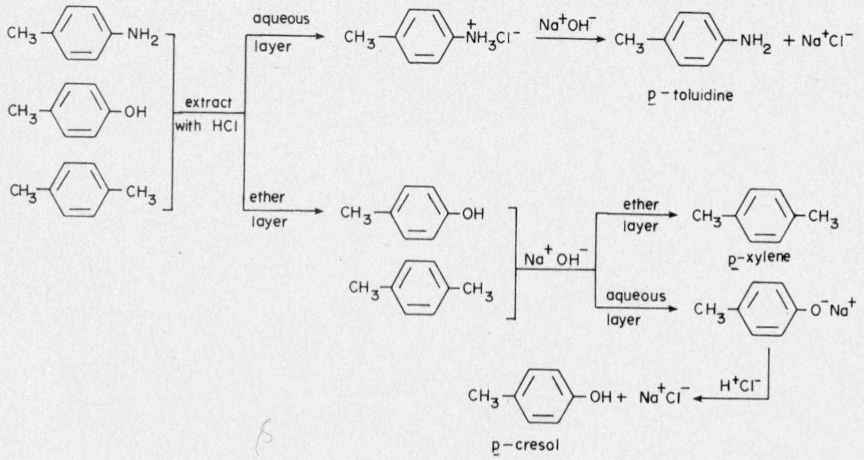

To recover the *p*-xylene, the ether is evaporated and the *p*-xylene distilled. In the case of *p*-toluidine and *p*-cresol, once the product is liberated from the corresponding salt, it is extracted from the water by ether. The ether is then evaporated and the desired product is distilled.

The order of extraction--acid first, then base--can be reversed.

12.33

The unshared electron pair on the amino group can be delocalized not only to the *ortho* and *para* carbons of the ring as with aniline (p. 326 of the text) but also to the oxygen of the nitro group.

12.34 Protonation of an amide on nitrogen would give a cation with the positive charge localized on the nitrogen:

Protonation on oxygen, however, gives a charge-delocalized cation:

This effect overrides the normally stronger basicity of nitrogen compared to that of oxygen.

12.35 a. Compare with eq. 12.17.

b. $(CH_3CH_2)_3N$ + H—OSO_3H $\longrightarrow$ $(CH_3CH_2)_3\overset{+}{N}H$ + $\overset{-}{O}SO_3H$

c. $(CH_3CH_2)_2\overset{+}{N}H_2Cl^-$ + Na^+OH^- $\longrightarrow$ $(CH_3CH_2)_2NH$ + Na^+Cl^- + H_2O

d. [benzene ring]—$\overset{\cdot\cdot}{N}(CH_3)_2$ + CH_3I ⟶ [benzene ring]—$\overset{+}{N}(CH_3)_3$ + I^-

e. [cyclohexane ring]—NH_2 + $CH_3\overset{\overset{O}{\|}}{C}$—O—$\overset{\overset{O}{\|}}{C}CH_3$ ⟶ [cyclohexane ring]—$\overset{\overset{O}{\|}}{NHCCH_3}$ + CH_3COOH

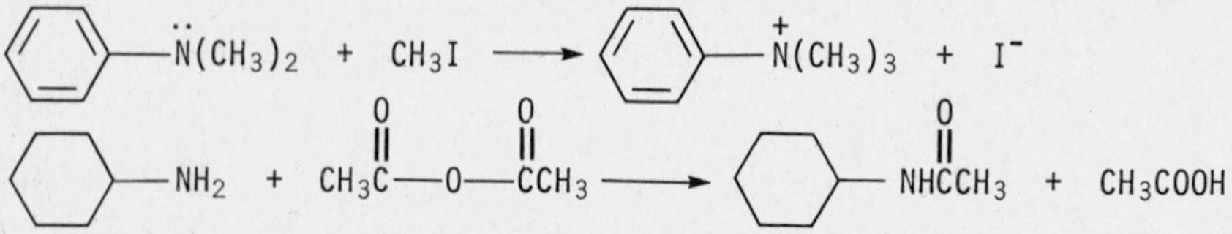

12.36 The reaction begins with nucleophilic attack by the amine on the carbonyl group of the anhydride.

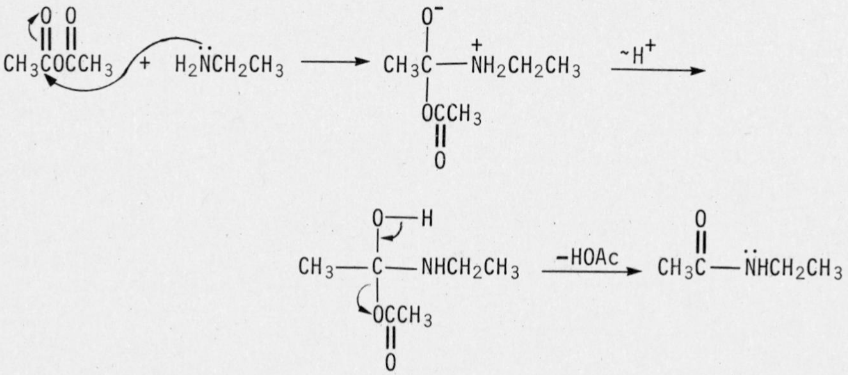

$CH_3\overset{\overset{O}{\|}}{C}O\overset{\overset{O}{\|}}{C}CH_3$ + $H_2\overset{\cdot\cdot}{N}CH_2CH_3$ ⟶ $CH_3\overset{\overset{O^-}{|}}{\underset{\underset{O}{\overset{\overset{OCCH_3}{|}}{\|}}}{C}}\overset{+}{N}H_2CH_2CH_3$ $\xrightarrow{\sim H^+}$

$CH_3—\overset{\overset{O—H}{|}}{\underset{\underset{O}{\overset{\overset{OCCH_3}{|}}{\|}}}{C}}—NHCH_2CH_3$ $\xrightarrow{-HOAc}$ $CH_3\overset{\overset{O}{\|}}{C}—\overset{\cdot\cdot}{N}HCH_2CH_3$

Even though the resulting amide has an unshared electron pair on nitrogen, it does not react with a second mole of acetic anhydride to become diacylated:

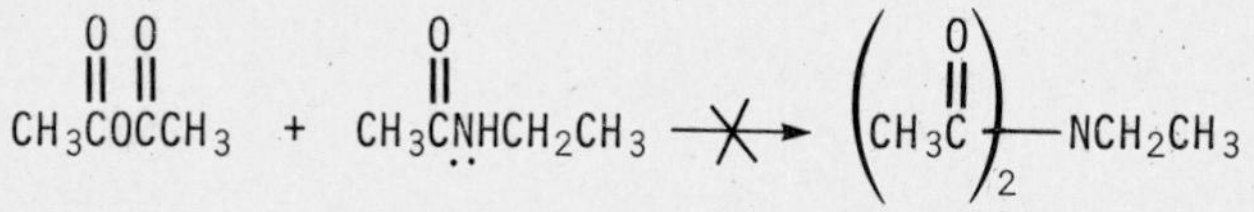

$CH_3\overset{\overset{O}{\|}}{C}O\overset{\overset{O}{\|}}{C}CH_3$ + $CH_3\overset{\overset{O}{\|}}{C}\overset{\cdot\cdot}{N}HCH_2CH_3$ $\xcancel{\longrightarrow}$ $\left(CH_3\overset{\overset{O}{\|}}{C}\right)_2—NCH_2CH_3$

The reason is that amides are poor nucleophiles because the unshared electron pair on the nitrogen is delocalized through resonance:

$$\left[CH_3\overset{\overset{O}{\|}}{C}—\overset{\cdot\cdot}{N}HCH_2CH_3 \longleftrightarrow CH_3\overset{\overset{O^-}{|}}{C}=\overset{+}{N}HCH_2CH_3 \right]$$

The amide is ineffective with respect to nucleophilic attack on the carbonyl group of acetic anhydride.

12.37 The negative charge in the carbanion resulting from loss of a proton from a barbiturate can be delocalized over *two* carbonyl groups:

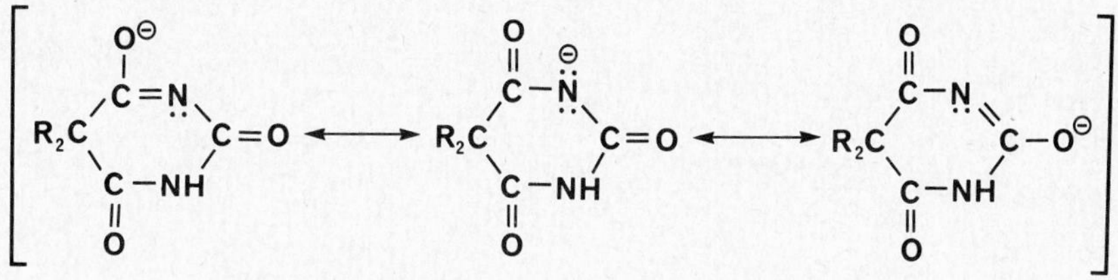

Compare this with eq. 12.16 for simple amides.

12.38 The mechanism involves nucleophilic attack by the hydroxyl group of naphthol on the C=O bond of the isocyanate. It follows the general mechanism for carbamate formation on p. 332 of the text:

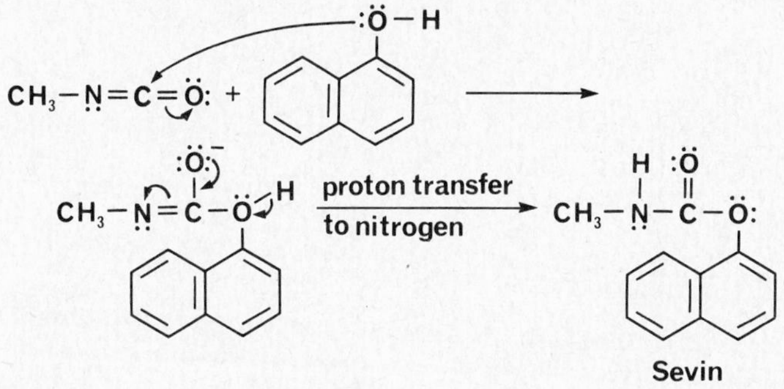

proton transfer
to nitrogen

Sevin

12.39 The reaction involves an S_N2 ring opening of ethylene oxide.

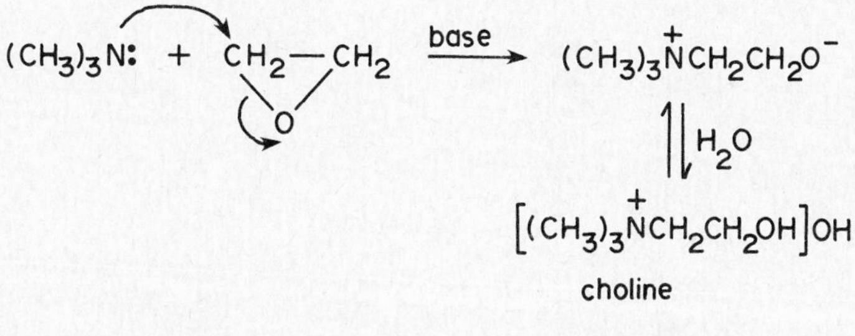

$(CH_3)_3N: + CH_2-CH_2 \xrightarrow{base} (CH_3)_3\overset{+}{N}CH_2CH_2O^-$

$\big\downarrow\big\uparrow H_2O$

$[(CH_3)_3\overset{+}{N}CH_2CH_2OH]OH^-$

choline

12.40 The amines are primary, secondary, and tertiary, respectively. The reactions of amines with nitrous acid are described in Sec. 12.12.

$$(CH_3)_2CHNH_2 \ + \ HONO \ \xrightarrow{0°C} \ (CH_3)_2CHOH \ + \ N_2\uparrow \ + \ H_2O \quad \text{(eq. 12.29)}$$

A gas will be evolved.

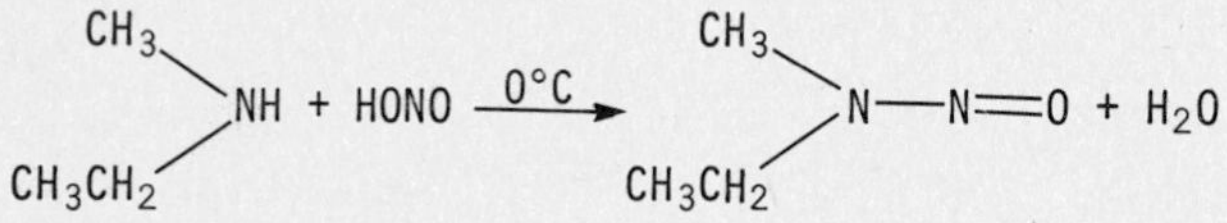

$$\begin{array}{c} CH_3 \\ {} \\ CH_3CH_2 \end{array}\!\!NH \ + \ HONO \ \xrightarrow{0°C} \ \begin{array}{c} CH_3 \\ {} \\ CH_3CH_2 \end{array}\!\!N{-}N{=}O \ + \ H_2O \qquad \text{(eq. 12.27)}$$

A yellow, oily upper layer will appear.

$$(CH_3)_3N \ + \ HONO \ \xrightarrow{0°C} \ \text{no observable reaction}$$

12.41 Loss of nitrogen from an aryldiazonium ion would produce an aryl carbocation:

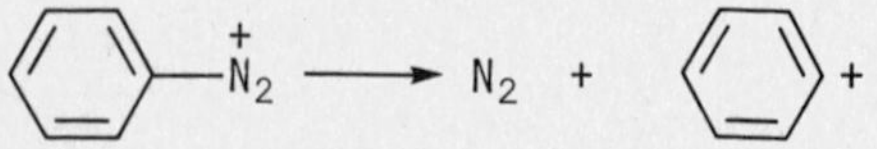

Aryl cations are particularly unstable and therefore difficult to form. The positive carbon in such cations has only two groups attached and should be sp-hybridized (in contrast to ordinary carbocations, which are sp^2-hybridized). The six-membered ring of course precludes the linear geometry expected for sp hybridization. Another reason for the stability of aryl diazonium ions is that the positive charge can be delocalized to the aromatic ring:

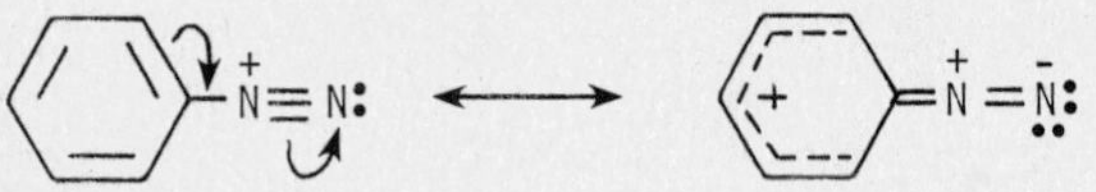

Therefore, aryldiazonium ions are appreciably more stable than aliphatic diazonium ions.

12.42 These equations illustrate the reactions in Secs. 12.13 and 12.14.

a. 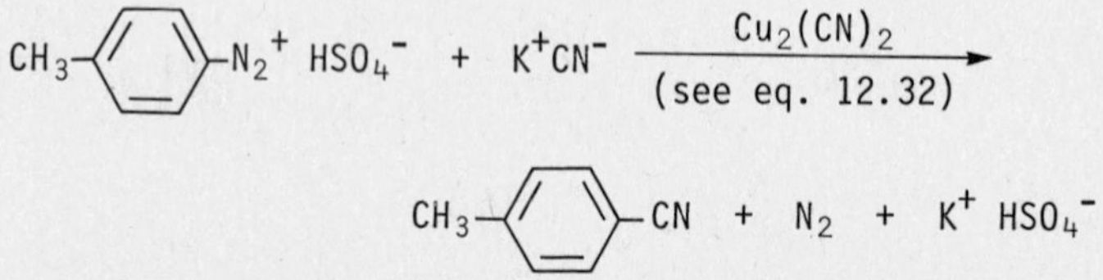 $\quad CH_3{-}\!\!\bigcirc\!\!{-}N_2^+ \ HSO_4^- \ + \ K^+CN^- \ \xrightarrow[\text{(see eq. 12.32)}]{Cu_2(CN)_2}$

$$CH_3{-}\!\!\bigcirc\!\!{-}CN \ + \ N_2 \ + \ K^+ \ HSO_4^-$$

b. 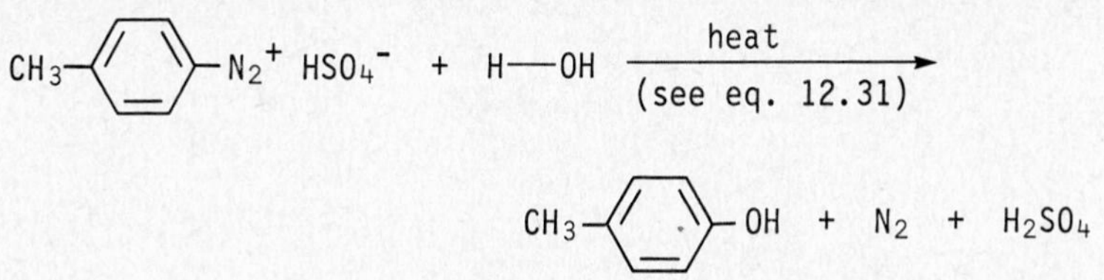 $CH_3-\langle\bigcirc\rangle-N_2^+ HSO_4^-$ + $H-OH$ $\xrightarrow[\text{(see eq. 12.31)}]{\text{heat}}$

$$CH_3-\langle\bigcirc\rangle-OH \ + \ N_2 \ + \ H_2SO_4$$

c. 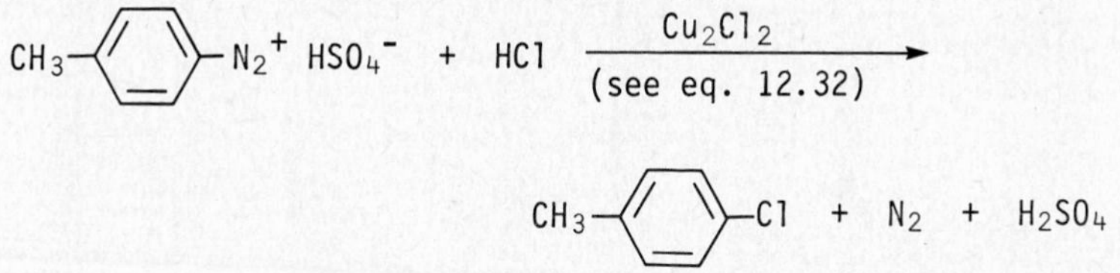 $CH_3-\langle\bigcirc\rangle-N_2^+ HSO_4^-$ + HCl $\xrightarrow[\text{(see eq. 12.32)}]{Cu_2Cl_2}$

$$CH_3-\langle\bigcirc\rangle-Cl \ + \ N_2 \ + \ H_2SO_4$$

d. 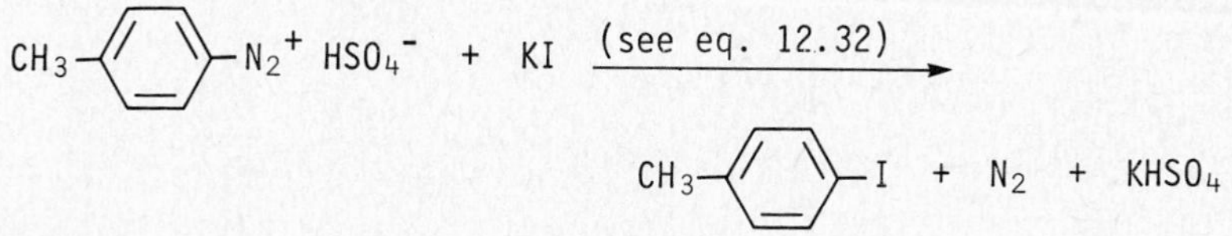 $CH_3-\langle\bigcirc\rangle-N_2^+ HSO_4^-$ + KI $\xrightarrow{\text{(see eq. 12.32)}}$

$$CH_3-\langle\bigcirc\rangle-I \ + \ N_2 \ + \ KHSO_4$$

e. Since *para* coupling is blocked by the methyl substituent, *ortho* coupling occurs:

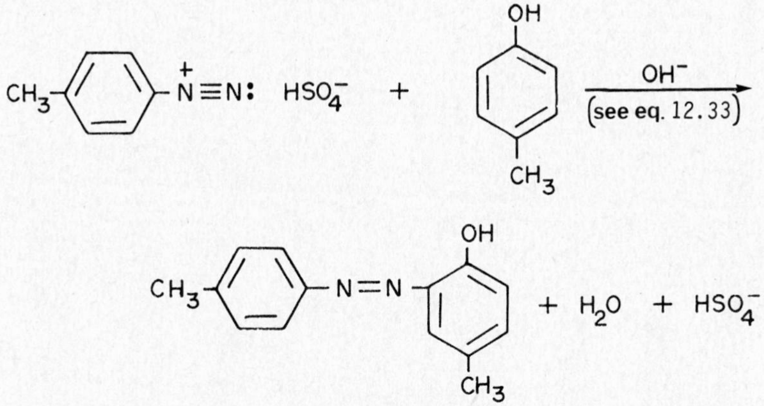

$CH_3-\langle\bigcirc\rangle-\overset{+}{N}\equiv N\text{:} \ HSO_4^-$ + $\overset{\displaystyle OH}{\underset{\displaystyle CH_3}{\langle\bigcirc\rangle}}$ $\xrightarrow[\text{(see eq. 12.33)}]{OH^-}$

$$CH_3-\langle\bigcirc\rangle-N=N-\overset{\displaystyle OH}{\underset{\displaystyle CH_3}{\langle\bigcirc\rangle}} \ + \ H_2O \ + \ HSO_4^-$$

f. 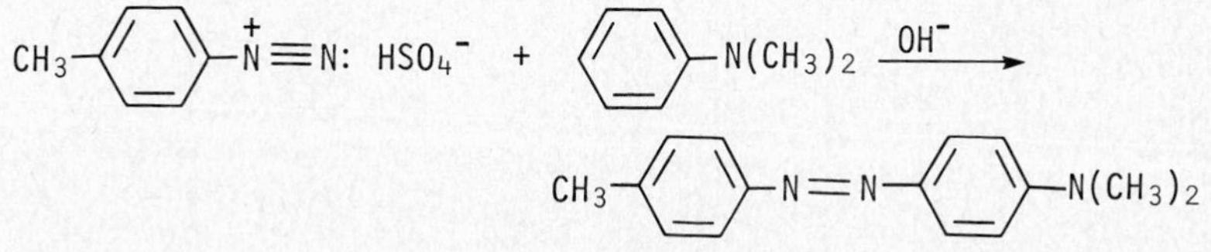 $CH_3-\langle\bigcirc\rangle-\overset{+}{N}\equiv N\text{:} \ HSO_4^-$ + $\langle\bigcirc\rangle-N(CH_3)_2$ $\xrightarrow{OH^-}$

$$CH_3-\langle\bigcirc\rangle-N=N-\langle\bigcirc\rangle-N(CH_3)_2$$

12.43 a.

b.

Note that the order of each step in the sequence is important. The benzene must be nitrated first, then brominated to attain *meta* orientation. Bromination of iodobenzene would not give *meta* product, so this indirect route must be used.

12.44 Benzidine can be diazotized at each amino group. It can then couple with two equivalents of the aminosulfonic acid. Coupling occurs *ortho* to the amino group, since the *para* position is blocked by the sulfonic acid group.

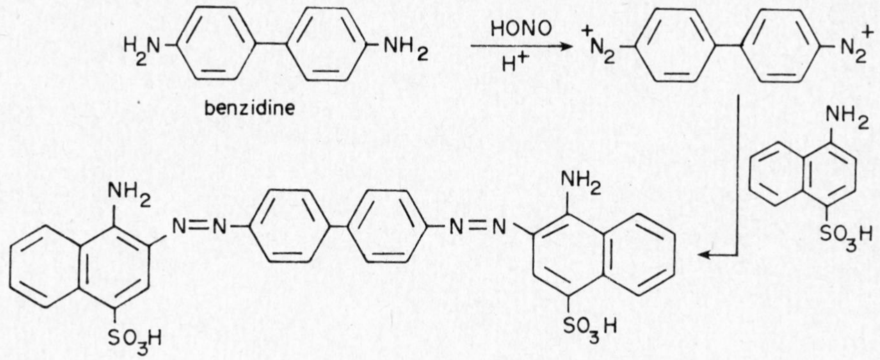

benzidine

12.45 1,4-Addition of chlorine to 1,3-butadiene gives a dihalide in which both chlorines are allylic and therefore quite easily displaced by nucleophiles. Reaction with sodium cyanide gives the dinitrile.

$$CH_2 \!\!=\!\! CH \!\!-\!\! CH \!\!=\!\! CH_2 \;+\; Cl_2 \longrightarrow \underset{Cl}{CH_2} \!\!-\!\! CH \!\!=\!\! CH \!\!-\!\! \underset{Cl}{CH_2}$$

$$NCCH_2CH \!\!=\!\! CHCH_2CN \xleftarrow{\;NaCN\;}$$

The double bond can then be hydrogenated to give the saturated nitrile. Catalytic hydrogenation gives hexamethylenediamine.

$$NCCH_2CH \!\!=\!\! CHCH_2CN \xrightarrow[Ni]{H_2} NCCH_2CH_2CH_2CH_2CN$$

$$\xrightarrow[Ni]{4\ H_2} H_2N \!\!-\!\! (CH_2)_6 \!\!-\!\! NH_2$$

hexamethylenediamine

12.46 The first steps are as follows:

$$HOOC(CH_2)_4COOH \xrightarrow{NH_3} NH_4^+ \;{}^-OOC(CH_2)_4COO^- \; NH_4^+$$

$$\text{heat} \;\Big|\; -2\ H_2O$$

$$H_2N \!\!-\!\! \overset{O}{\overset{\|}{C}}(CH_2)_4\overset{O}{\overset{\|}{C}}NH_2$$

Dehydration of the diamide gives the dinitrile, which can then be reduced to hexamethylenediamine.

$$H_2NC(CH_2)_4CNH_2 \xrightarrow{-2\ H_2O} N\equiv C(CH_2)_4C\equiv N \xrightarrow[Ni]{4\ H_2} H_2N(CH_2)_6NH_2$$

12.47 $\left(NH(CH_2)_5NHC(CH_2)_8C\right)_n$

12.48 a.

b.

c.

d.

e.

f.

g.

h.

i.

j. $(CH_3CH_2)_2N-C$

(See p. 343 in the text for the structure of lysergic acid.)

12.49 The reactions of heterocyclic amines with acid, alkyl halides, nitrous acid, or acylating agents are entirely analogous to those of aliphatic or aromatic amines.

a.

b.

c.

d.

e.

The nitrogens in pyrimidine are identical; thus either one can be protonated.

f.

The acyclic nitrogen is more basic than the indole-type
nitrogen.

12.50   The exocyclic nitrogen is present as an amide. The heterocyclic
nitrogen is present in the four-membered ring as a lactam (a cyclic
amide; see Sec. 12.15).

CHAPTER THIRTEEN   CARBOHYDRATES

CHAPTER SUMMARY

   Carbohydrates are polyhydroxy aldehydes or ketones, or substances that give such compounds on hydrolysis. They are classified as polysaccharides, oligosaccharides, or monosaccharides.

   Monosaccharides, also called simple sugars, are classified by the number of carbon atoms (triose, tetrose, pentose, etc.) and by the nature of the carbonyl group (aldose or ketose).

   R-(+)-**Glyceraldehyde** is an aldotriose designated by the following **Fischer projection formula:**

$$CH{=}O$$
$$H{-}\!\!-\!\!-OH$$
$$CH_2OH$$

$$\equiv$$

$$CH{=}O$$
$$HOCH_2\overset{|\!|\!|\!\cdot}{\underset{H}{C}}OH$$

Horizontal groups are assumed to come out of the plane of the paper toward the viewer and vertical groups to recede behind the plane of the paper. This configuration is designated D, whereas the enantiomer with the H and OH positions reversed is L. In larger monosaccharides, the letters D and L are used to designate the configuration of the chiral center with the *highest* number, the chiral carbon most remote from the carbonyl group. The D-aldoses through the hexoses are listed in Table 13.1.

   Epimers are stereoisomers that differ in configuration at *only one* chiral center.

   Monosaccharides with more than four carbons usually exist in a cyclic hemiacetal form, in which a hydroxyl group on C-4 or C-5 reacts with the carbonyl group (at C-1 in aldoses) to form a hemiacetal. In this way, C-1 also becomes chiral and is called the **anomeric carbon**. **Anomers** differ in configuration only at this chiral center and are designated α or β. The common ring sizes for the cyclic hemiacetals are six-membered (called **pyranoses**) or five-membered (called **furanoses**). The rings contain one oxygen atom and five or four carbon atoms, respectively.

Anomers usually interconvert in solution, resulting in a gradual change in optical rotation from that of the pure anomer to an equilibrium value for the mixture. This rotational change is called **mutarotation**.

Haworth formulas are a useful way of representing the cyclic forms of monosaccharides. The rings are depicted as flat, with hydroxyl groups or other substituents above or below the ring plane.

Monosaccharides can be oxidized at the aldehyde carbon to give carboxylic acids called **aldonic acids**. Oxidation at both ends of the carbon chain gives **aldaric acids**. Reduction of the carbonyl group to an alcohol gives polyols called **alditols**. The ──OH groups in sugars, like those in simpler alcohols, can be esterified or etherified.

Monosaccharides react with alcohols (H^+ catalyst) to give **glycosides**. The ──OH group at the anomeric carbon is replaced by an ──OR group; the product is an acetal. Alcohols and phenols often occur in nature combined with sugars as glycosides; this renders them water-soluble.

Disaccharides consist of two monosaccharides linked by a glycosidic bond between the anomeric carbon of one unit and a hydroxyl group (often on C-4) of the other unit. Examples include **maltose** and **cellobiose** (formed from two glucose units joined by a 1,4-linkage and differing only in configuration at the anomeric carbon, being α and β, respectively), **lactose** (from a galactose and glucose unit linked 1,4 and β), and **sucrose**, or cane sugar (from a fructose and glucose unit, linked at the anomeric carbon of each, or 1,2).

Sugars such as fructose, glucose, and sucrose are sweet, but others (for example, lactose and galactose) are not. Some noncarbohydrates, such as **saccharin** and **aspartame**, also taste sweet.

Polysaccharides have many monosaccharide units linked by glycosidic bonds. **Starch** and **glycogen** are polymers of D-glucose, mainly linked 1,4 or 1,6 and α. **Cellulose** consists of D-glucose units linked 1,4 and β.

Monosaccharides with modified structures are often biologically important. Examples include **sugar phosphates**, **deoxy sugars**, **amino sugars**, and **ascorbic acid** (vitamin C).

REACTION SUMMARY

Hydrolysis

Polysaccharide $\xrightarrow{H_3O^+}$ oligosaccharide $\xrightarrow{H_3O^+}$ monosaccharide

Acyclic and Cyclic Equilibration

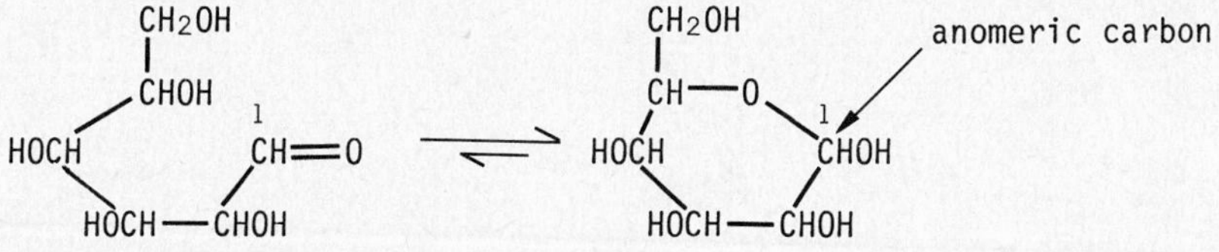

acyclic (aldehyde)   cyclic (hemiacetal)

Oxidation

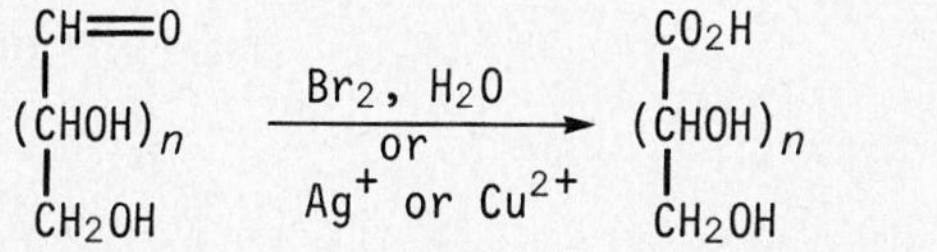

aldose   aldonic acid

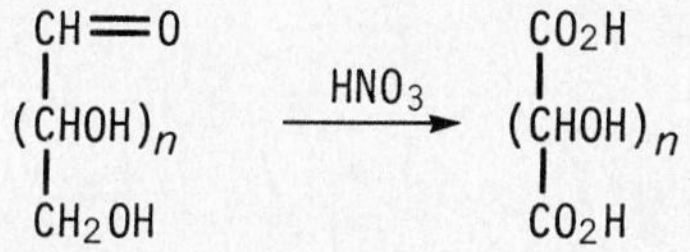

aldose   aldaric acid

Reduction

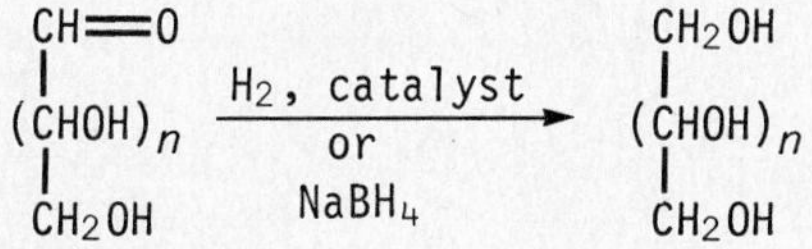

aldose   alditol

Esterification and Etherification

$$HOCH_2 \text{ sugar} \xrightarrow{Ac_2O} AcOCH_2 \text{ sugar} \qquad Ac = CH_3\overset{O}{\overset{\|}{C}}-$$

$$\xrightarrow[\substack{\text{or} \\ CH_3I, Ag_2O}]{NaOH, (CH_3)_2SO_4} CH_3OCH_2 \text{ sugar}$$

Formation of Glycosides

$$HOCH_2 \text{ sugar-OH} + ROH \underset{}{\overset{H^+}{\rightleftharpoons}} HOCH_2 \text{ sugar-OR} + H_2O$$

Hydrolysis of Glycosides

$$HOCH_2 \text{ sugar-OR} + H_2O \underset{}{\overset{H^+}{\rightleftharpoons}} HOCH_2 \text{ sugar-OH} + ROH$$

LEARNING OBJECTIVES

1. Know the meaning of: carbohydrate, monosaccharide, oligosaccharide, polysaccharide, disaccharide, trisaccharide.

2. Know the meaning of: aldose, ketose, triose, tetrose, pentose, hexose, glyceraldehyde, dihydroxyacetone.

3. Know the meaning of: D- or L-sugar, Fischer projection formula, Haworth formula, epimer.

4. Know the meaning of: α and β configurations, anomer, furanose and pyranose forms, mutarotation.

5. Know the meaning of: glycosidic bond, reducing and nonreducing sugar, aldaric acid, aldonic acid, alditol.

6. Learn the formulas for some common monosaccharides, especially the D forms of glucose, mannose, galactose, and fructose.

7. Draw the Fischer projection formula for a simple monosaccharide.

8. Convert the Fischer projection formula for a tetrose to a sawhorse or Newman projection formula, and vice versa.

9. Tell whether two structures are epimers or anomers.

10. Given the acyclic formula for a monosaccharide, draw its cyclic structure in either the pyranose or furanose form, and either α or β configuration.

11. Given the rotations of two pure anomers and of their equilibrium mixture, calculate the percentage of each anomer present at equilibrium.

12. Given the formula for a monosaccharide, draw the formula for its glycoside with a given alcohol, or with a given additional monosaccharide.

13. Draw the cyclic structures (Haworth projection and conformational structure) for α-D- and β-D-glucose and the corresponding methyl glucosides.

14. Write all the steps in the mechanism for the formation of a glycoside from a given sugar and alcohol.

15. Write all the steps in the mechanism for the hydrolysis of a given glycoside to the corresponding sugar and alcohol.

16. Write all the steps in the mechanism for the hydrolysis of a given disaccharide to the component monosaccharides.

17. Given the structure of a sugar, write equations for its reaction with each of the following reagents: acetic anhydride, bromine water, nitric acid, sodium borohydride, and Tollens' or Fehling's reagent.

18. Know the structures of: maltose, cellobiose, lactose, sucrose.

19. Write the structures for the repeating units in starch and cellulose.

20. Know the meaning of: sugar phosphate, deoxy sugar, amino sugar, ascorbic acid (vitamin C).

ANSWERS TO PROBLEMS

Problems Within the Chapter

13.1 The L isomer is the mirror image of the D isomer; the configuration at *every* chiral center is reversed.

a.

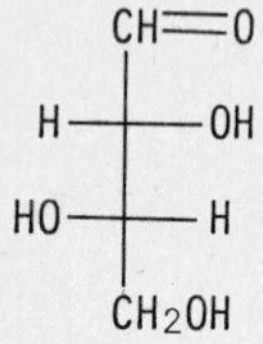

b.

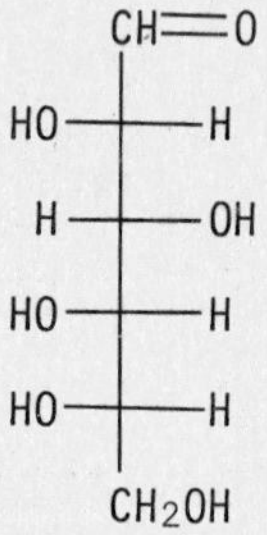

13.2 Follow Example 13.3 as a guide.

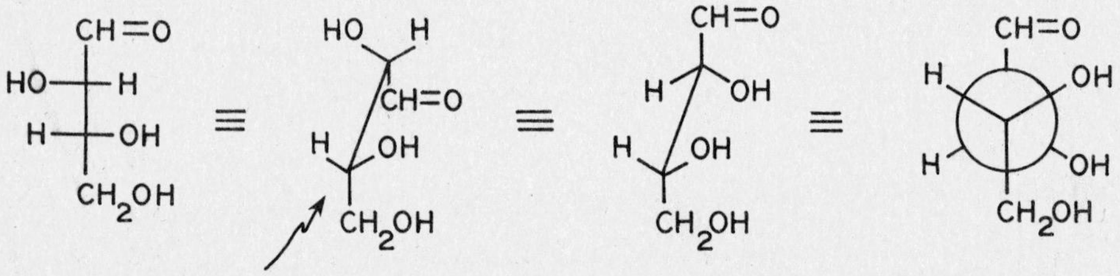

If you view this formula from the "top", you will see that it is just a three-dimensional representation of the Fischer projection; horizontal groups at each chiral center come up toward you, and vertical groups recede away from you.

13.3 For each D-aldehexose there will be, as one more chiral center is added, two aldoheptoses. Altogether, there are sixteen D-aldoheptoses.

13.4 D-Ribose and D-xylose are identical except for the configuration at C-3. So too are D-arabinose and D-lyxose.

13.5 Because the D refers only to the configuration at the chiral center most remote from the aldehyde group (C-5 in D-talose).

13.6 Follow Example 13.4 as a guide.

$$\% \text{ of } \beta = \frac{151 - 84}{151 - (-53)} \times 100 = \frac{67}{204} \times 100 = 32.8\%$$

$$\% \text{ of } \alpha = 100 - 32.8 = 67.2\%$$

13.7 D-Galactose differs from D-glucose only in configuration at C-4, so the structures for D-glucose can be used as a starting point, and a change is made only at C-4.

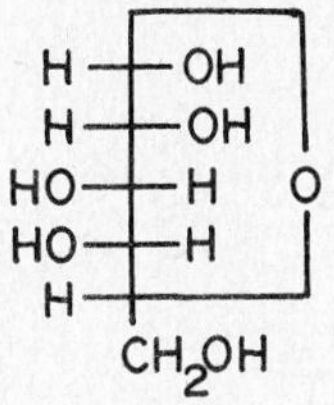

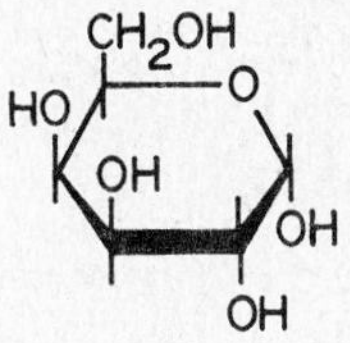

13.8

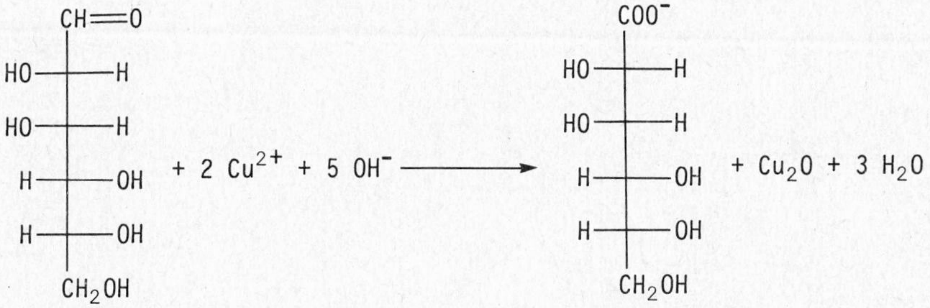

$$+ 2 \ Cu^{2+} + 5 \ OH^- \longrightarrow \qquad + Cu_2O + 3 \ H_2O$$

13.9

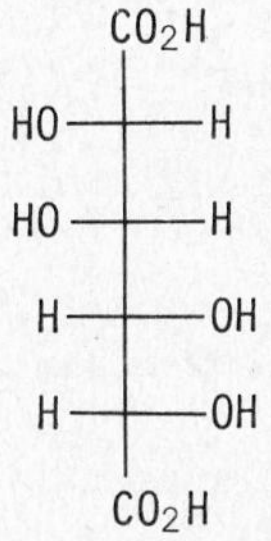

13.10

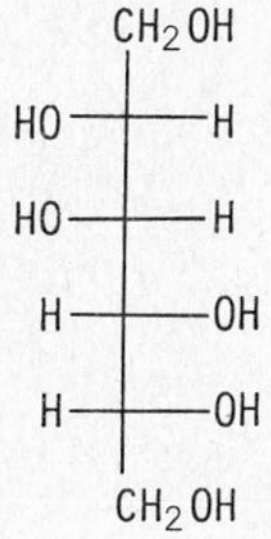

13.11 Using Haworth formulas, we have the following:

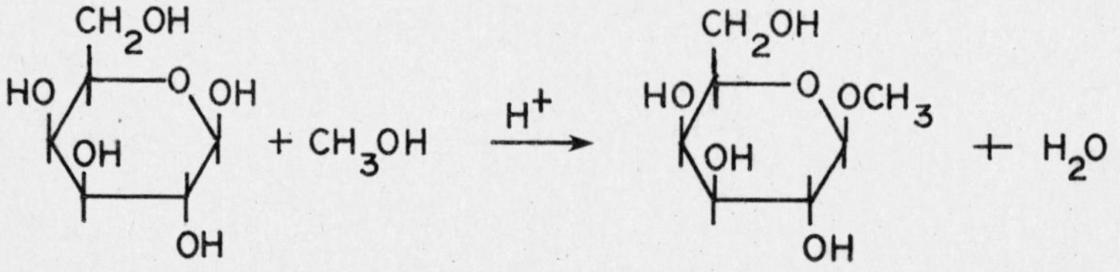

Although we show only the methyl β-D-galactoside as a product, some of the α epimer will also be formed, because the carbocation inter-mediate can be attacked at either face by the methanol.

13.12

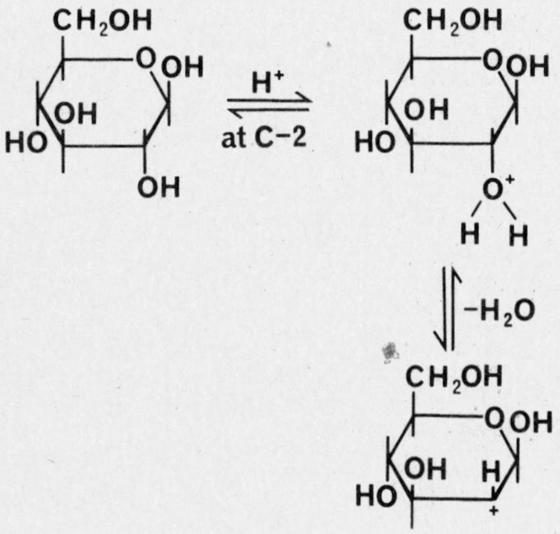

The resulting carbocation is secondary, because two carbons and a hydrogen are attached to the positive carbon. The same would be true of a carbocation with the positive charge on C-3 or C-4. The carbocation with the positive charge on C-6 would be primary. In contrast, the C-1 carbocation (see Example 13.7) has an *oxygen*, a carbon, and a hydrogen attached to the positive carbon.

13.13 Carbon-1 of the second (right-hand) glucose unit in maltose is a hemiacetal carbon. Thus the α and β forms *at this carbon atom* can equilibrate via the open-chain aldehyde form. Mutarotation is therefore possible. Note that carbon-1 of the first (left-hand) glucose unit is an acetal carbon. Its configuration is therefore fixed (as α).

13.14 Carbon-1 of the glucose unit in lactose is a hemiacetal carbon and will be in equilibrium with the open-chain aldehyde form. Therefore, lactose will be oxidized by Fehling's solution and will mutarotate.

Additional Problems

13.15 If you have difficulty with any part of this problem, consult the indicated section in the text, where the term is defined and/or illustrated.

a. Sec. 13.3   b. Sec. 13.3
c. Secs. 13.2 and 13.3   d. Secs. 13.2 and 13.13
e. Secs. 13.2 and 13.14   f. Sec. 13.8
g. Sec. 13.8   h. Sec. 13.12
i. Sec. 13.5

13.16 D-Sugars have the same configuration at the carbon atom adjacent to the primary alcohol function as D-(+)-glyceraldehyde (which is R).

<pre>
    CH=O   CHO   CH₂OH
     | | |
H────┼──OH   (CHOH)ₙ   C=O
     | | |
    CH₂OH   H─────┼──OH   (CHOH)ₙ
                         | |
                        CH₂OH   H─────┼──OH
                                             |
                                            CH₂OH

D-(+)-glyceraldehyde   a D-aldose   a D-ketose
</pre>

L-Sugars have the opposite configuration at the chiral center adjacent to the primary alcohol function.

<pre>
    CH=O   CHO   CH₂OH
     | | |
HO───┼──H   (CHOH)ₙ   C=O
     | | |
    CH₂OH   HO────┼──H   (CHOH)ₙ
                         | |
                        CH₂OH   HO────┼──H
                                             |
                                            CH₂OH

L-(-)-glyceraldehyde   an L-aldose   an L-ketose
</pre>

13.17

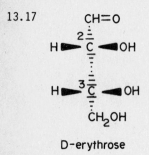

D-erythrose

At C-2, the priority order of groups is
$OH > CH{=}O > CH(OH)CH_2OH > H$
The configuration is *R*.

At C-3, the priority order of groups is
$OH > CH(OH)CHO > CH_2OH > H$
The configuration is *R*.

13.18 Use the answer to Problem 13.17 to help you set up the priority orders.

C2 : R
C3 : S
C4 : R
C5 : R

C1 : R

13.19 D-Gulose and D-idose differ only in the configuration at C-2. They
are epimers at C-2.

13.20 <u>The D-ketoses:</u>

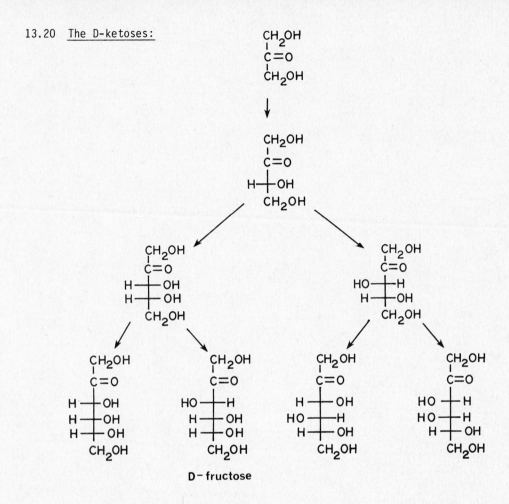

D-fructose

13.21

Compare the final structure (which is obtained by rotating the third structure around the C—CHO bond) with the formula for *S* (or *L*) glyceraldehyde shown in Sec. 13.4 and in Example 13.1.

CARBOHYDRATES

13.22 a.

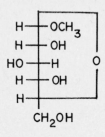

b.

c.

d. The structure is the enantiomer of the structure shown in part a above.

13.23 a. Consult Figure 13.1 for the Fischer projection formula of
           D-mannose. The L isomer is enantiomeric with the D isomer.

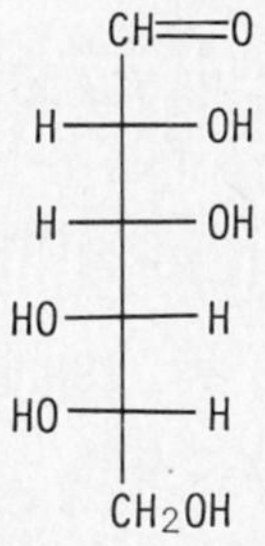

$$
\begin{array}{c}
\mathrm{CH{=}O} \\
\mathrm{H} \!-\!\!\!\!-\! \mathrm{OH} \\
\mathrm{H} \!-\!\!\!\!-\! \mathrm{OH} \\
\mathrm{HO} \!-\!\!\!\!-\! \mathrm{H} \\
\mathrm{HO} \!-\!\!\!\!-\! \mathrm{H} \\
\mathrm{CH_2OH}
\end{array}
$$

L-mannose

   b. See the answer to Problem 13.20 for the Fischer projection
      formula of D-fructose.

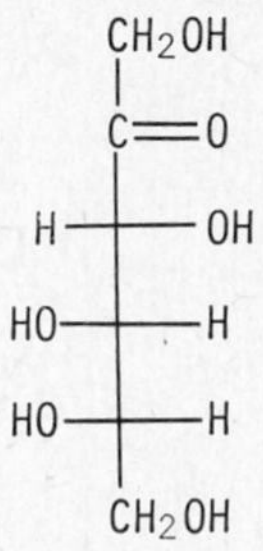

$$
\begin{array}{c}
\mathrm{CH_2OH} \\
\mathrm{C{=}O} \\
\mathrm{H} \!-\!\!\!\!-\! \mathrm{OH} \\
\mathrm{HO} \!-\!\!\!\!-\! \mathrm{H} \\
\mathrm{HO} \!-\!\!\!\!-\! \mathrm{H} \\
\mathrm{CH_2OH}
\end{array}
$$

L-fructose

13.24

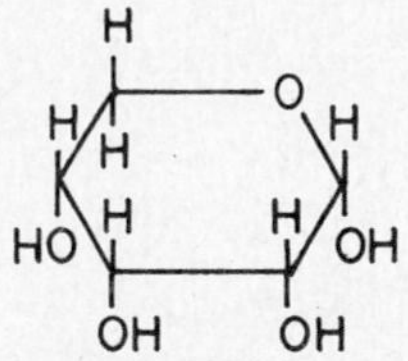

α-pyranose

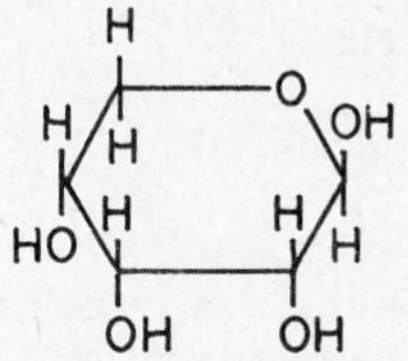

β-pyranose

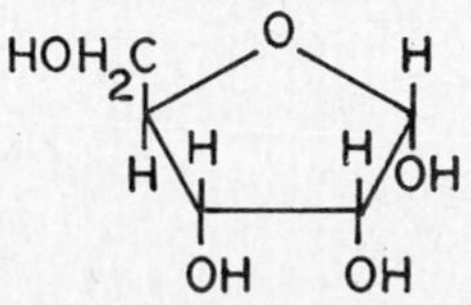

α-furanose

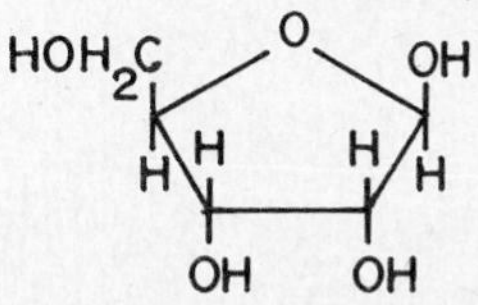

β-furanose

13.25

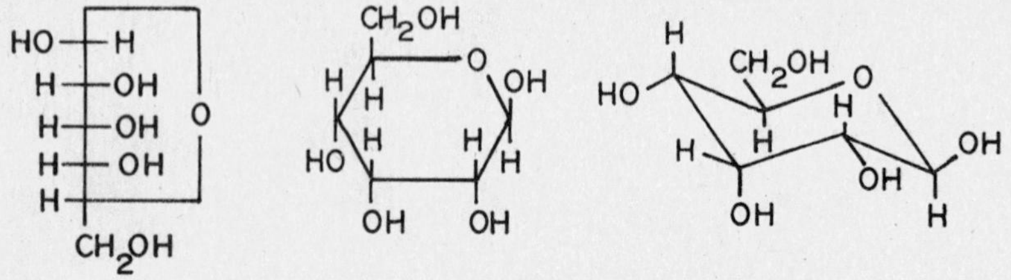

See Figure 13.1 for the Fischer projection of the acyclic form.

13.26 The compounds are diastereomers; that is, they are stereoisomers but not mirror images. They are therefore expected to differ in all properties, including water solubility.

13.27 Use Example 13.4 as a guide.

$$\% \text{ of } \beta = \frac{21 - (-92)}{21 - (-133)} \times 100 = \frac{113}{154} \times 100 = 73.4\%$$

13.28

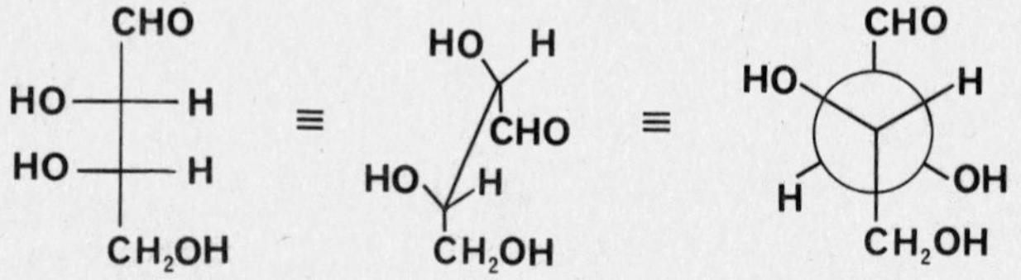

L—erythrose

13.29

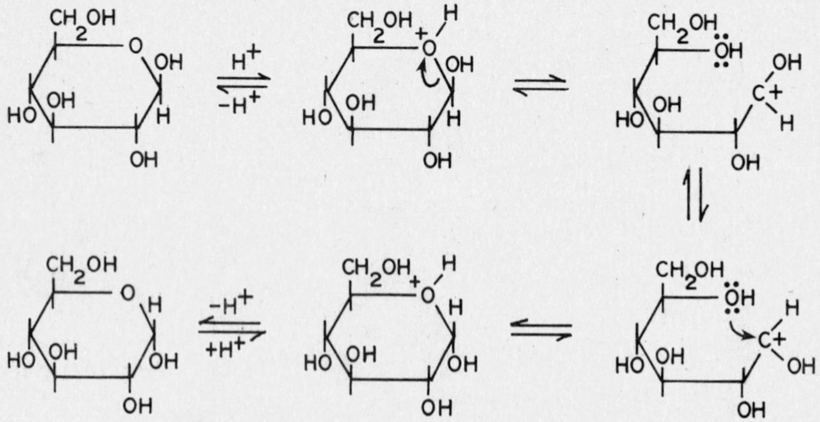

13.30 The structures are shown in Figure 13.1. Oxidation of D-erythrose gives *meso*-tartaric acid:

```
   CHO                      CO2H
H ─┼─ OH        ──────►   H ─┼─ OH
H ─┼─ OH                  H ─┼─ OH
   CH2OH                    CO2H
```

D-erythrose   (optically inactive)
                         meso-tartaric acid

Analogous oxidation of D-threose gives an optically active tartaric acid:

```
    CHO                     CO2H
HO ─┼─ H        ──────►  HO ─┼─ H
 H ─┼─ OH                 H ─┼─ OH
    CH2OH                   CO2H
```

D-threose   (optically active)
                        S,S-tartaric acid

In this way, we can readily assign structures to the two tetroses.

13.31 Consult Sec. 13.9 as a guide.

```
a.     COOH           b.     COOH
    H ─┼─ OH             H ─┼─ OH
   HO ─┼─ H            HO ─┼─ H
   HO ─┼─ H            HO ─┼─ H
    H ─┼─ OH             H ─┼─ OH
       CH2OH                COOH
```

13.32 a.

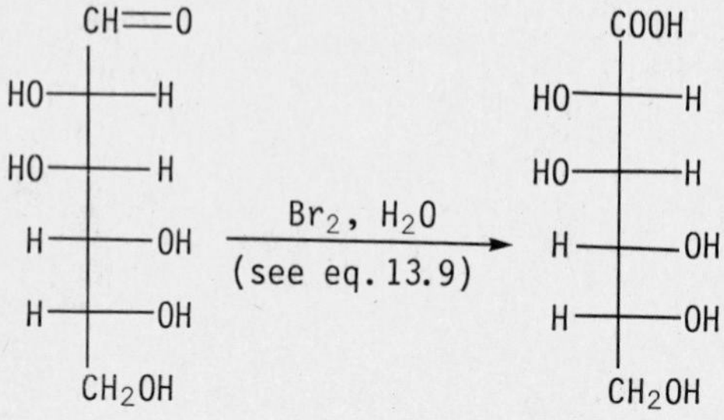

$$\xrightarrow[\text{(see eq. 13.9)}]{Br_2, H_2O}$$

D-mannonic acid

b.

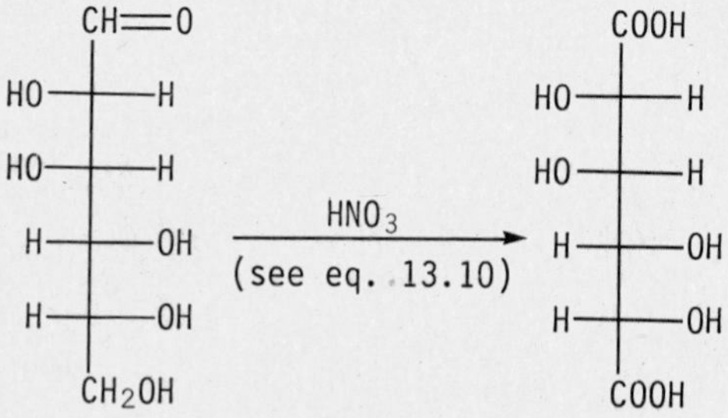

$$\xrightarrow[\text{(see eq. 13.10)}]{HNO_3}$$

D-mannaric acid

c.

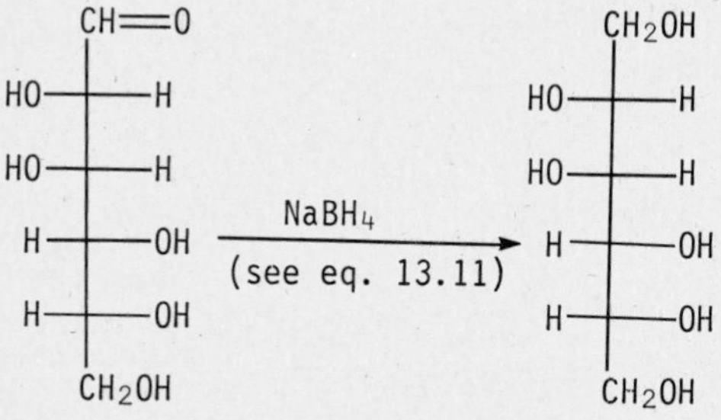

$$\xrightarrow[\text{(see eq. 13.11)}]{NaBH_4}$$

D-mannitol

d.

$$\xrightarrow[\text{(see eq. 13.12)}]{\left(CH_3\overset{O}{\overset{\|}{C}}\right)_2O}$$

$Ac = CH_3\overset{O}{\overset{\|}{C}}-$

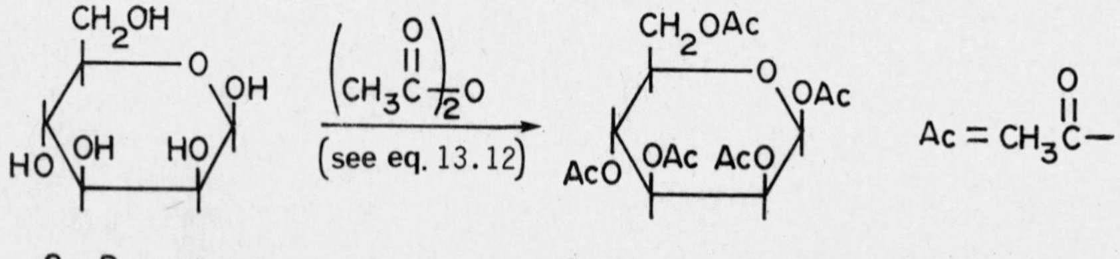

β − D − mannose   β − D − mannose pentaacetate

13.33 Using Fischer projection formulas, we can write the structures of D-glucitol (eq. 13.11) and D-mannitol (see the answer to Problem 13.10).

```
     CH2OH              CH2OH
  H —––— OH        HO —––— H
 HO —––— H         HO —––— H
  H —––— OH         H —––— OH
  H —––— OH         H —––— OH
     CH2OH              CH2OH
  D-glucitol        D-mannitol
```

Note that the configurations are identical at C-3, C-4, and C-5. The fact that D-fructose gives both of these polyols on reduction tells us that it must also have the same configuration as they do at C-3, C-4, and C-5. Thus the keto group of D-fructose must be at C-2.

```
     CH2OH                  CH2OH              CH2OH
     |                   H —––— OH         HO —––— H
     ==O                                              
 HO —––— H    NaBH4    HO —––— H    +     HO —––— H
  H —––— OH    ——–→     H —––— OH          H —––— OH
  H —––— OH             H —––— OH          H —––— OH
     CH2OH                 CH2OH              CH2OH
```

13.34 If glucose had the acyclic aldehyde structure, one might expect the aldehyde to react with *two moles* of methanol to give an acetal:

```
     CH==O                   CH(OCH3)2
  H —––— OH                H —––— OH
 HO —––— H                HO —––— H
  H —––— OH   2 CH3OH      H —––— OH    or C8H18O7
  H —––— OH   ——––→         H —––— OH
     CH2OH      H+            CH2OH
```

This reaction does *not* occur; only *one mole* of methanol is consumed, showing that glucose is already in the hemiacetal form (eq. 13.7). The two products are the α- and β-glucosides (eq. 13.14).

CARBOHYDRATES

13.35 See eq. 13.10.

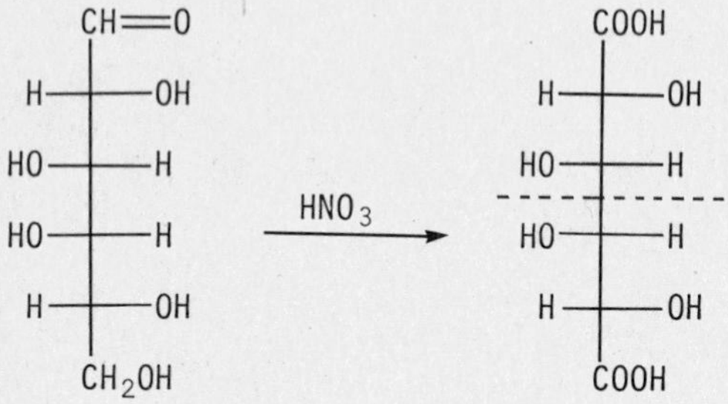

The dicarboxylic acid obtained as the product is a *meso* form,
with a plane of symmetry shown by the dashed line. Therefore, this
acid is achiral. (Experiments such as this were helpful in assigning
configurations to the various monosaccharides.)

13.36 a. For the structure of maltose, see Sec. 13.13a.

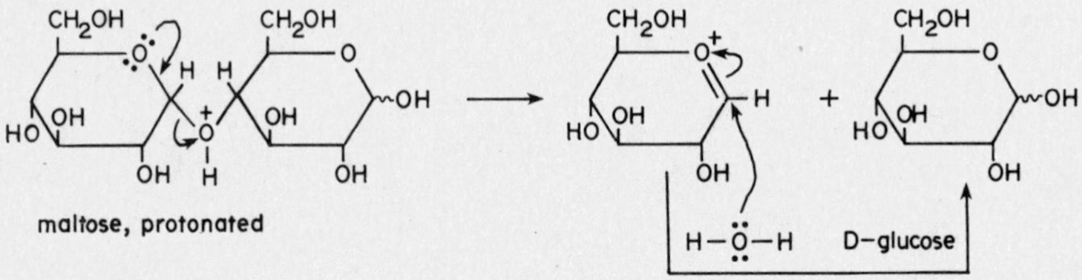

maltose, protonated

b. For the structure of lactose, see Sec. 13.13c.

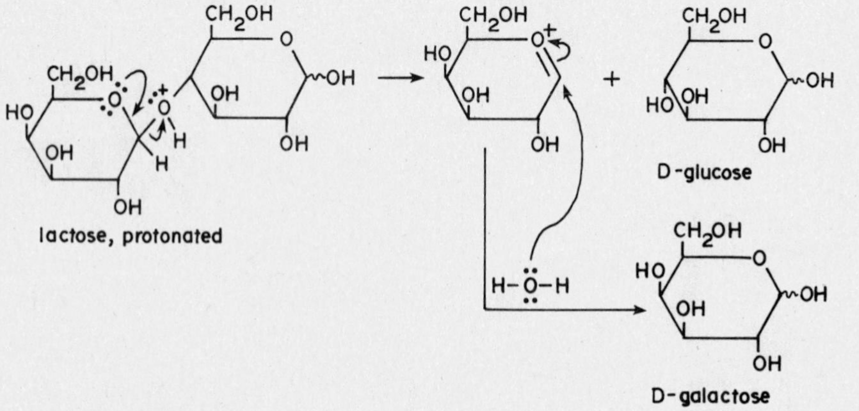

lactose, protonated

300

c. For the structure of sucrose, see Sec. 13.13d.

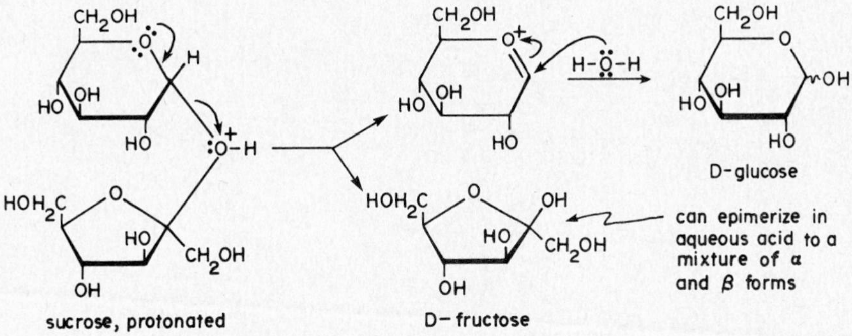

sucrose, protonated   D-fructose   D-glucose

can epimerize in aqueous acid to a mixture of α and β forms

The mechanism shown is one of two that are possible. The alternative mechanism would break the other glycoside bond to give glucose and the carbocation from the fructose unit. Both mechanisms undoubtedly occur simultaneously. The products, of course, are the same from both paths, namely a mixture of the α and β forms of D-glucose and D-fructose.

13.37 The formula for maltose is given in Sec. 13.13a.

a.

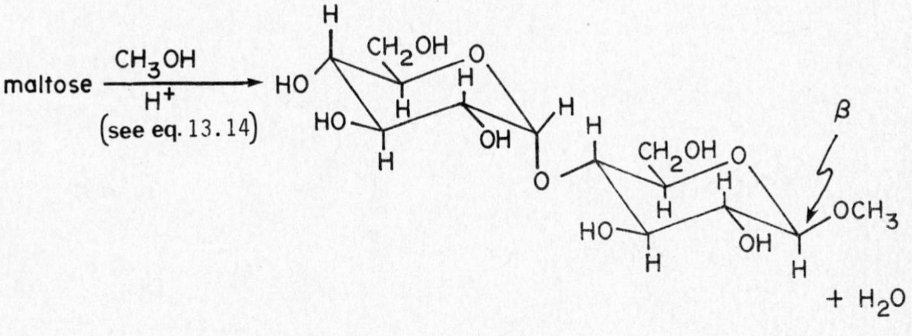

$$\text{maltose} \xrightarrow[\substack{H^+ \\ \text{(see eq. 13.14)}}]{CH_3OH}$$

+ H_2O

Only the β isomer is shown, but the α isomer will also be formed.

b.

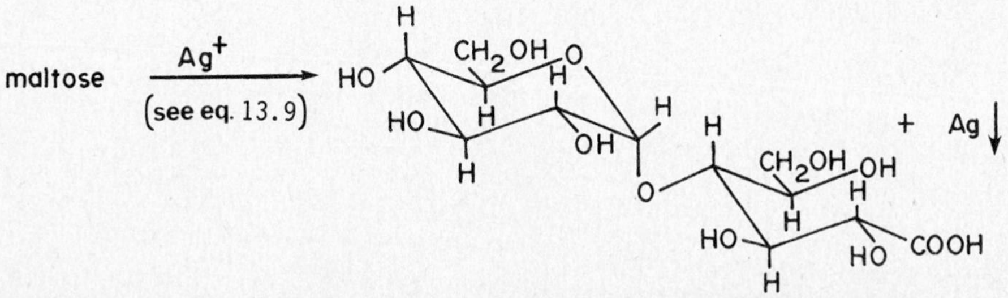

$$\text{maltose} \xrightarrow[\text{(see eq. 13.9)}]{Ag^+}$$

+ Ag↓

c. The reaction of maltose with bromine water will yield the same product as in part b (see eq. 13.9).

d.

maltose $\xrightarrow[\text{(see eq. 13.12)}]{\left(CH_3\overset{\overset{O}{\|}}{C}\right)_2 O}$

Eight equivalents of acetic anhydride are required to acetylate all eight hydroxyl groups of maltose.

13.38 a. Hydrolysis of trehalose gives two moles of D-glucose. From the "left" portion, we get

which is easily recognizable as D-glucose. From the "right" portion, we get

Turning this structure 180° in the plane of the paper to put the ring oxygen in the customary position gives

which is now also recognizable as D-glucose.

b. Since both anomeric carbons in trehalose are tied up in the glucosidic bond, no hemiacetal group remains in the structure. Therefore, equilibration with the aldehyde form and oxidation by Fehling's reagent are not possible. The test, as with sucrose, will be negative.

13.39 a. For the structure, see Sec. 13.13c (p. 369). The α form is shown. The β form is identical with it, except for the configuration at C-1 in the "right-hand" unit, where the OH group is equatorial instead of axial.

b. Follow example 13.4 as a guide for the calculation.

$$\% \text{ of } \beta = \frac{92.6 - 52}{92.6 - 34} \times 100 = \frac{40.6}{58.6} \times 100 = 69.3\%$$

The % of α is 100 - 69.3 = 30.7%

13.40 a. See eq. 13.12.
b. See eq. 13.9.
c. See eq. 13.11.

d.

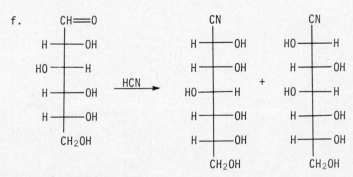

Compare with eq. 9.26.

e. See eq. 13.14.

f.

Compare with eq. 9.21.

g.

$$\underset{\text{(Fischer projection: CH=O, H—OH, HO—H, H—OH, H—OH, CH}_2\text{OH)}}{} + 2\ Cu^{2+} + 5\ OH^- \longrightarrow \underset{\text{(Fischer projection: CO}_2^-,\ H—OH, HO—H, H—OH, H—OH, CH}_2\text{OH)}}{} + Cu_2O + 3\ H_2O$$

Compare with eq. 9.34.

13.41 In sucrose, both anomeric carbons are involved in the glycosidic
bond. No hemiacetal function is present; both anomeric carbons are
in the acetal (ketal) form. Therefore, equilibration with an acyclic
aldehyde form is not possible, and the sugar cannot reduce Tollens',
Fehling's, or Benedict's reagent. In maltose (Sec. 13.13a), on the
other hand, carbon-1 of the "right-hand" glucose unit is a hemiacetal
carbon, in equilibrium with the open-chain aldehyde form, which can
reduce those same reagents.

13.42 Two strongly electron-withdrawing substituents are attached to the

amide nitrogen. They are the $\text{—C}{=}\text{O}$ and $\text{—S}\overset{\displaystyle O}{\underset{\displaystyle O}{\diagup\!\!\!\diagdown}}$ groups. Both

act to enhance saccharin's acidity.

The negative charge is not only delocalized to the carbonyl oxygen
(as with ordinary amide anions; see eq. 12.16), but is also stabilized
by the partial positive charge on the adjacent sulfur atom.

13.43

The enolate anion of 3-methyl-2-cyclohexenone attacks the carbonyl group of the acyclic unsaturated ketone to give the product.

13.44 a. Use the formula for cellulose (Fig. 13.5) as a guide, and replace the OH at C-2 by $NHCOCH_3$:

b.

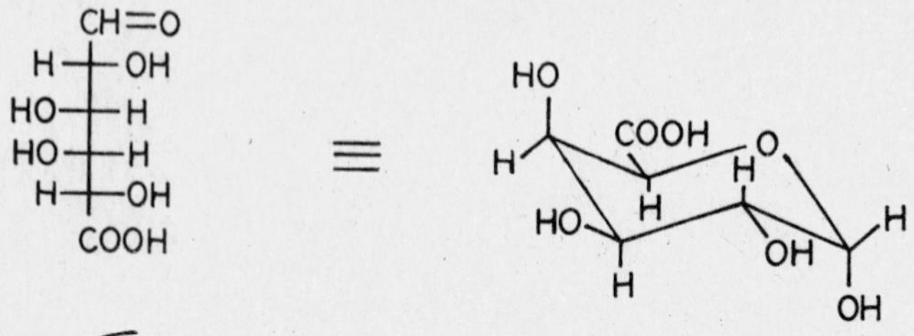

α–D–galacturonic acid

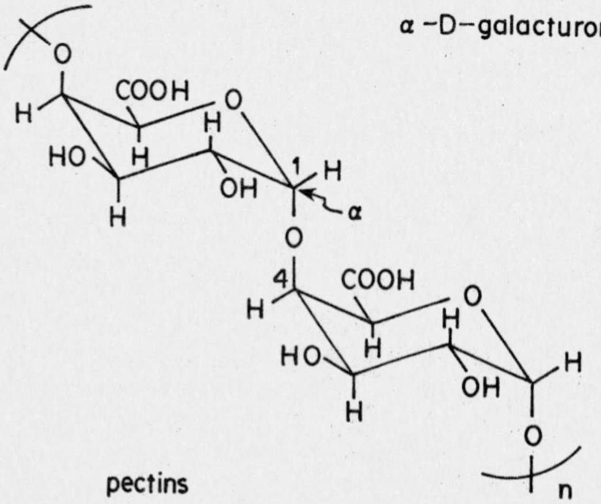

pectins

13.45 The formula of L-galactose can be obtained from that of D-galactose, which is shown in Figure 13.1. Deoxy sugars are discussed in Sec. 13.16.

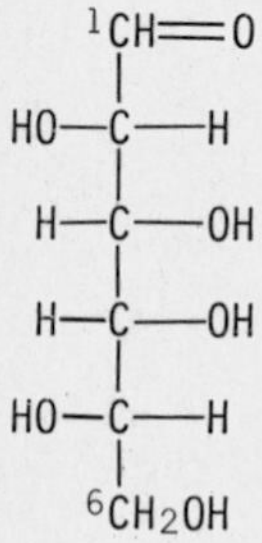

L-galactose

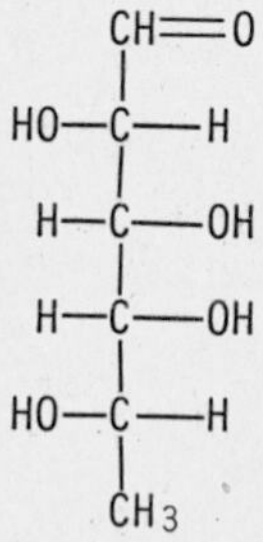

L-fucose
(6-deoxy-L-galactose)

13.46 See eq. 13.15 for the formula of ascorbic acid.

The negative charge can be almost equally spread over the two oxygen atoms.

13.47 The formula for β-D-xylopyranose, derived from the Fischer projection of D-xylose in Figure 13.1, is

Since the xylans have these units linked 1,4, their structure is

13.48 Be systematic in deducing all the structures.

The system to use in writing these formulas is to begin with all hydroxyl groups *cis*. Then we move one hydroxyl to the opposite face, two hydroxyls to the opposite face (1,2; 1,3; 1,4), and finally three hydroxyls to the opposite face (1,2,3; 1,2,4; 1,3,5).

CHAPTER FOURTEEN   AMINO ACIDS, PEPTIDES, AND PROTEINS

CHAPTER SUMMARY

Proteins are natural polymers composed of α-**amino acids** linked by **amide (peptide) bonds**. Except for **glycine** (aminoacetic acid), protein-derived amino acids are chiral and have the L-configuration. Table 14.1 lists the names, three-letter abbreviations, and structures of the twenty common amino acids. Of these, eight (the **essential amino acids**) cannot be synthesized in the bodies of adult humans and must be ingested in food. A measurement of the gradual change in optical activity of amino acids that remain in human or animal bones and teeth after death can be used in archaeological dating.

Amino acids with one amino and one carboxyl group exist as **dipolar ions**. Amino acids are **amphoteric**: in strong acid, they are protonated and become positively charged (ammonium carboxylic acids); in base, they lose a proton and become negatively charged (amine carboxylates). If placed in an electric field, amino acids migrate toward the cathode (negative electrode) at low pH and toward the anode (positive electrode) at high pH. The intermediate pH at which they do not migrate toward either electrode is the **isoelectric point**. (The isoelectric points for the twenty common amino acids are listed in Table 14.1). For amino acids with one amino group and one carboxyl group, the isoelectric pH is about 6. Amino acids with two carboxyl groups and one amino group have isoelectric points at a low pH (about 3), whereas amino acids with two amino groups and one carboxyl group have isoelectric points at a high pH (about 9). **Electrophoresis** is a process that makes use of the dependence of charge on pH to separate amino acids and proteins.

Amino acids undergo reactions that correspond to each functional group. In addition to exhibiting both acidic and basic behavior, for example, the carboxyl group can be esterified and the amino group can be acylated.

Amino acids react with **ninhydrin** to give a violet dye; this reaction is useful for detection and quantitative analysis of amino acids.

An amide bond between the carboxyl group of one amino acid and the amino group of another is called a **peptide bond**. It links amino acids to one another in peptides and proteins. The amino acid (often abbreviated **aa**) at one end of the peptide chain will have a free amino group (the **N-terminal amino acid**), and the amino acid at the other end of the chain (the **C-terminal amino acid**) will have a free carboxyl group. By convention, we write these structures from left to right starting at the N-terminal end.

Another type of covalent bond in proteins is the **disulfide bond,** formed by oxidative coupling of the —SH groups in **cysteine.** **Oxytocin** is an example of a cyclic peptide with an S——S bond.

By the **primary structure** of a peptide or protein, we mean its **amino acid sequence.** Complete hydrolysis gives the amino acid content. The N-terminal amino acid can be identified by the **Sanger method,** using **2,4-dinitrofluorobenzene.** The **Edman degradation** uses **phenyl isothiocyanate** to clip off one amino acid at a time from the N-terminal. Other reagents selectively cleave peptide chains at certain amino acid links. A combination of these methods can "sequence" a protein, and the methods have been automated.

Polypeptides are usually synthesized by the **Merrifield solid-phase technique.** An N-protected amino acid is linked by an ester bond to a benzyl chloride-type unit in a polystyrene. The protecting group is then removed, and the next N-protected amino acid is linked to the polymer-bound one. The cycle is repeated until the peptide is assembled, after which it is detached from the polymer. The most common N-protecting group is the *t*-**butoxycarbonyl (Boc) group:** $(CH_3)_3COC$——. It is connected
$$\overset{\|}{O}$$
to the amino group using di-*t*-butyldicarbonate and can be removed from the amino group by mild acid hydrolysis. The reagent used to attach each N-protected amino acid to the growing peptide chain is **dicyclohexylcarbodiimide.** Detachment of the peptide chain from the polymer is accomplished using HBr in trifluoroacetic acid.

Two features that affect **secondary protein structure** (molecular shape) include the rigid, planar geometry and restricted rotation of the peptide bond, and interchain or intrachain hydrogen bonding of the type $C{=}O\cdots H$——N. The α **helix** and the **pleated sheet** are common protein shapes.

Proteins may be **fibrous** or **globular.** The structure and polarity of the particular amino acid R groups and their sequence affect the solubility properties and **tertiary structure** of proteins. **Quaternary structure** refers to the aggregation of similar protein subunits.

REACTION SUMMARY

REACTION SUMMARY

Dissociation of Amino Acids

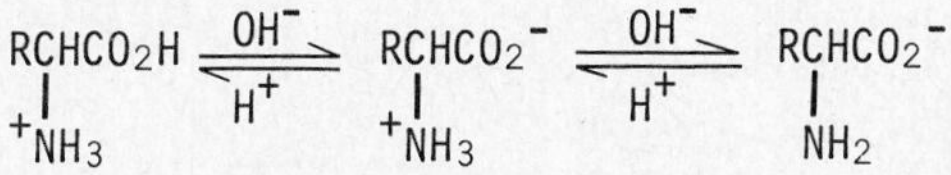

$$\underset{\substack{| \\ {}^{+}NH_3}}{RCHCO_2H} \underset{H^+}{\overset{OH^-}{\rightleftharpoons}} \underset{\substack{| \\ {}^{+}NH_3}}{RCHCO_2^-} \underset{H^+}{\overset{OH^-}{\rightleftharpoons}} \underset{\substack{| \\ NH_2}}{RCHCO_2^-}$$

dipolar ion

Esterification

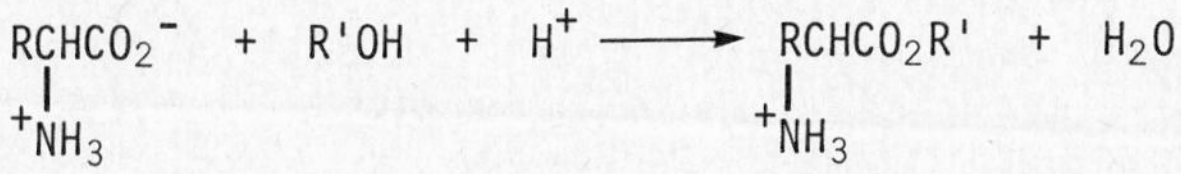

$$\underset{\substack{| \\ {}^{+}NH_3}}{RCHCO_2^-} + R'OH + H^+ \longrightarrow \underset{\substack{| \\ {}^{+}NH_3}}{RCHCO_2R'} + H_2O$$

Acylation

$$\underset{\substack{| \\ {}^{+}NH_3}}{RCHCO_2^-} + \underset{\substack{\| \\ O}}{R'C}\!-\!Cl \xrightarrow{2\ OH^-} \underset{\substack{| \\ R'CNH \\ \| \\ O}}{RCHCO_2^-} + 2\ H_2O + Cl^-$$

Ninhydrin Reaction

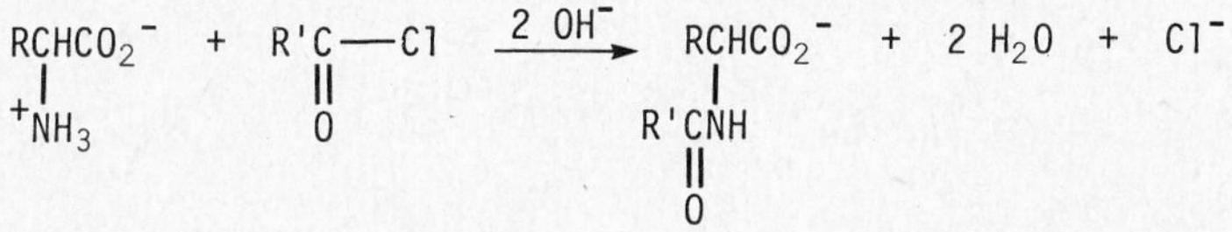

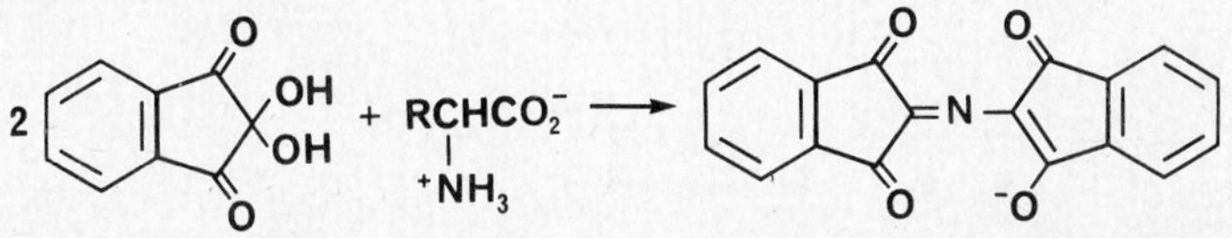

$$+\ RCHO + CO_2 + 3\ H_2O + H^+$$

Sanger's Reagent

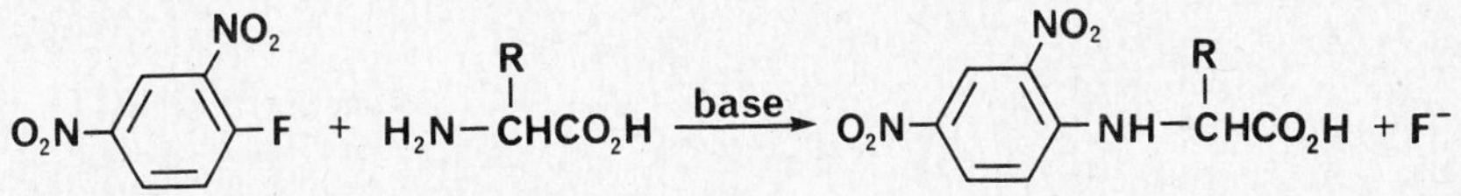

Edman Degradation

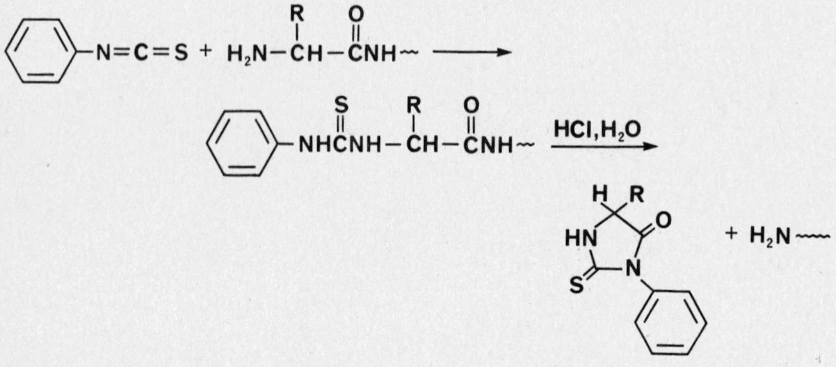
HCl,H₂O

Peptide Synthesis

a. N-protection:

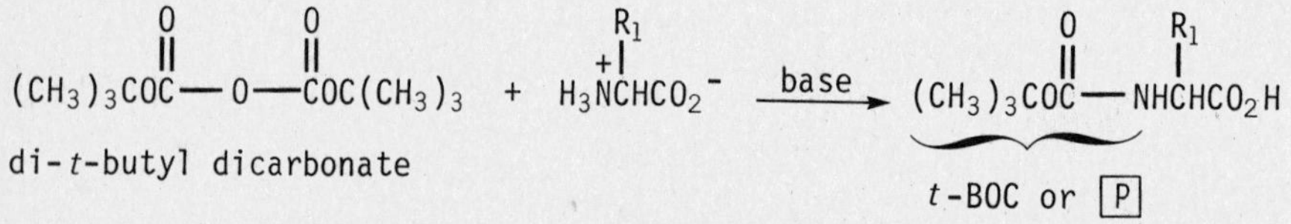
base
di-t-butyl dicarbonate
t-BOC or P

b. Polymer attachment:

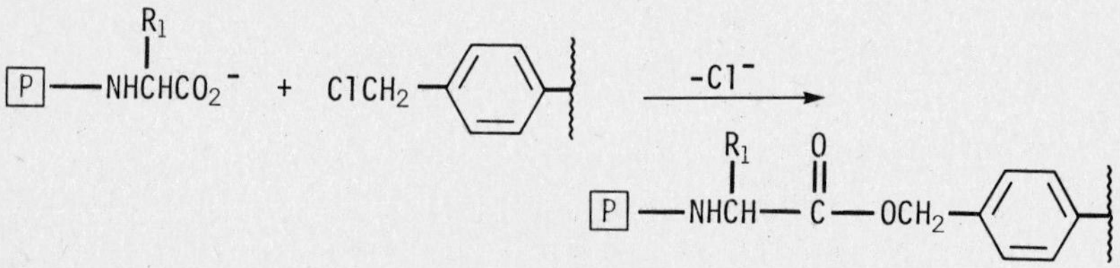

c. Deprotection (removal of the protecting group):

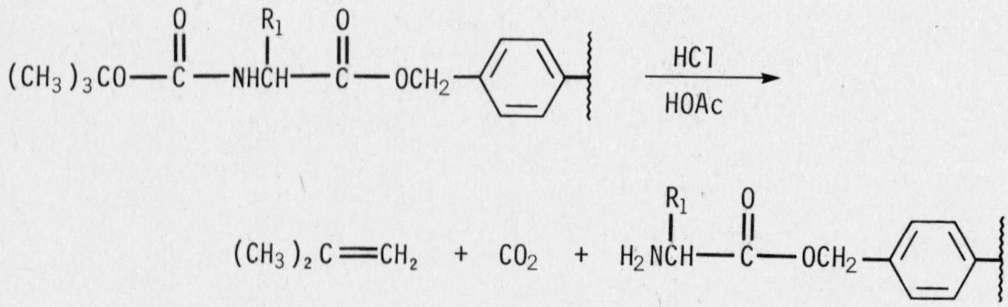
HCl
HOAc

d. <u>Amino acid coupling with DDC:</u>

$$\boxed{P}-NHC\overset{\overset{O}{\|}}{}-\overset{\overset{R_2}{|}}{CH}-CO_2^- \; + \; H_2NCH\overset{\overset{R_1}{|}}{}-\overset{\overset{O}{\|}}{C}-OCH_2-\bigcirc\!\!\dashv \qquad \bigcirc\!-N\!\!=\!\!C\!\!=\!\!N-\bigcirc \longrightarrow$$

$$\boxed{P}-NHC\overset{\overset{O}{\|}}{}-\overset{\overset{R_2}{|}}{CH}-\overset{\overset{O}{\|}}{C}-NHCH\overset{\overset{R_1}{|}}{}-\overset{\overset{O}{\|}}{C}-OCH_2-\bigcirc\!\!\dashv \; + \; \bigcirc\!-NHCNH\overset{\overset{O}{\|}}{}-\bigcirc$$

e. <u>Detachment from the polymer:</u>

$$H_2NC\overset{\overset{O}{\|}}{}-\overset{\overset{R_2}{|}}{CH}-\overset{\overset{O}{\|}}{C}-NHCH\overset{\overset{R_1}{|}}{}-\overset{\overset{O}{\|}}{C}-OCH_2-\bigcirc\!\!\dashv \xrightarrow[\text{CF}_3\text{CO}_2\text{H}]{\text{HBr}}$$

$$H_2NC\overset{\overset{O}{\|}}{}-\overset{\overset{R_2}{|}}{CH}-\overset{\overset{O}{\|}}{C}-NHCH\overset{\overset{R_1}{|}}{}-CO_2H \; + \; BrCH_2-\bigcirc\!\!\dashv$$

MECHANISM SUMMARY

<u>Nucleophilic Aromatic Substitution (Addition-Elimination)</u>

$$\bigcirc\!\!-X \; + \; Nu{:}^- \longrightarrow \left[\bigcirc\!\!\overset{X}{\underset{Nu}{\diagdown}} \right] \xrightarrow{-X^-} \bigcirc\!\!-Nu$$

Exemplified by Sanger's reaction (eq. 14.8) and facilitated by electron-withdrawing (carbanion-stabilizing) groups *ortho* or *para* to the leaving group X.

LEARNING OBJECTIVES

1. Know the meaning of: α amino acid, essential amino acid, dipolar ion, amphoteric, L configuration, isoelectric point, electrophoresis.

2. Learn the names, structures, and abbreviations of the amino acids listed in Table 14.1.

3. Write the structure of a given amino acid as a function of the pH of the solution.

4. Write an equation for the reaction of an amino acid (dipolar form) with strong acid or strong base.

313

5. Write the form of a given amino acid that is likely to be predominant at high or low pH, and at the isoelectric pH.

6. Given the isoelectric pH's of several amino acids, predict the directions of their migration during electrophoresis.

7. Given the isoelectric pH's of several amino acids, select a pH at which they can be separated by electrophoresis.

8. Write the equation for the reaction of a given amino acid with ninhydrin reagent.

9. Write the equations for the reaction of a given amino acid with (a) a given alcohol and H$^+$, and (b) acetic anhydride or another activated acyl derivative.

10. Know the meaning of: peptide bond; peptide, dipeptide, tripeptide, and so on; N-terminal and C-terminal amino acid; cysteine unit; disulfide bond.

11. Given the structures of the component amino acids and the name of a di-, tri-, or polypeptide, draw its structure.

12. Identify the N-terminal and C-terminal amino acid for a given peptide.

13. Given the three-letter abbreviated name for a peptide, write its structure.

14. Write the equation for the hydrolysis of a given di- or polypeptide.

15. Know the meaning of: amino acid sequence, Sanger's reagent, Edman degradation, selective peptide cleavage.

16. Write the equation for the reaction of an amino acid with 2,4-dinitrofluorobenzene.

17. Write the equations for the Edman degradation of a given tri- or polypeptide.

18. Given information on selective peptide cleavage and the sequences of fragment peptides, deduce the sequence of the original polypeptide.

19. Know the meaning of: protecting group, solid-phase technique, t-butoxycarbonyl (Boc) group, dicyclohexylcarbodiimide (DCC).

20. Write an equation for the protection of a given amino acid using di-t-butyl dicarbonate and for its deprotection using acid.

21. Write an equation for the linking of two protected amino acids with dicyclohexylcarbodiimide (DCC).

22. Write an equation for the linking of the N-protected C-terminal amino acid to a polymer for peptide synthesis, and write an equation for detachment of the peptide from the polymer.

23. Describe the geometry of the peptide bond.

24. Know the meaning of: primary, secondary, tertiary, and quaternary protein structure; α-helix; pleated sheet; fibrous and globular protein.

ANSWERS TO PROBLEMS

Problems Within the Chapter

14.1   These reactions occur if acid is added to the product of eq. 14.3. That is, they are essentially the reverse of the process shown in eqs. 14.3 and 14.2.

$$CH_3\underset{\underset{NH_2}{|}}{C}HCO_2^-Na^+ + HCl \longrightarrow CH_3\underset{\underset{^+NH_3}{|}}{C}HCO_2^- + Na^+Cl^-$$

$$CH_3\underset{\underset{^+NH_3}{|}}{C}HCO_2^- + HCl \longrightarrow CH_3\underset{\underset{^+NH_3}{|}}{C}HCO_2H + Cl^-$$

14.2   The $-CO_2H$ group is more acidic than the $-NH_3^+$ group. Thus, on treatment with base, a proton is removed from the $-CO_2H$ group (see eq. 14.2); then, with a second equivalent of base, a proton is removed from the $-NH_3^+$ group (see eq. 14.3).

14.3   The $-NH_2$ group is more basic than the $-CO_2^-$ group. Thus, on treatment with acid (as in Problem 14.1), a proton adds first to the $-NH_2$ group; then, with a second equivalent of acid, a proton adds to the $-CO_2^-$ group.

14.4 $-CO_2H > -NH_3^+ > -CO_2^- > -NH_2$
(See the answers to Problems 14.2 and 14.3.)

14.5 a.

$$\text{(phenyl)}-CH_2CHCO_2^-$$
$$\overset{|}{\underset{+NH_3}{}}$$

The dipolar ion will not migrate toward either electrode.

b. $CH_3S\!\!-\!\!CH_2CH_2\!\!-\!\!CHCO_2H$
$$\overset{|}{\underset{+NH_3}{}}$$

The ion will move toward the cathode (negative electrode).

c. $CH_2\!\!-\!\!CH\!\!-\!\!CO_2^-$
$$\overset{|}{OH}\quad\overset{|}{NH_2}$$

The ion will migrate toward the anode (positive electrode).

14.6 The other possible form, with a net charge of -1, is

$$^-O_2CCH_2CHCO_2^-$$
$$\overset{|}{\underset{+NH_3}{}}$$

It is the species formed when the dipolar ion form is treated with base.

14.7 pH 1: $CH_2CH_2CH_2CH_2CHCO_2H$
$$\overset{|}{\underset{+NH_3}{}}\qquad\overset{|}{\underset{+NH_3}{}}$$

pH 9.7:   Two dipolar forms can be drawn:

$$CH_2CH_2CH_2CH_2CHCO_2^-$$
$$\overset{|}{NH_2}\qquad\overset{|}{\underset{+NH_3}{}}$$

$$CH_2CH_2CH_2CH_2CHCO_2^-$$
$$\overset{|}{\underset{+NH_3}{}}\qquad\overset{|}{NH_2}$$

The first of these two forms predominates. The amino group at C-2 is more basic than the amino group at C-6, because of the electron-donating effect of the adjacent $-CO_2^-$ group.

pH 12: $CH_2CH_2CH_2CH_2CHCO_2^-$
$$\overset{|}{NH_2}\qquad\overset{|}{NH_2}$$

14.8   a. At pH 7.0, glycine (isoelectric point = 6.0) is present mainly
in the form $CH_2CO_2^-$ with some $CH_2CO_2^-$ and will migrate toward

$$\underset{^+NH_3}{|} \qquad\qquad \underset{NH_2}{|}$$

the anode (positive electrode), whereas lysine (isoelectric
point = 9.7) is present as

$$\underset{^+NH_3}{CH_2}\underset{}{CH_2}\underset{}{CH_2}\underset{^+NH_3}{CH_2}\underset{}{CHCO_2H} \quad \text{and} \quad \underset{^+NH_3}{CH_2}\underset{}{CH_2}\underset{}{CH_2}\underset{^+NH_3}{CH_2}\underset{}{CHCO_2^-}$$

and will migrate toward the cathode (negative electrode).

   b. At pH 6.0, phenylalanine (isoelectric point = 5.5) will migrate
toward the anode. Its structure is

$$\text{C}_6\text{H}_5\text{—CH}_2\underset{\underset{NH_2}{|}}{\text{CHCO}_2^-}$$

Leucine (isoelectric point = 6.0) will not migrate. Its
structure is

$$(CH_3)_2 CHCH_2 \underset{\underset{^+NH_3}{|}}{CHCO_2^-}$$

Proline (isoelectric point = 6.3) will migrate toward the
cathode. Its structure is

$$\begin{array}{c} CH_2 \text{—} CH \text{—} CO_2H \\ |\qquad | \\ CH_2 \quad {}^+NH_2 \\ \diagdown\;\diagup \\ CH_2 \end{array}$$

To generalize, if the solution pH is less than the isoelectric pH,
the net charge on the amino acid will be positive and it will migrate
toward the cathode; if the solution pH is greater than the isoelectric
pH, the net charge on the amino acid will be negative and it will
migrate toward the anode.

14.9   The isoelectric pH's are as follows: valine, 6.0; glutamic acid,
3.2; lysine, 9.7. At pH 2.0, all three migrate toward the cathode
(not necessarily at the same rate, however). At pH 6.0, glutamic
acid migrates toward the anode, lysine migrates toward the cathode,
and valine does not migrate. At pH 12.0, all three migrate toward
the anode (again, not necessarily at the same rate).

14.10 a.

C_6H_5—$\text{CH}_2\text{CHCO}_2^-$ + CH_3OH + HCl $\longrightarrow$
$\quad\quad\quad\quad\quad\quad\quad\quad\underset{^+\text{NH}_3}{\overset{|}{}}$

$\quad\quad\quad\quad\quad\quad$ C_6H_5—$\text{CH}_2\text{CHCO}_2\text{CH}_3$ + H_2O + Cl^-
$\quad\quad\quad\quad\quad\quad\quad\quad\quad\quad\quad\underset{^+\text{NH}_3}{\overset{|}{}}$

b. $(\text{CH}_3)_2\text{CHCHCO}_2^-$ + C_6H_5—COCl + $2\,\text{OH}^-$ $\longrightarrow$
$\quad\quad\quad\quad\underset{^+\text{NH}_3}{\overset{|}{}}$

$\quad\quad\quad\quad\quad\quad$ $(\text{CH}_3)_2\text{CHCHCO}_2^-$ + $2\,\text{H}_2\text{O}$ + Cl^-
$\quad\quad\quad\quad\quad\quad\quad\quad\quad\quad\overset{|}{}$
$\quad\quad\quad\quad\quad\quad\quad\quad\quad\quad\underset{\overset{||}{\text{O}}}{\text{HNC}}$—$\text{C}_6\text{H}_5$

c. $\underset{^+\text{NH}_3}{\overset{|}{\text{CH}_2\text{CO}_2^-}}$ + $\text{CH}_3\overset{\overset{\text{O}}{||}}{\text{C}}$—$\text{O}$—$\overset{\overset{\text{O}}{||}}{\text{C}}\text{CH}_3$ $\longrightarrow$ $\text{CH}_3\overset{\overset{\text{O}}{||}}{\text{C}}$—$\text{NHCH}_2\text{CO}_2\text{H}$ + $\text{CH}_3\text{CO}_2\text{H}$

14.11 Rewrite eq. 14.6, with $R = CH_3$.

14.12 a. $(\text{CH}_3)_2\text{CHCH}\overset{\overset{|}{\text{NH}_2}}{}$—$\overset{\overset{\text{O}}{||}}{\text{C}}$—$\text{NHCH}\overset{\overset{\text{CH}_3}{|}}{}\text{COOH}$

b. $\text{CH}_3\text{CHC}\overset{\overset{|}{\text{NH}_2}}{}$—$\overset{\overset{\text{O}}{||}}{}$—$\text{NHCH}\overset{\overset{\text{CH}(\text{CH}_3)_2}{|}}{}\text{COOH}$

14.13 $\underset{\text{NH}_2}{\overset{\overset{\text{O}}{||}}{\text{CH}_2\text{C}}}$—$\text{NHCH}\overset{\overset{\text{CH}_3}{|}}{}$—$\overset{\overset{\text{O}}{||}}{\text{C}}$—$\text{NHCH}\overset{\overset{\text{CH}_2\text{OH}}{|}}{}\text{COOH}$

Note that, by convention, in writing peptide structures, each

peptide bond is written in the —$\overset{\overset{\text{O}}{||}}{\text{C}}$—$\text{NH}$— direction.

14.14 Gly-Ser-Ala
       Ala-Gly-Ser
       Ala-Ser-Gly
       Ser-Gly-Ala
       Ser-Ala-Gly

There are six isomeric structures altogether, including Gly-Ala-Ser.

14.15 For alanylglycine, the reactions are as follows:

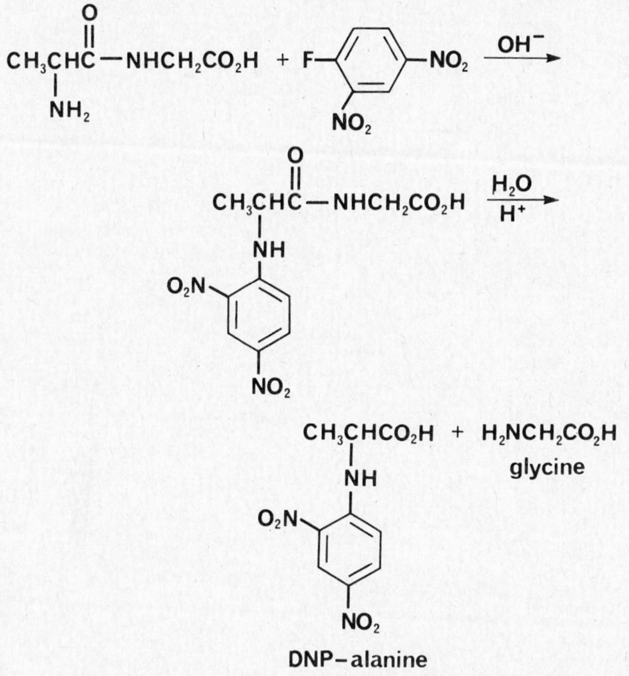

DNP–alanine

For glycylalanine, we have the following:

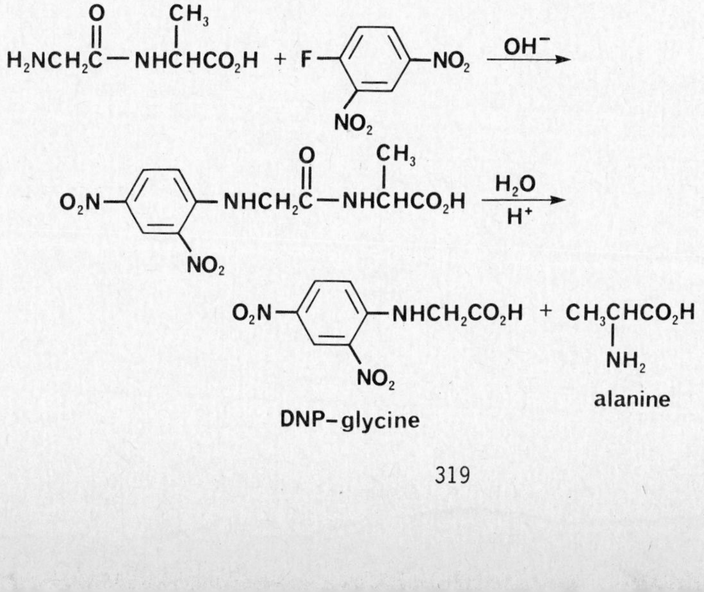

DNP–glycine

14.16 a. Trypsin cleaves peptides on the *carboxyl* side of lysine and
           arginine. Bradykinin contains no lysines and two arginines.
           Cleavage will occur after the first arginine (carboxyl side)
           but *not* in front of the second arginine (amino side). The
           cleavage products will be

           Arg and Pro-Pro-Gly-Phe-Ser-Pro-Phe-Arg

       b. Chymotrypsin will cleave bradykinin on the *carboxyl* sides of
           Phe to give three products:

           Arg-Pro-Pro-Gly-Phe and Ser-Pro-Phe and Arg

14.17 The A chain of insulin has two Tyr units (Fig. 14.7) and would be
       cleaved by chymotrypsin on the *carboxyl* side of these units, to give:

       Gly-Ile-Val-Glu-Gln-Cys-Cys-Ala-Ser-Val-Cys-Ser-Leu-Tyr
       and Gln-Leu-Glu-Asn-Tyr and Cys-Asn

Additional Problems

14.18 Definitions and/or examples are given in the text, where indicated.

       a. Sec. 14.7   b. Sec. 14.3
       c. Sec. 14.7   d. Sec. 14.2
       e. Sec. 14.2   f. Table 14.1, entries 1-5, 10
       g. Table 14.1, entries 6-9, 11-20 h. Sec. 14.3
       i. Sec. 14.3   j. Sec. 14.6

14.19

       The configuration is *S*.

14.20 Use Figure 14.1 and Table 14.1 as guides.

14.21 a. $CH_3CHCO_2^-$ + HCl $\longrightarrow$ CH_3CHCO_2H + Cl^-
       | |
       $^+NH_3$ $^+NH_3$

b. $CH_3CHCO_2^-$ + NaOH $\longrightarrow$ $CH_3CHCO_2^-Na^+$ + H_2O
   | |
   $^+NH_3$ NH_2

14.22 a. $(CH_3)_2CHCHCO_2^-$ b. $HOCH_2CHCO_2^-$
                  | |
                  $^+NH_3$ $^+NH_3$

c. CH_2—$CHCO_2^-$ d. HO—⟨⟩—$CH_2CHCO_2^-$
   | | |
   CH_2 $^+NH_2$ $^+NH_3$
      \   /
      CH_2

14.23 a. The most acidic proton is on the carboxyl group nearest the
         —$^+NH_3$ substituent.

b. The ammonium ion is more acidic than the alcohol function:

$HOCH_2CHCO_2^-$
       |
       $^+NH_3$ ⟵ most acidic

c. The carboxyl group is a stronger acid than the —$^+NH_3$ group.

d. The proton on the positive nitrogen is the only appreciably
   acidic proton.

14.24 a. The most basic group is CO_2^-; the product of protonation is

CH_3CH—$CHCO_2H$
     | |
     OH $^+NH_3$

b. The carboxylate ion remote from the —$^+NH_3$ group is most basic.
   The product of protonation is

$HO_2CCH_2CHCO_2^-$
            |
            $^+NH_3$

14.25 Because of its positive charge, the —$^+NH_3$ group is an electron-
      attracting substituent. As such, it enhances the acidity of the
      nearby carboxyl group (in protonated alanine).

14.26 $HO_2CCH_2CH_2CHCO_2H$ $\underset{\overset{pH}{2.19}}{\rightleftharpoons}$ $HO_2CCH_2CH_2CHCO_2^-$ $\underset{\overset{pH}{4.25}}{\rightleftharpoons}$

           $\overset{|}{+NH_3}$ $\overset{|}{+NH_3}$

          $^-O_2CCH_2CH_2CHCO_2^-$ $\underset{\overset{pH}{9.67}}{\rightleftharpoons}$ $^-O_2CCH_2CH_2CHCO_2^-$

                   $\overset{|}{+NH_3}$ $\overset{|}{NH_2}$

14.27   The proton adds to the imino nitrogen ($HN{=}C{<}$) to give a resonance-stabilized cation:

14.28

The order of basicities (and therefore the sequence of protonation of groups) is guanidine > $-NH_2$ > $-CO_2^-$.

14.29   In strong acid (pH 1), histidine has two resonance contributors:

14.30   a.   Use eq. 14.4 with R = CH_3 and R' = CH_3CH_2-.

     CH_3CHCO_2H + CH_3CH_2OH $\xrightarrow{H^+}$ $CH_3CHCO_2CH_2CH_3$ + H_2O

      $\overset{|}{NH_2}$ $\overset{|}{NH_2}$

b. Use eq. 14.5 with R = CH$_3$ and R' =

$$CH_3CHCO_2H \ + \ \text{[benzoyl chloride]} \xrightarrow{OH^-} CH_3CHCO_2H \ + \ H_2O + Cl^-$$

with the amino acid bearing NH$_2$ on the left, and NHC(=O)-phenyl on the product.

c.
$$CH_3CHCO_2H \ + \ (CH_3CO)_2O \longrightarrow CH_3CHCO_2H \ + \ CH_3CO_2H$$
left reactant has NH$_2$; product has NHCCH$_3$ with carbonyl O.

14.31 a. The amino group and the hydroxyl group react with acetic anhydride.

$$CH_2—CHCO_2H \ + \ 2 \ CH_3C—O—CCH_3 \longrightarrow$$
with OH and NH$_2$ substituents, and each acetyl group bearing O.

$$CH_3C—O \quad CH_2—CH—CO_2H \ + \ 2 \ CH_3CO_2H$$
with NHCCH$_3$ (amide) substituent and carbonyl O.

ester   amide

b. The phenolic ring is easily brominated (see Sec. 7.16, eq. 7.39).

$$HO-\text{[benzene ring]}-CH_2CHCO_2H \xrightarrow{2 \ Br_2}$$
with NH$_2$ substituent.

$$HO-\text{[dibromo benzene ring]}-CH_2CHCO_2H \ + \ 2HBr$$
with Br at two ring positions and NH$_2$ substituent.

c. Both the hydroxyl and amino groups react with benzoyl chloride.

$$CH_3CH-CHCO_2H \;+\; 2 \;\; \underset{}{\text{C}_6\text{H}_5}-\overset{\overset{\text{O}}{\|}}{C}Cl \longrightarrow$$

with OH and NH$_2$ substituents

$$CH_3CH-CHCO_2H \;+\; 2\,HCl$$

d. Both carboxyl groups are esterified.

$$HOOC-CH_2CH_2-\underset{\underset{NH_2}{|}}{CH}COOH \;+\; 2\,CH_3OH \xrightarrow{\;H^+\;}$$

$$CH_3O-\overset{\overset{\text{O}}{\|}}{C}-CH_2CH_2\underset{\underset{NH_2}{|}}{CH}-\overset{\overset{\text{O}}{\|}}{C}-OCH_3 \;+\; 2\,H_2O$$

e. The amide group is hydrolyzed, and the carboxyl group is converted to its salt.

$$H_2N-\overset{\overset{\text{O}}{\|}}{C}-CH_2CH_2\underset{\underset{NH_2}{|}}{CH}COOH \xrightarrow[\text{heat}]{NaOH} {}^-O-\overset{\overset{\text{O}}{\|}}{C}-CH_2CH_2\underset{\underset{NH_2}{|}}{CH}\overset{\overset{\text{O}}{\|}}{C}-O^- \;+\; NH_3$$

14.32 Follow eq. 14.6, with R = $C_6H_5-CH_2-$.

14.33 Review Sec. 14.7 if necessary, and use Table 14.1 for the structures of the amino acids. Write the structures in neutral form, recognizing of course that dipolar ion structures are possible and that the exact form and degree of ionization depend on the pH of the solution.

a. $\quad H_2N\underset{\underset{CH_3}{|}}{CH}\overset{\overset{\text{O}}{\|}}{C}-NH\underset{\underset{CH_3}{|}}{CH}CO_2H$

b.

$$
\underset{\underset{CH(CH_3)_2}{|}}{H_2NCHC} \overset{\overset{O}{\|}}{\underset{}{}} \!\!-\!\! \underset{\underset{CH_2-\text{(indole)}}{|}}{NHCHCO_2H}
$$

c.

$$
\underset{\underset{CH_2-\text{(indole)}}{|}}{H_2NCHC} \overset{\overset{O}{\|}}{\underset{}{}} \!\!-\!\! \underset{\underset{CH(CH_3)_2}{|}}{NHCHCO_2H}
$$

d. $\underset{}{H_2NCH_2C} \overset{\overset{O}{\|}}{} \!-\! \underset{\underset{CH_3}{|}}{NHCHC} \overset{\overset{O}{\|}}{} \!-\! NHCH_2CO_2H$

e. $\underset{\underset{CH_2OH}{|}}{H_2NCHC} \overset{\overset{O}{\|}}{} \!-\! \underset{\underset{CH_2}{|}\ \underset{CH(CH_3)_2}{|}}{NHCHC} \overset{\overset{O}{\|}}{} \!-\! \underset{\underset{CH_2CH_2CH_2NHC=\!\!=NH}{|}\ \underset{NH_2}{|}}{NHCHCO_2H}$

f.

$$
\underset{\underset{CH_2-\text{(imidazole, HN\ N)}}{|}}{H_2NCHC} \overset{\overset{O}{\|}}{} \!-\! NHCH_2C \overset{\overset{O}{\|}}{} \!-\! NHCH_2C \overset{\overset{O}{\|}}{} \!-\! \underset{\underset{CH_2CH_2CO_2H}{|}}{NHCHCO_2H}
$$

14.34 a. $\underset{\underset{NH_2}{|}}{(CH_3)_2CHCH_2CHC} \overset{\overset{O}{\|}}{} \!-\! \underset{\underset{CH_2OH}{|}}{NHCHCO_2H} \quad \xrightarrow[\;H^+\;]{\;H_2O\;} \quad \underset{\underset{NH_2}{|}}{(CH_3)_2CHCH_2CHCO_2H} \;+\; \underset{\underset{CH_2OH}{|}}{H_2NCHCO_2H}$

$\qquad\qquad\qquad\qquad\qquad\qquad\qquad\qquad\qquad\qquad\qquad\qquad\qquad\qquad\qquad\quad$ Leu $\qquad\qquad\qquad\qquad$ Ser

b.

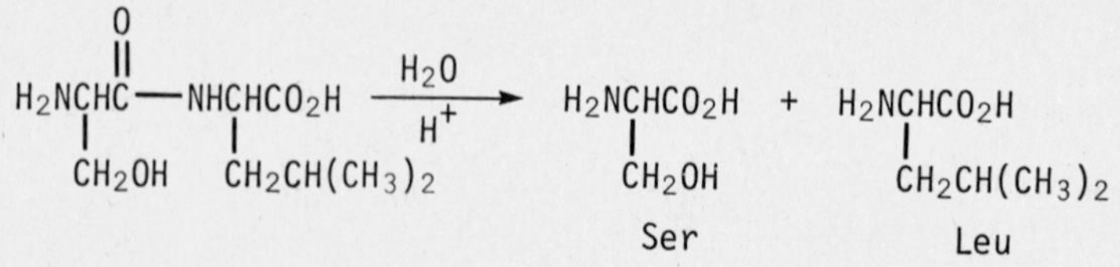

$$\underset{\substack{|\\CH_2OH}}{H_2NCHC}\overset{\overset{O}{\|}}{-}\underset{\substack{|\\CH_2CH(CH_3)_2}}{NHCHCO_2H} \xrightarrow[H^+]{H_2O} \underset{\substack{|\\CH_2OH}}{H_2NCHCO_2H} + \underset{\substack{|\\CH_2CH(CH_3)_2}}{H_2NCHCO_2H}$$

Ser   Leu

The hydrolysis products in parts a and b are identical.

c.

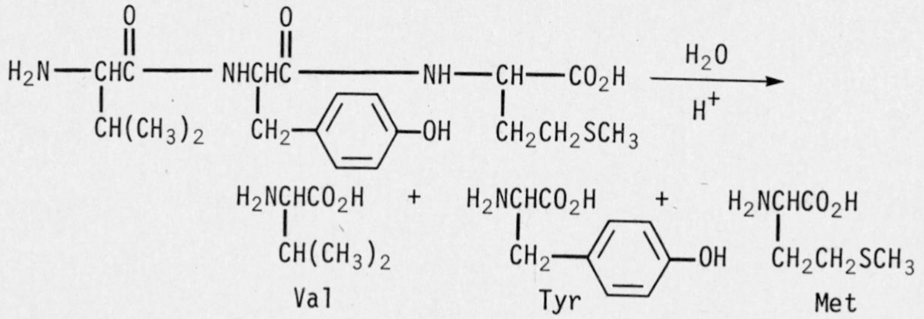

$$H_2N\underset{\substack{|\\CH(CH_3)_2}}{-CHC}\overset{\overset{O}{\|}}{-}\underset{\substack{|\\CH_2-\bigcirc-OH}}{NHCHC}\overset{\overset{O}{\|}}{-}\underset{\substack{|\\CH_2CH_2SCH_3}}{NH-CH-CO_2H} \xrightarrow[H^+]{H_2O}$$

$$\underset{\substack{|\\CH(CH_3)_2}}{H_2NCHCO_2H} + \underset{\substack{|\\CH_2-\bigcirc-OH}}{H_2NCHCO_2H} + \underset{\substack{|\\CH_2CH_2SCH_3}}{H_2NCHCO_2H}$$

Val   Tyr   Met

14.35

$$\underset{\text{most acidic (pH 1)}}{\overset{+}{N}H_3\underset{\substack{|\\CH_3}}{-CH}\overset{\overset{O}{\|}}{-C}-NHCH_2CO_2H} \xrightarrow{^-OH} \underset{\text{about pH 6 (dipolar ion)}}{\overset{+}{N}H_3\underset{\substack{|\\CH_3}}{-CH}\overset{\overset{O}{\|}}{-C}-NHCH_2CO_2^-}$$

$$\text{OH}^-$$

$$\underset{\text{most basic (pH 10)}}{NH_2\underset{\substack{|\\CH_3}}{-CH}\overset{\overset{O}{\|}}{-C}-NHCH_2CO_2^-} \longleftarrow$$

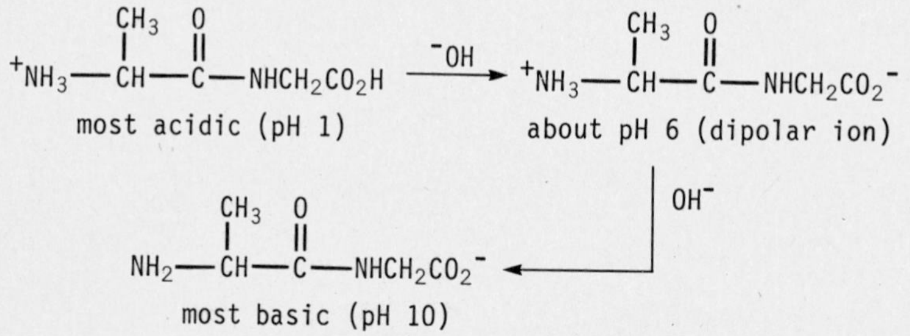

The behavior of a simple dipeptide like this, with no acidic or basic functions in the R group of each unit, is very much like that of a simple amino acid.

14.36 With four different amino acids there are 4!, or 4 x 3 x 2 x 1 = 24, possible structures. They are as follows:

Gly-Ala-Val-Leu   Ala-Gly-Val-Leu
Gly-Ala-Leu-Val   Ala-Gly-Leu-Val
Gly-Val-Ala-Leu   Ala-Val-Gly-Leu
Gly-Val-Leu-Ala   Ala-Val-Leu-Gly
Gly-Leu-Ala-Val   Ala-Leu-Gly-Val
Gly-Leu-Val-Ala   Ala-Leu-Val-Gly

Val-Gly-Ala-Leu	Leu-Gly-Ala-Val
Val-Gly-Leu-Ala	Leu-Gly-Val-Ala
Val-Ala-Gly-Leu	Leu-Ala-Gly-Val
Val-Ala-Leu-Gly	Leu-Ala-Val-Gly
Val-Leu-Gly-Ala	Leu-Val-Gly-Ala
Val-Leu-Ala-Gly	Leu-Val-Ala-Gly

14.37 Hydrogen peroxide oxidizes thiols to disulfides (Sec. 7.18, eq. 7.44). See also eq. 14.7.

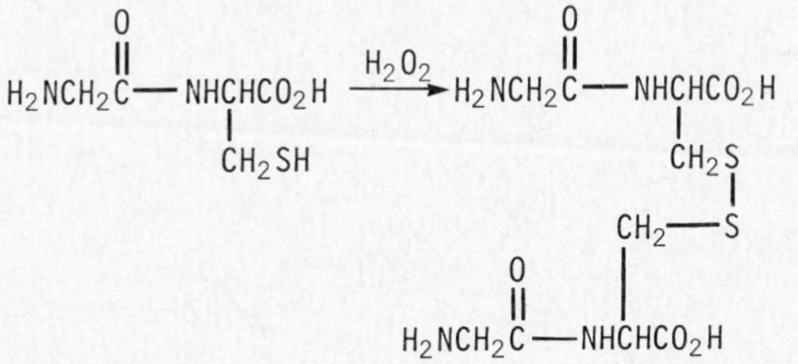

$$H_2NCH_2\overset{O}{\overset{\|}{C}}-NHCHCO_2H \xrightarrow{H_2O_2} H_2NCH_2\overset{O}{\overset{\|}{C}}-NHCHCO_2H$$

$$\underset{CH_2SH}{} \quad \underset{\underset{CH_2-S}{CH_2S}}{}$$

$$H_2NCH_2\overset{O}{\overset{\|}{C}}-NHCHCO_2H$$

14.38 a. Use eq. 14.8, with R = H.

b. Lysine contains two primary amino groups. Each group can react with 2,4-dinitrofluorobenzene:

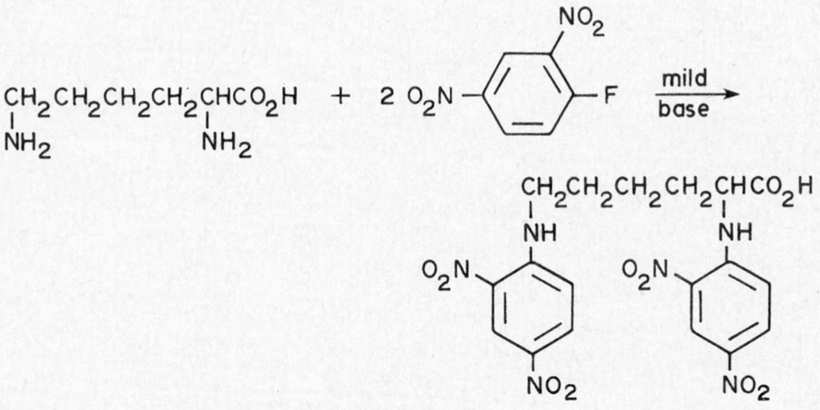

$$\underset{NH_2}{CH_2}CH_2CH_2CH_2\underset{NH_2}{CHCO_2H} + 2\ O_2N-\underset{}{\bigcirc}-F \xrightarrow[base]{mild}$$

14.39 The first result tells us that the N-terminal amino acid is methionine. The structure at this point is

Met(2 Met, Ser, Gly)

From dipeptide D, we learn that one sequence is

Ser-Met

From dipeptide C, we learn that one sequence is
Met—Met

From tripeptide B, we learn that one sequence is
Met(Met, Ser)

and in view of the result from dipeptide D, tripeptide B must be
Met-Ser-Met

Since two methionines must be adjacent (dipeptide C), the possibilities are
Met-Met-Ser-Met-Gly and Met-Ser-Met-Met-Gly

Tripeptide A allows a decision, since two methionines and one glycine must be joined together. Therefore, the correct structure is

Met-Ser-Met-Met-Gly

14.40 The first step in an Edman degradation involves nucleophilic addition to the C$\equiv$S bond of phenyl isothiocyanate. The reaction is analogous to the addition of alcohols to isocyanates discussed in A Word About 24, p. 332.

$$R-NH_2 + C_6H_5-N=C=S \longrightarrow C_6H_5-N=C-NH_2R \; (S^-)$$

terminal amino group of peptide

$$C_6H_5-N-C(=S)-NH_2R \xrightarrow{-H^+} C_6H_5-NH-C(=S)-NHR$$

The product is a thiourea (compare its structure with the structure of urea on p. 280 of the text).

14.41 Follow the scheme in Figure 14.6 (p. 396), with R_1 = CH_3, R_2 = H, and R_3 = $(CH_3)_2CH$, and with an —OH attached to the carbonyl group at the right. The final products are the phenylthiohydantoin of alanine and the dipeptide glycylvaline.

14.42

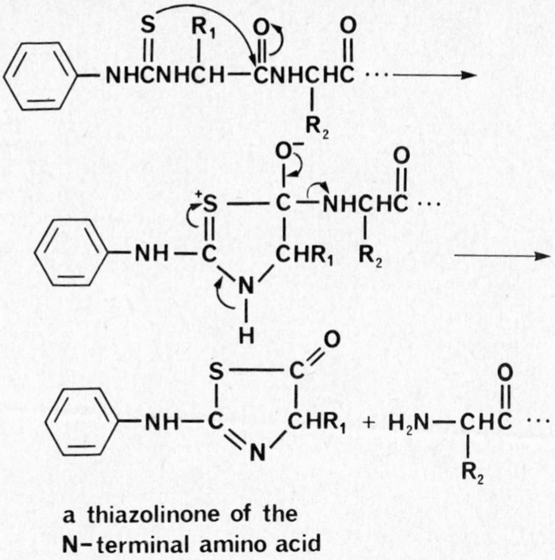

**a thiazolinone of the
N-terminal amino acid**

One possible mechanism for rearrangement to the hydantoin is the
following:

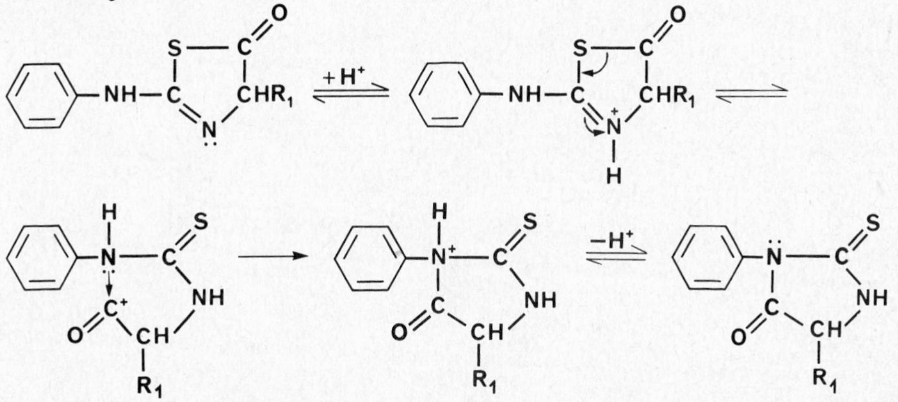

14.43 The hydrolysis products must overlap as shown below:

```
Ala-Gly
    Gly-Val
    Gly-Val-Tyr
        Val-Tyr-Cys
            Tyr-Cys-Phe
                Cys-Phe-Leu
                    Phe-Leu-Try
```

The peptide must be Ala-Gly-Val-Tyr-Cys-Phe-Leu-Try, an octapeptide.
The N-terminal amino acid is alanine, the C-terminal one is
tryptophan, and the name is alanylglycylvalyltyrosylcystylphenyl-
alanylleucyltryptophan.

14.44 The point of this problem is to illustrate that it is deceptively easy to use the three-letter abbreviations in designating peptide structures. Although Try-Gly-Gly-Phe-Met appears quite simple, the full structure is not. With a molecular formula of $C_{29}H_{36}O_6N_6S$, this pentapeptide has a molecular weight of 596 and is quite a complex polyfunctional molecule.

14.45 Trypsin cleaves peptides at the carboxyl side of Lys and Arg. The only fragment that does not end in Lys is the dipeptide. Therefore, it must come at the C-terminal end of endorphin, and the C-terminal amino acid is Gln.

Cyanogen bromide cleaves peptides at the carboxyl side of Met. There is only one Met in the structure (see the fifth trypsin fragment), and the cyanogen bromide cleavage gives a hexapeptide consisting of the first six amino acids of this fragment. The N-terminal amino acid must be Tyr. We now know the first ten and the last two amino acids of β-endorphin.

Chymotrypsin cleaves peptides at the carboxyl side of Phe, Tyr, and Trp. Working backwards in the 15-unit fragment, we see that it starts at the ninth amino acid in the last fragment from the trypsin digestion. We see that the first amino acid in this fragment (Ser) must be connected to the Lys at the end of the fifth fragment from the trypsin cleavage. Since this fragment begins with the N-terminal amino acid, we now know the sequence of the first twenty and the last two amino acids. The partial structure of β-endorphin must be

Tyr-Gly-Gly-Phe-Leu-Met-Thr-Ser-Glu-Lys-Ser-Gln-Thr-Pro┐

┌ Leu-Val-Thr-Leu-Phe-Lys-(Lys)(Asn-Ala-His-Lys)(Asn-Ala┐

┌ Ile-Val-Lys)-Gly-Gln

We cannot deduce, from the data given, the sequence of the amino acids in parentheses. To put them in order we must seek peptides from partial hydrolysis that contain overlapping sequences of these amino acids.

14.46 The nucleophile is the carboxylate anion; the leaving group is chloride ion from the benzyl chloride. Benzyl halides react rapidly in S_N2 displacements.

14.47 The protonated peptide is the leaving group; the nucleophile is bromide ion.

14.48 Base removes a proton from the —$\overset{+}{N}H_3$ group, thus generating the necessary nucleophile:

$$H_3\overset{+}{N}-\underset{\underset{R}{|}}{CH}-CO_2^- \;+\; OH^- \;\rightleftharpoons\; H_2\ddot{N}-\underset{\underset{R}{|}}{CH}-CO_2^- \;+\; H_2O$$

$$(CH_3)_3CO-\overset{O}{\overset{||}{C}}-O-\overset{O}{\overset{||}{C}}-OC(CH_3)_3 \;+\; H_2\ddot{N}-\underset{\underset{R}{|}}{CH}-CO_2^- \;\longrightarrow$$

$$(CH_3)_3CO-\overset{O}{\overset{||}{C}}-O-\underset{\underset{OC(CH_3)_3}{|}}{\overset{\overset{O^-}{|}}{C}}-\overset{+}{N}H_2-CHCO_2^- \;\longrightarrow\; (CH_3)_3COC-\overset{O}{\overset{||}{}}\overset{+}{N}H_2\underset{\underset{R}{|}}{CHCO_2^-}$$

The other fragment of this step is

$(CH_3)_3CO-\overset{O}{\overset{||}{C}}-OH$, which is decarboxylated giving CO_2 and $(CH_3)_3COH$.

$$(CH_3)_3C\overset{O}{\overset{||}{O}}C-NHCHCO_2H \quad \overset{R}{}$$
$$\underbrace{\qquad\qquad\qquad}_{\boxed{P}}$$

14.49 First, protect the C-terminal amino acid (alanine):

$$(CH_3)_3CO-\overset{O}{\overset{||}{C}}\diagdown\atop(CH_3)_3CO-\overset{O}{\underset{||}{C}}\diagup O \;+\; H_3\overset{+}{N}-\underset{\underset{CH_3}{|}}{CH}-CO_2^- \quad\xrightarrow[\text{(see eq. 14.14)}]{\text{base}}$$

$$(CH_3)_3COCNHCHCO_2H \quad \overset{O\;\;CH_3}{\overset{||\;\;|}{}}$$

Then attach it to the polymer:

$$(CH_3)_3COCNHCHCO_2^- \overset{O\;\;CH_3}{\overset{||\;\;|}{}} \;+\; ClCH_2-\!\!\left\langle\!\!\bigcirc\!\!\right\rangle\!\!\sim \quad\xrightarrow[\text{(see eq. 14.11)}]{-Cl^-}$$

$$(CH_3)_3COCNHCHCOCH_2-\!\!\left\langle\!\!\bigcirc\!\!\right\rangle\!\!\sim \overset{O\;\;CH_3}{\overset{||\;\;|}{}}\underset{O}{\overset{||}{}}$$

Next, deprotect the attached alanine:

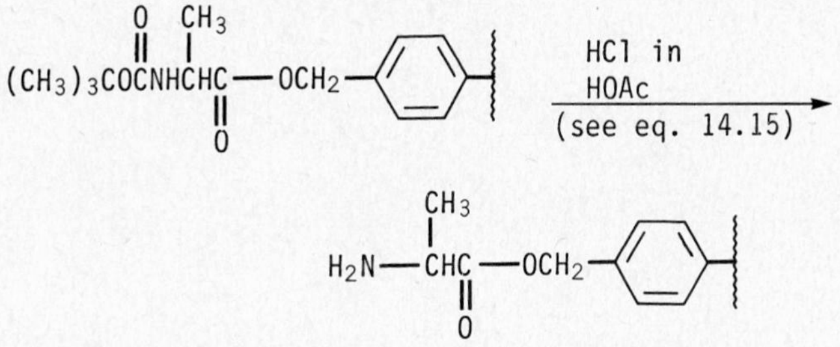
HCl in
HOAc
(see eq. 14.15)

Couple the protected (as in eq. 14.14) glycine to the deprotected, polymer-bound alanine:

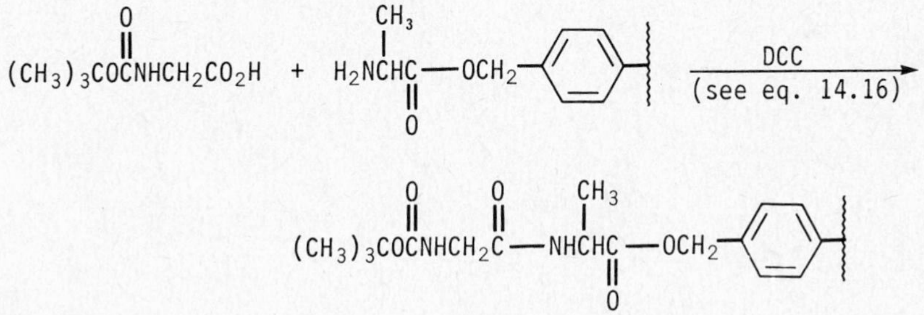
DCC
(see eq. 14.16)

Then deprotect again, couple the protected alanine, and deprotect a final time:

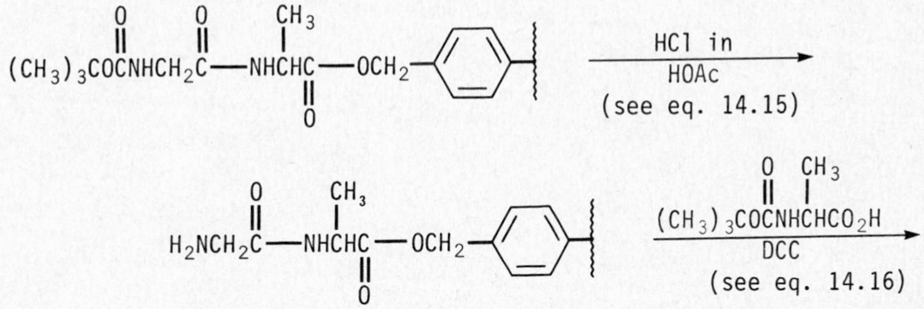
HCl in
HOAc
(see eq. 14.15)
DCC
(see eq. 14.16)

$$(CH_3)_3COC\overset{O}{\underset{\parallel}{}}-NHCH\overset{CH_3}{\underset{|}{}}C\overset{O}{\underset{\parallel}{}}-NHCH_2C\overset{O}{\underset{\parallel}{}}-NHCH\overset{CH_3}{\underset{|}{}}C\overset{O}{\underset{\parallel}{}}-OCH_2-\underset{}{\bigcirc}\!\!\!\xrightarrow[\substack{HOAc \\ (see\ eq.\ 14.15)}]{HCl\ in}$$

$$H_2NCH\overset{CH_3}{\underset{|}{}}C\overset{O}{\underset{\parallel}{}}-NHCH_2C\overset{O}{\underset{\parallel}{}}-NHCH\overset{CH_3}{\underset{|}{}}C\overset{O}{\underset{\parallel}{}}-OCH_2-\bigcirc$$

At this stage, the tripeptide is constructed, but it must be detached
from the polymer:

$$H_2NCH\overset{CH_3}{\underset{|}{}}C\overset{O}{\underset{\parallel}{}}-NHCH_2C\overset{O}{\underset{\parallel}{}}-NHCH\overset{CH_3}{\underset{|}{}}C\overset{O}{\underset{\parallel}{}}-OCH_2-\bigcirc\!\!\!\xrightarrow[\substack{CF_3CO_2H \\ (See\ eq.\ 14.13)}]{HBr}$$

$$H_2NCH\overset{CH_3}{\underset{|}{}}C\overset{O}{\underset{\parallel}{}}-NHCH_2C\overset{O}{\underset{\parallel}{}}-NHCH\overset{CH_3}{\underset{|}{}}-C\overset{O}{\underset{\parallel}{}}-OH\ +\ BrCH_2-\bigcirc$$

Ala-Gly-Ala

This problem, which involves only the construction of a simple
tripeptide, should give you some sense of the number of operations
required in solid-phase peptide syntheses.

14.50
$$\left[\ H_2NCH_2C\overset{O}{\underset{\parallel}{}}-NHCH_2CO_2H\ \longleftrightarrow\ H_2NCH_2C\overset{O^-}{\underset{+}{}}=NHCH_2CO_2H\ \right]$$

Rotation is restricted at this bond.

The six atoms that lie in one plane are

$$\underset{\underset{C}{\overset{|}{N}}\overset{\diagup H}{}}{\overset{\displaystyle C\overset{O}{\overset{\parallel}{}}}{}}$$

14.51 Use Table 14.2 as a guide.

    a. With trypsin, we expect four fragments:

        His-Ser-Glu-Gly-Thr-Phe-Thr-Ser-Asp-Tyr-Ser-Lys
        Tyr-Leu-Asp-Ser-Arg
        Arg
        Ala-Gln-Asp-Phe-Val-Gln-Trp-Leu-Met-Asn-Thr

    b. With chymotrypsin, we expect six fragments:

        His-Ser-Glu-Gly-Thr-Phe
        Thr-Ser-Asp-Tyr
        Ser-Lys-Tyr
        Leu-Asp-Ser-Arg-Arg-Ala-Gln-Asp-Phe
        Val-Gln-Trp
        Leu-Met-Asn-Thr

14.52 Amino acids with nonpolar R groups will have those groups pointing toward the center of a globular protein. These are phenylalanine (b) and isoleucine (c). The other amino acids listed (a, d, e, f) have polar R groups, which will point toward the surface and hydrogen-bond with water, thus helping to solubilize the protein.

14.53 One possibility is the slow racemization via the enol of the carboxylic acid:

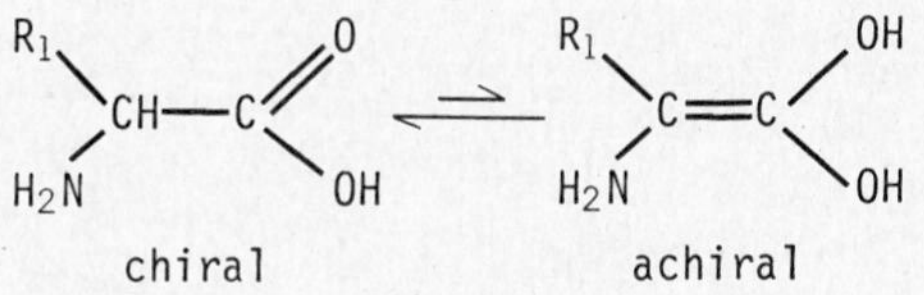

      chiral   achiral

CHAPTER SUMMARY

Nucleic acids, the carriers of genetic information, are macromolecules that are composed of and can be hydrolyzed to **nucleotide units**. Hydrolysis of a nucleotide gives one equivalent each of a **nucleoside** and phosphoric acid. Further hydrolysis of a nucleoside gives one equivalent each of a sugar and a heterocyclic base.

The DNA sugar is **2-deoxy-D-ribose**. The four heterocyclic bases in DNA are **cytosine, thymine, adenine**, and **guanine**. The first two are **pyrimidines**, and the latter two are **purines**. In nucleosides, the bases are attached to the anomeric carbon (C-1) of the sugar as β-**N-glycosides**. In nucleotides, the ——OH at C-3 or C-5 of the sugar is present as a phosphate ester.

The primary structure of DNA consists of nucleotides linked by a phosphodiester bond between the 5' ——OH of one unit and the 3' ——OH of the next unit. To fully describe a DNA molecule, the **base sequence** must be known. Methods for sequencing are rapidly being developed, and, at present, over 150 bases can be sequenced per day. The counterpart of sequencing, the synthesis of oligonucleotides having known base sequences, is also rapidly progressing.

The secondary structure of DNA is a **double helix**; two helical poly-nucleotide chains coil around a common axis. Each helix is right-handed, and the two run in opposite directions with respect to their 3' and 5' ends. The bases are located inside the double helix, in planes perpendicular to the helix axis. They are paired (A-T and G-C) by hydrogen bonds, which hold the two chains together. The sugar-phosphate backbones form the exterior surface of the double helix. Genetic information is passed on when the double helix uncoils and each strand acts as a template for binding and linking nucleotides to form the next generation.

RNA differs from DNA in three ways: the sugar is D-ribose, the pyrimidine **uracil** replaces thymine (the other three bases are the same), and the molecules are mainly single-stranded. There are three principal types of RNA: **messenger RNA** (involved in transcribing the genetic code), **transfer RNA** (which carries a specific amino acid to the site of protein synthesis), and **ribosomal RNA**.

CHAPTER SUMMARY

The **genetic code** involves sequences of three bases called **codons**, each of which translates to a specific amino acid. The code is degenerate (that is, there is more than one codon per amino acid), and some codons are "stop" signals that terminate synthesis. **Protein biosynthesis** is the process by which the message carried in the base sequence is transformed into an amino acid sequence in a peptide or protein.

In addition to their role in genetics, nucleotides play other important roles in biochemistry. Key enzymes and coenzymes such as **nicotinamide adenine dinucleotide** (NAD), **flavin adenine dinucleotide** (FAD), and vitamin B_{12} also include nucleotides as part of their structures.

REACTION SUMMARY

Hydrolysis of Nucleic Acids

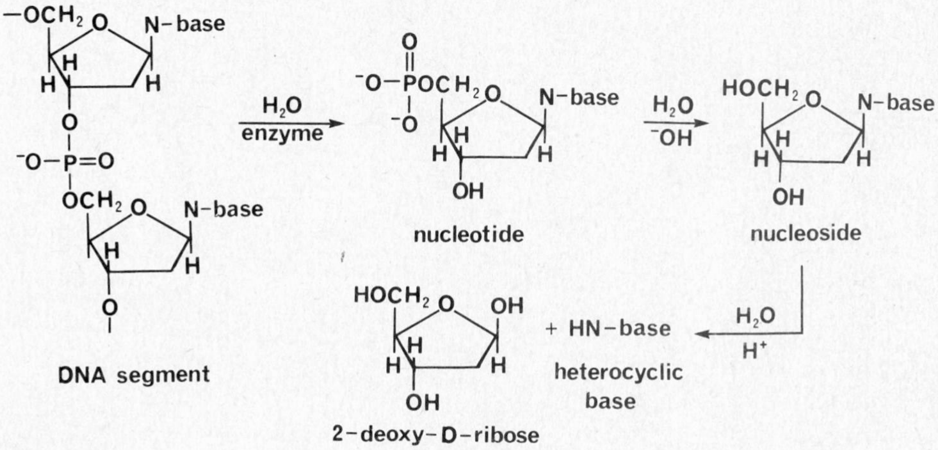

DNA segment
N-base
H₂O
enzyme
nucleotide
H₂O
⁻OH
nucleoside
H₂O
H⁺
+ HN-base
heterocyclic base
2-deoxy-D-ribose

LEARNING OBJECTIVES

1. Know the meaning of: nucleic acid, nucleotide, nucleoside.
2. Know the structures of: cytosine, thymine, adenine, guanine, uracil, 2-deoxy-D-ribose, D-ribose.
3. Know the meaning of: DNA, RNA, N-glycoside, pyrimidine base, purine base.
4. Given the name, draw the structure of a specific nucleoside.

5. Write an equation for the hydrolysis of a specific nucleoside by aqueous acid. Write the steps in the reaction mechanism.

6. Given the name, draw the structure of a specific nucleotide.

7. Write an equation for the hydrolysis of a specific nucleotide by aqueous base.

8. Draw the structure of an N-glycoside and an O-glycoside.

9. Given the name or abbreviation for a DNA or RNA nucleotide or nucleoside, draw its structure.

10. Draw the primary structure of a segment of an RNA or DNA chain.

11. Explain why only pyrimidine-purine base pairing is permissible in the double helix structure.

12. Describe the main features of the secondary structure of DNA.

13. Explain, with the aid of structures, the role of hydrogen bonding in nucleic acid structures.

14. Describe the main features of DNA replication.

15. Given the base sequence in one strand of a DNA molecule, write the base sequence in the other strand, or in the derived mRNA. Conversely, given a base sequence for mRNA, write the base sequence in one strand of the corresponding DNA.

16. Given a synthetic polyribonucleotide and the peptide sequence in the resulting polypeptide, deduce the codons for the amino acids.

17. Describe the main features of protein biosynthesis.

18. Explain the different functions of messenger, ribosomal, and transfer RNA.

19. Know the meaning of: codon, anticodon, genetic code, transcription.

20. Draw the structure of: AMP, ADP, ATP, and cAMP.

ANSWERS TO PROBLEMS

Problems Within the Chapter

15.1

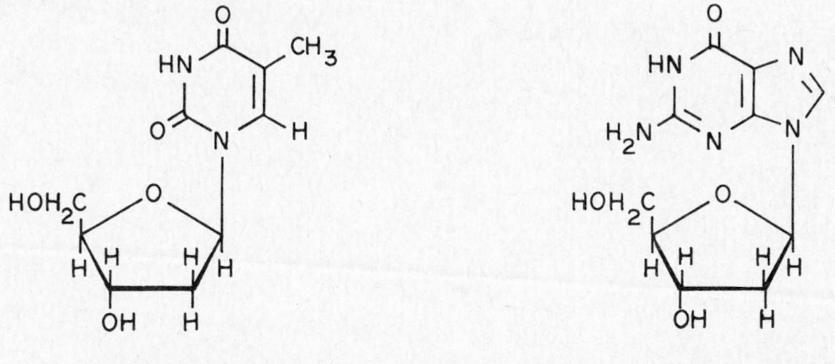

2'-deoxythymidine   2'-deoxyguanosine

15.2

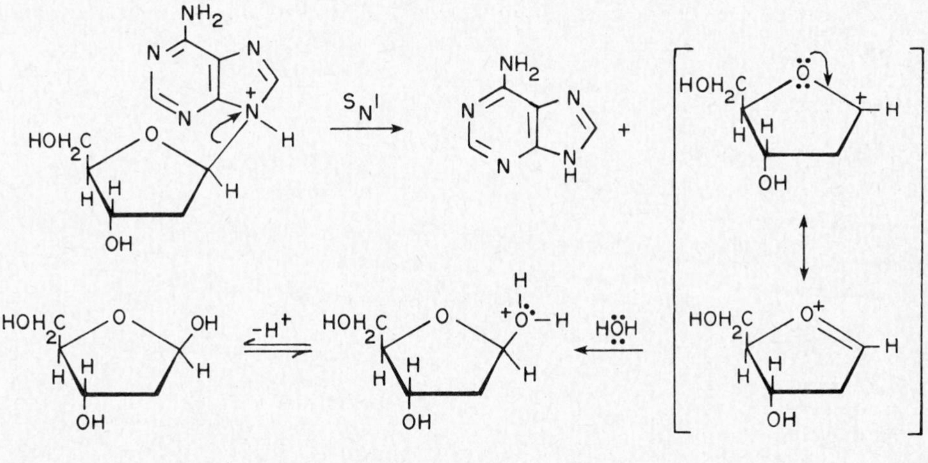

After protonation of the nitrogen, the bond between the anomeric carbon of the sugar and the nitrogen of the heterocyclic base is cleaved by an S_N1 mechanism. The resulting carbocation is stabilized by resonance as shown. Reaction with water as a nucleophile (followed by proton loss) then gives 2-deoxyribose, either the form as shown or the α form; that is, the water can attack from the top or bottom face of the furanose ring. This mechanism is exactly analogous to that of the hydrolysis of an acetal (the reverse of eq. 9.10, p. 237; see also the answer to Problem 9.9).

15.3 a.

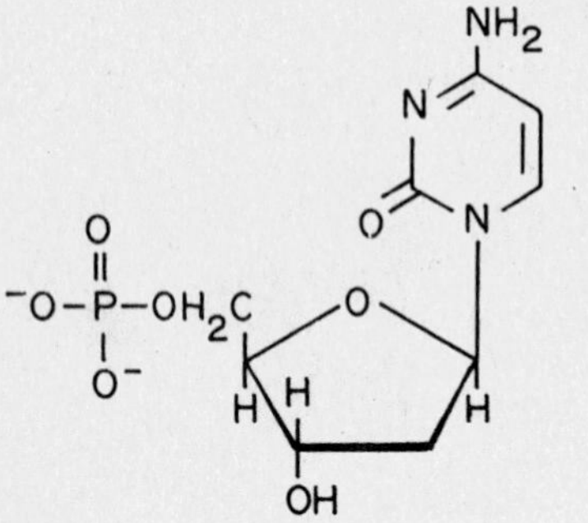

b.

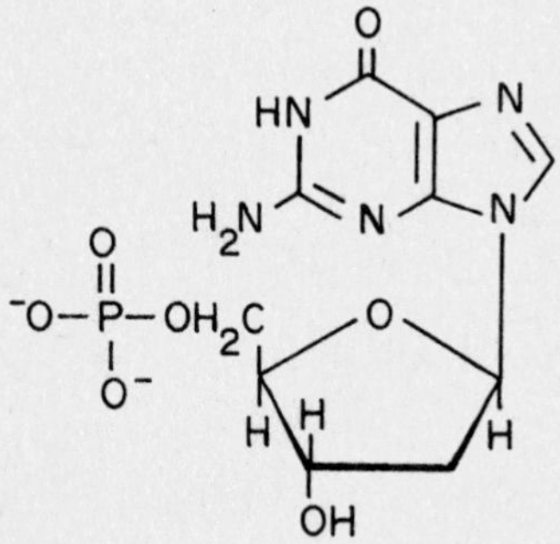

15.4

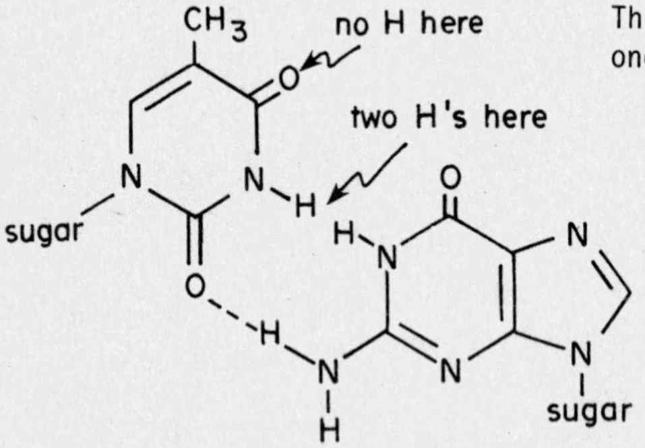

no H here

two H's here

sugar

This structure has only one hydrogen bond.

Although this structure has two hydrogen bonds, the distance between the sugar groups is substantially increased, and also the guanine amino group is left with no stabilizing hydrogen bond.

15.5   —TCGGTACA— (written from the 3' end to the 5' end).

15.6   a.

Note that C-2' has a hydroxyl group; AMP is an RNA mononucleotide.

b.

Note that there is a hydroxyl group at C-2' of each unit, and that the 5' and 3' ends do not have phosphate groups attached.

15.7　A polynucleotide made from UA will have the codons UAU and AUA:

    U A U A U A U A . . .

UAU is the codon for Tyr, and AUA must be a codon for Ile. Note that this codon differs from the codon in Example 15.4 only in the last letter. Such differences are common; the first two letters of codons are frequently more important than the third letter.

Additional Problems

15.8　a.　See the formulas for cytosine and thymine (Figure 15.1) and for uracil (Sec. 15.11).
     b.　See the formulas for adenine and guanine (Figure 15.1).
     c.　See Sec. 15.4.
     d.　See Sec. 15.5.

15.9

15.10 All the ring atoms in adenine but one (the NH nitrogen in the five-membered ring) are sp^2-hybridized and planar. Since both rings are aromatic, we expect the molecule to be planar or nearly so. In guanine, the ring atoms are also sp^2-hybridized (except for one NH nitrogen in each ring). The amide group in the six-membered ring (N-1, C-6) is also planar. Therefore, once again we expect the molecule to be planar, or very nearly so.

15.11 In this case, the rings will also be essentially planar, for the same reasons given in the answer to Problem 15.10.

15.12 a.

b. See the starting material in eq. 15.2, p. 421.

c. See Sec. 15.11 as a guide.

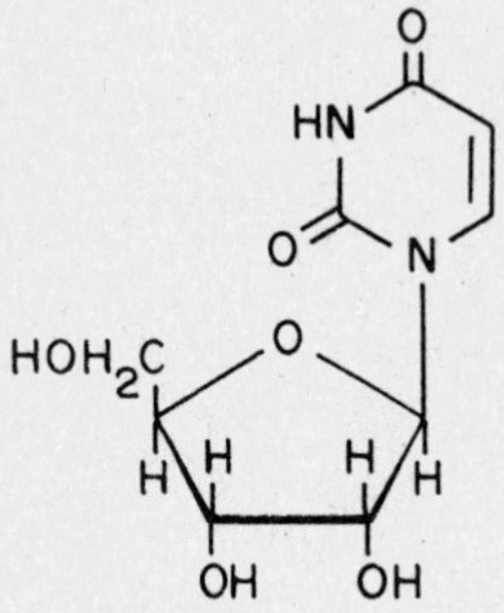

d. See the answer to Problem 15.1.

15.13

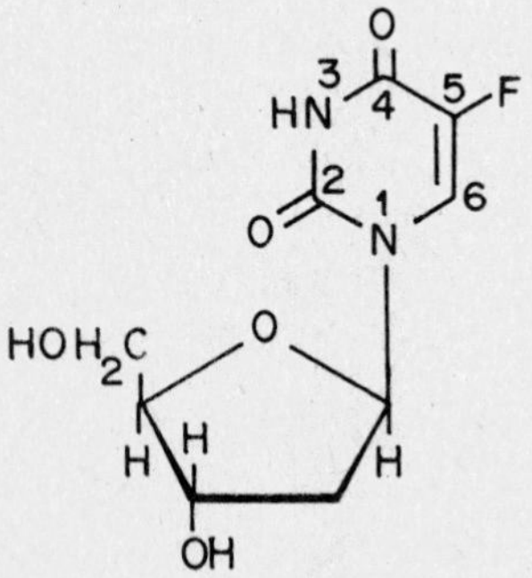

15.14

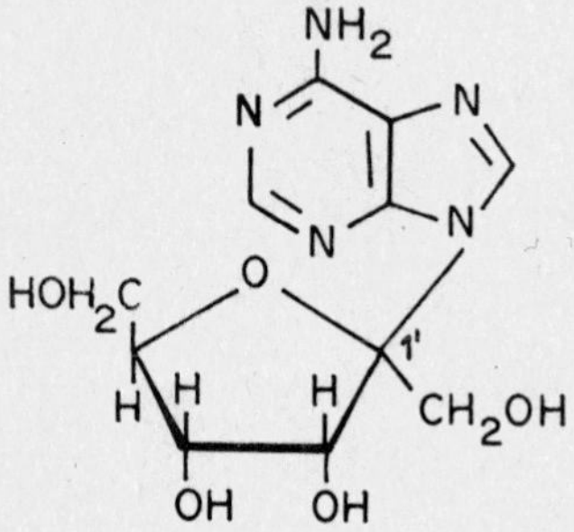

15.15

The products are adenine, ribose, and inorganic phosphate.

15.16 a.

The sugar part of the molecule has a hydroxyl group at C-2'
(D-ribose).

b.

15.17 Hydroxide ion attacks the primary C-5' carbon atom in an S_N2
       displacement. The leaving group is phosphate ion.

15.18 a.

b.

c.

15.19 a.

b.

c.

15.20 a. The products are two equivalents of 2'-deoxyadenosine-3'-
monophosphate, one equivalent of 2'-deoxythymidine-3'-
monophosphate, one equivalent of 2'-deoxycytidine.

b. The products are two equivalents of 2'-deoxyadenosine, one
equivalent of 2'-deoxythymidine, one equivalent of
2'-deoxycytidine, and three equivalents of inorganic phosphate.

15.21 a.

b.

c.

15.22 The structure is identical to that of the T-A base pair shown on p. 427 of the text, except that the methyl group in the thymine unit is replaced by a hydrogen.

15.23 3' T-T-C-G-A-C-A-T-G 5'

15.24 3' U-U-C-G-A-C-A-U-G 5'

The only difference between this sequence and the answer to Problem 15.23 is that each T is replaced by U.

15.25 Given mRNA sequence: 5' A-G-C-U-G-C-U-C-A 3'

DNA strand from which mRNA was transcribed:

    3' T-C-G-A-C-G-A-G-T 5' ⎫
       | | | | | | | | | ⎬ DNA double helix
    5' A-G-C-T-G-C-T-C-A 3' ⎭

Note that the DNA strand that was *not* transcribed (the last segment shown above) is identical with the given mRNA segment, except that each U is replaced by T.

15.26 For each T there must be an A; for each G there must be a C. As a consequence of this base pairing, the mole percentages of T in any sample of DNA will be equal to the mole percentages of A, and the same will be true for G and C. It is *not* necessary, however, that there be any special relationship between the percentages of the two pyrimidines (T and C) or of the two purines (A and G).

15.27 The code reads from the 5' to the 3' end of mRNA.

    5' C-A-U 3'   mRNA
    3' G-T-A 5'   DNA-transcribed chain
    5' C-A-T 3'   DNA complement

15.28

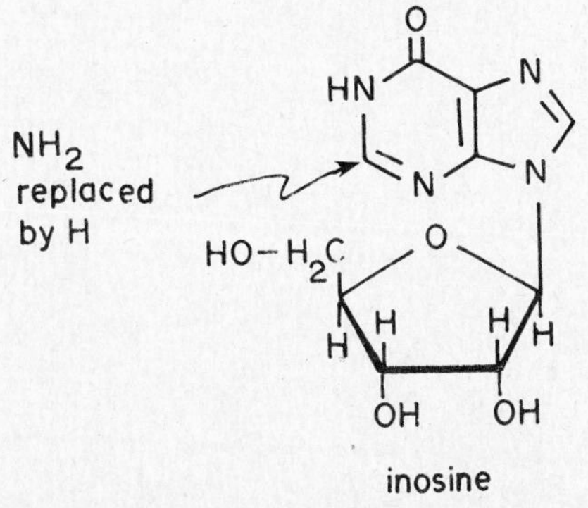

NH$_2$ replaced by H

inosine

15.29 Each of the three base pairs has two hydrogen bonds.

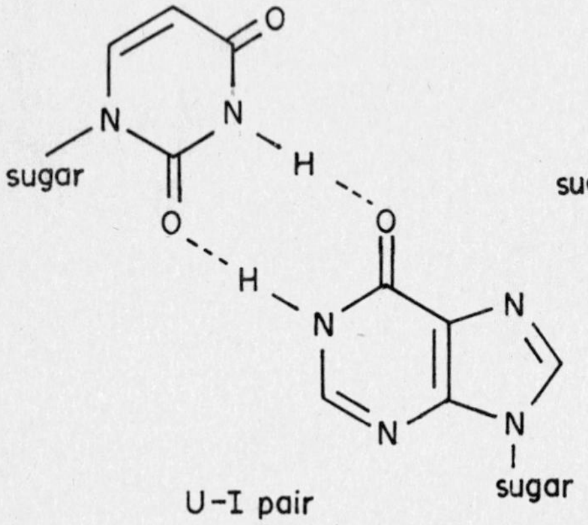

U-I pair

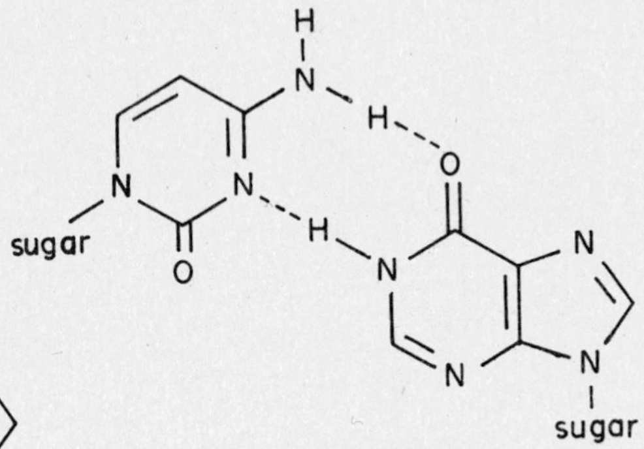

C-I pair

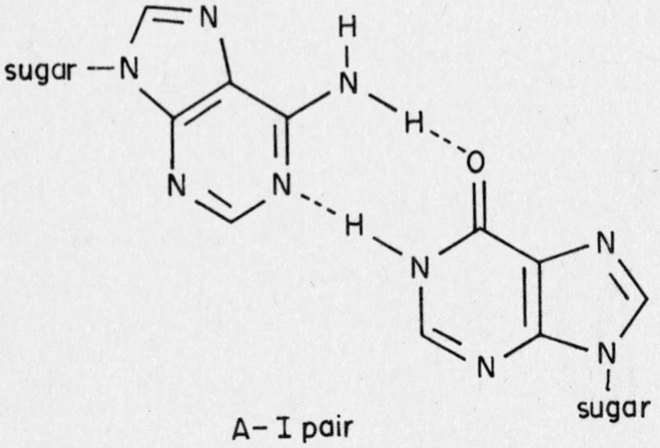

A-I pair

Although the distance between sugar parts of these pairs varies, this factor is not important in *m*RNA-*t*RNA interactions, because there is no helical structure involved.

15.30 Hydrolysis occurs at all glycosidic, ester, and amide linkages. The products are as follows:

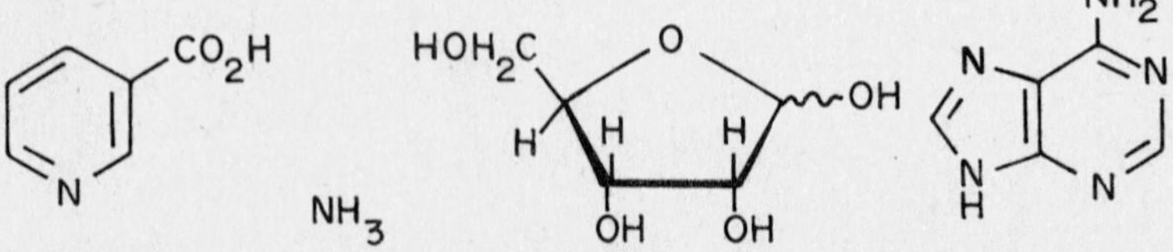

nicotinic acid ammonia D-ribose (2 mol) adenine and phosphoric acid (2 mol)

15.31

15.32 The formula for uridine monophosphate is shown on p. 429 of the text, and the formula for α-D-glucose is shown in eq. 13.7, p. 362. The structure of UDP-glucose is

15.33 There are no N—H bonds in caffeine; therefore (unlike adenine and guanine), it cannot form N-glycosides.

CHAPTER SIXTEEN   SPECTROSCOPY, POLYMERS, AND PHARMACEUTICALS

CHAPTER SUMMARY

Spectroscopy

Spectroscopic methods provide rapid, nondestructive ways to determine molecular structures. One of the most powerful of these methods is **nuclear magnetic resonance (NMR) spectroscopy.** It involves the excitation of nuclei from lower to higher energy spin states while they are placed between the poles of a powerful magnet. In organic chemistry, the most important nuclei measured are ^1H and ^{13}C.

Protons in different chemical environments have different **chemical shifts,** measured in δ (delta) units from the reference peak of tetramethyl-silane [TMS, $(CH_3)_4Si$]. **Peak areas** are proportional to the number of protons. Peaks may be split (**spin-spin splitting**) depending on the number of nearby protons. Proton NMR gives at least three types of structural information: (1) the number of signals and their chemical shifts can be used to identify the kinds of chemically different protons in the molecules; (2) peak areas tell how many protons of each kind are present; (3) spin-spin splitting patterns identify the number of near-neighbor protons.

^{13}C NMR spectroscopy can tell how many different "kinds" of carbon atoms are present, and ^1H-^{13}C splitting can help identify particular carbons.

Infrared spectroscopy is mainly used to tell what types of bonds are present in a molecule (using the **functional group region,** 1500-5000 cm^{-1}) and whether two substances are identical or different (using the **fingerprint region,** 700-1500 cm^{-1}).

Visible and ultraviolet spectroscopy employs radiation with wavelengths of 200-800 nm. This radiation corresponds to energies that are associated with **electronic transitions,** in which an electron "jumps" from a filled orbital to a vacant orbital with higher energy. Visible-ultraviolet spectra are most commonly used to detect conjugation. In general, the greater the degree of the conjugation, the longer the wavelength of energy absorbed.

Mass spectra are used to determine molecular weights and molecular composition (from the **parent** or **molecular ion**) and to obtain structural information from the fragmentation of the molecular ion into **daughter ions.**

Polymers

   Polymers are macromolecules built of smaller units called **monomers;** the process by which they are formed is called **polymerization. Homo-polymers** are made from a single monomer; **copolymers** are made from two or more monomers. Polymers may be **linear, branched,** or **cross-linked,** depending on how the monomer units are arranged. These details of structure affect polymers' properties.

   Addition, or **chain-growth, polymers** are made by adding one monomer unit at a time to the growing polymer chain. The reaction requires initiation to produce some sort of reactive intermediate, which may be a **free radical,** a **cation,** or an **anion.** The intermediate adds to the monomer, giving a new intermediate, and the process continues until the chain is terminated in some way. **Polystyrene** is a typical free-radical chain-growth polymer.

   Chiral centers can be generated when a substituted vinyl compound is polymerized. The resulting polymers are classified as **atactic, isotactic,** or **syndiotactic,** depending on whether the chiral centers are **random, identical,** or **alternating** in configuration as one proceeds down the polymer chain. Ziegler-Natta catalysts (one example is a mixture of trialkylaluminum and titanium tetrachloride) usually produce stereoregular polymers, whereas free-radical catalysts generally give stereorandom polymers.

   Condensation, or **step-growth, polymers** are usually formed in a reaction between two monomers, each of which is at least difunctional. Polyesters and polyamides are typical examples of step-growth polymers. These polymers grow by steps or leaps rather than one monomer unit at a time.

Pharmaceuticals

   Pharmaceuticals may be classified according to biological activity. Often substances with quite different chemical structures can be effective against similar diseases.

   Antibiotics are used to kill or arrest the growth of bacteria. Among the first to see large-scale use were the **sulfa drugs** (for example, sulfanilamide). More generally effective are the **penicillins** and **cephalosporins,** both of which are β-**lactams** (amides with a four-membered ring). Other important antibiotics include the **tetracyclines, erythro-mycin, streptomycin,** and **chloramphenicol.**

Drugs that affect the central nervous system include tranquilizers (**Valium** and **Librium**), hypnotics or sleep-inducing drugs (**barbiturates**), and stimulants (**caffeine** and **amphetamines**), among others.

Cardiovascular drugs are used to treat various heart diseases. They include the antihypertensive drug **L-methyldopa**, the diuretics **hydrochlorthiazide**, **triamterine**, and **furosemide**, the vasodilator **nitroglycerine**, and the antiarrhythmic agents **digoxin** and **propranolol**.

Other important drugs include the **steroids** (sex hormones, oral contraceptives, antiinflammatory agents, diuretics, etc.), **aspirin** (an analgesic), antihistamines to treat allergies (for example, **diphenhydramine**), antiviral agents (**amantidine** and **idoxuridine**); and anticancer agents (*cis*-**platin**, **5-fluorouracil**, and **doxorubicin**).

REACTION SUMMARY

Free-Radical Chain-Growth Polymerization

Initiation:

$$I \xrightarrow[\text{or light}]{\text{heat}} 2\ R\cdot$$

Propagation:

$$R\cdot + \overset{\frown}{C}=C \longrightarrow R-C-\overset{\cdot}{C} \xrightarrow{\ \ C=C\ \ } \text{etc.}$$

Termination:

radical coupling

radical disproportionation

Chain Transfer (Hydrogen Abstraction)

REACTION SUMMARY

Cationic Chain-Growth Polymerization

$$R^+ \ + \ CH_2\!=\!\!\overset{\displaystyle |}{\underset{\displaystyle X}{CH}} \longrightarrow RCH_2\!-\!\overset{+}{\underset{\displaystyle X}{\overset{\displaystyle |}{CH}}} \quad \xrightarrow{\ CH_2\!=\!\!\overset{\displaystyle |}{\underset{\displaystyle X}{CH}}\ } \ etc.$$

Anionic Chain-Growth Polymerization

$$R\!:^- \ + \ CH_2\!=\!\!\overset{\displaystyle |}{\underset{\displaystyle X}{CH}} \longrightarrow RCH_2\!-\!\overset{..}{\underset{\displaystyle X}{\overset{\displaystyle |}{CH}}} \quad \xrightarrow{\ CH_2\!=\!\!\overset{\displaystyle |}{\underset{\displaystyle X}{CH}}\ } \ etc.$$

Step-Growth Polymerization (illustrated for polyesters)

$$HO\!\rightsquigarrow\!OH \ + \ HO\overset{O}{\overset{||}{C}}\rightsquigarrow\overset{O}{\overset{||}{C}}OH \longrightarrow HO\rightsquigarrow O\overset{O}{\overset{||}{C}}\rightsquigarrow\overset{O}{\overset{||}{C}}OH \quad \xrightarrow[\text{or}]{HO\rightsquigarrow OH} \quad \text{three units}$$

(two units)

$$\overset{O\quad O}{\overset{||\quad ||}{HOC\rightsquigarrow COH}}$$

$$HO\rightsquigarrow O\overset{O}{\overset{||}{C}}\rightsquigarrow\overset{O}{\overset{||}{C}}OH$$

$$HO\rightsquigarrow O\overset{O}{\overset{||}{C}}\rightsquigarrow\overset{O}{\overset{||}{C}}O\rightsquigarrow O\overset{O}{\overset{||}{C}}\rightsquigarrow\overset{O}{\overset{||}{C}}OH$$

(four units)

LEARNING OBJECTIVES

Spectroscopy

1. Know the meaning of: NMR, applied magnetic field, spin state, chemical shift, δ value, TMS (tetramethylsilane), peak area, spin-spin splitting, $n + 1$ rule.

2. Given a structure, tell how many different "kinds" of protons are present.

3. Given a simple structure, use the data in Table 16.1 to predict the appearance of its ^1H NMR spectrum.

4. Given a ^{1}H NMR spectrum and other data such as molecular formula, use Table 16.1 to deduce a possible structure.

5. Use spin-spin splitting patterns on a spectrum to help assign a structure.

6. Use ^{1}H and ^{13}C NMR spectra with other data or spectra to deduce a structure.

7. Know the meaning of the functional group region and fingerprint region of an infrared spectrum, and tell what kind of information can be obtained from each.

8. Use the information in Table 16.3 to distinguish between different classes of organic compounds.

9. In connection with visible-ultraviolet spectroscopy, know the meaning of: nanometer, electronic transition, Beer's law, molar absorptivity or extinction coefficient.

10. Know the relationship between conjugation and visible-ultraviolet absorption, and use this relationship to distinguish between closely related structures.

11. Know the meaning of: mass spectrum, *m/e* ratio, molecular ion, parent ion, fragmentation, daughter ion.

12. Given the structure of a simple compound and its mass spectrum, deduce possible structures for the daughter ions.

13. Use all spectroscopic methods in conjunction to deduce a structure.

Polymers

1. Know the meaning of: monomer, polymer, macromolecule, polymerization, average molecular weight, degree of polymerization.

2. Know the meaning of: homopolymer, copolymer, linear, branched, and cross-linked polymer. For copolymers, know the meaning of: alternating, random, block, graft.

3. Know the meaning of and illustrate the difference between chain-growth (addition) and step-growth (condensation) polymerization.

4. Write the mechanism for an addition polymerization via a radical, cationic, or anionic intermediate. In each case, predict the direction of addition to the monomer if it is an unsymmetric alkene.

5. Write resonance structures for the reactive intermediate in addition polymerization, to show how it is stabilized by the alkene substituent.

6. Know the meaning of: radical coupling, radical disproportionation, and chain transfer, and illustrate each with examples.

7. Know the meaning of and illustrate with examples: Ziegler-Natta polymerization, atactic, isotactic, and syndiotactic polymers.

8. Write a mechanism for a step-growth polymerization, as in the formation of a polyester, polyamide, polyurethane, or phenol-formaldehyde polymer.

Pharmaceuticals

1. Know the meaning of: antibiotic, antihypertensive drug, diuretic, vasodilator, antihistamine, antiviral agent, and antineoplastic agent. Give an example of each.

2. Know the structure of and synthetic route to sulfanilamide. Know how its structure is modified in other sulfa drugs.

3. Know the meaning of lactam and β-lactam, and draw the structure of an example of each. Know the structure of penicillin.

4. Be familiar with the general structural features and uses of tetracyclines, macrolides, streptomycin, and chloramphenicol.

5. Be familiar with the structures and uses of adrenalin, amphetamine, L-methyldopa, furosemide, and propranolol.

6. Be familiar with the structures and uses of diphenhydramine, amantidine, *cis*-platin, and 5-fluorouracil.

ANSWERS TO PROBLEMS

Problems Within the Chapter

16.1 All the protons in the structures in parts a and c are equivalent and appear as a sharp, single peak. One way to test whether protons are equivalent is to replace any one of them by some group X. If the same product is obtained regardless of which proton is replaced, then the protons must be equivalent. Try this test with the compounds in parts a and c. Note that diethyl ether, the compound in part b, has two sets of equivalent protons (CH_2 and CH_3).

16.2   a. CH_3OH 3:1

b. $CH_3\overset{\overset{\textstyle O}{\|}}{C}OCH_3$ 1:1 (or 3:3)

c. $CH_3CH_2\overset{\overset{\textstyle O}{\|}}{C}CH_2CH_3$ 2:3 (or 4:6)

16.3   a. $CH_3\overset{\overset{\textstyle O}{\|}}{C}OH$ There will be two peaks, at approximately $\delta2.1-2.6$ and $\delta10-13$, with relative areas 3:1.

b. $(CH_3)_2C{=\!=}CH_2$ There will be two peaks, at approximately $\delta1.6-1.9$ and $\delta4.6-5.0$, with relative areas 6:2 (or 3:1).

16.4   The compound is $(CH_3)_3C\!-\!\overset{\overset{\textstyle O}{\|}}{C}\!-\!OCH_3$

$\delta0.9$ $\delta3.6$
9 H   3 H

The expected spectrum for its isomer is

$CH_3\overset{\overset{\textstyle O}{\|}}{C}\!-\!OC(CH_3)_3$

$\delta2.1-2.6$ $\delta0.9$
3 H   9 H

The only difference is the chemical shift of the methyl protons.

16.5 $C\underline{H}_3CHCl_2$ doublet, $\delta0.85-0.95$, area 3

$CH_3C\underline{H}Cl_2$ quartet, $\delta5.8-5.9$, area 1

16.6   a. $BrC\underline{H}_2CH_2Cl$ triplet, $\delta3.2-3.3$, area 2

$BrCH_2C\underline{H}_2Cl$ triplet, $\delta3.4-3.5$, area 2

Actually the spectrum will be more complex than this, because the chemical shifts of the two types of protons are close in value.

b. $ClCH_2CH_2Cl$ The protons are all equivalent, and the spectrum will consist of a sharp singlet at $\delta 3.4-3.5$.

16.7 $CH_3CH_2CH_2OH$

The spectrum *without* ^1H coupling will appear as three peaks:

C-1 δ ~ 65
C-2 δ ~ 32
C-3 δ ~ 11

In the ^1H-coupled spectra, the peaks for C-1 and C-2 would be triplets, and the C-3 peak would be a quartet.

16.8   We use Beer's law: $A = \varepsilon\, cl$
Rearranging, we get:

$$c = \frac{A}{\varepsilon\, l}$$

$$= \frac{2}{12,600 \times 1} = 1.587 \times 10^{-4} \text{ mol/L}$$

Note that ultraviolet spectra are often obtained on very dilute solutions.

16.9   Conjugation is possible between the two rings in biphenyl, but in diphenylmethane the $—CH_2—$ group interrupts this conjugation. Thus for comparable electronic transitions biphenyl is expected to absorb at longer wavelengths.

16.10   The molecular ion peak ($C_7H_{14}O^+$) will appear at $m/e = 114$ (14 mass units, or one $—CH_2—$ group less than for 4-octanone). Since the ketone is symmetric, it should fragment to yield a $C_3H_7CO^+$ peak ($m/e = 71$) and a $C_3H_7^+$ peak ($m/e = 43$); only one set of daughter ions (instead of the two seen in Figure 16.8) will be observed.

16.11   The molecular weight of a polyethylene molecule with 1000 units plus one benzoyl unit is 28,000 + 105 = 28,105. The percentage that is due to the initiator radical is

$$\frac{105}{28,105} \times 100 = 0.37\%$$

For polystyrene, the molecular weight is 104,000 + 105 = 104,105, and the percentage that is due to initiator is

$$\frac{105}{104,105} \times 100 = 0.10\%$$

16.12

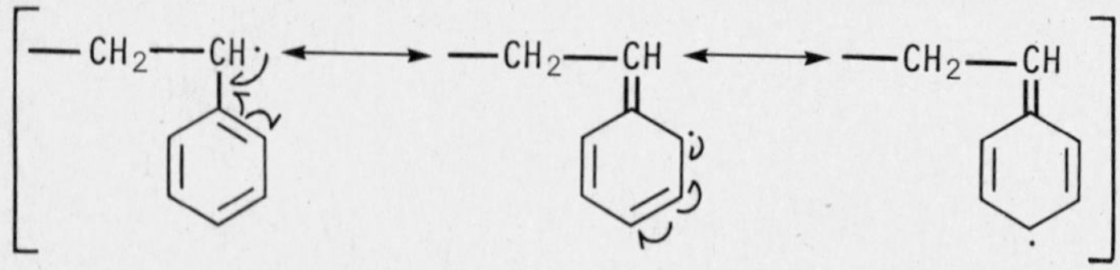

$$\left[\begin{array}{ccc} -CH_2-CH\cdot & \longleftrightarrow & -CH_2-CH & \longleftrightarrow & -CH_2-CH \end{array} \right]$$

The odd electron can be delocalized to the *ortho* and *para* carbons of the benzene ring.

16.13 $R^-Li^+ + CH_2 = C - CO_2CH_3 \longrightarrow RCH_2 - \overset{Li^+}{\underset{CH_3}{\overset{\ominus}{C}}} - CO_2CH_3$

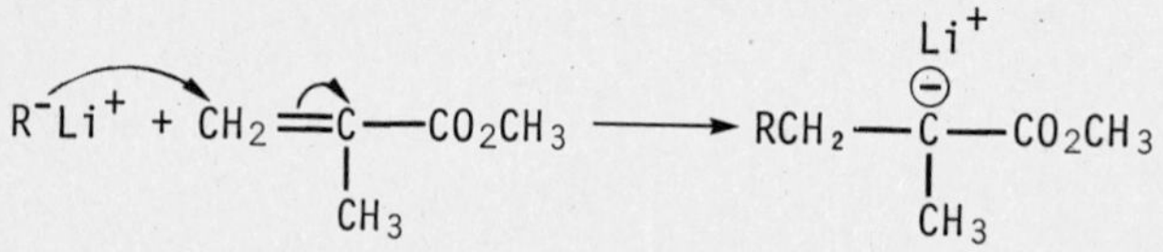

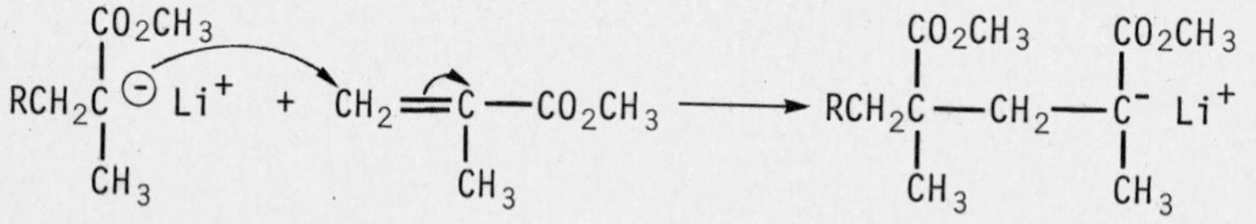

$\underset{CH_3}{\overset{CO_2CH_3}{RCH_2C}} \ominus Li^+ \; + \; CH_2 = \underset{CH_3}{C} - CO_2CH_3 \longrightarrow RCH_2\overset{CO_2CH_3}{\underset{CH_3}{C}} - CH_2 - \overset{CO_2CH_3}{\underset{CH_3}{C^-}} Li^+$ and so on

$(R = CH_3CH_2CH_2CH_2-)$

The negative charge in the intermediate carbanion can be delocalized to the carbonyl oxygen of the ester group:

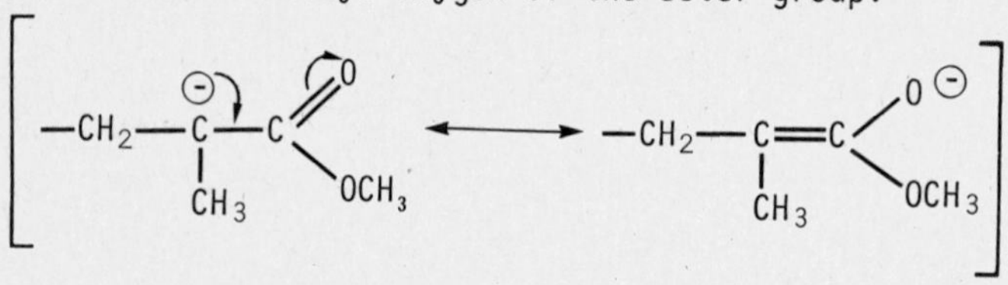

$$\left[\begin{array}{ccc} -CH_2-\overset{\ominus}{\underset{CH_3}{C}}-C\overset{O}{\underset{OCH_3}{\diagdown}} & \longleftrightarrow & -CH_2-\underset{CH_3}{C}=C\overset{O^\ominus}{\underset{OCH_3}{\diagdown}} \end{array} \right]$$

16.14. a.

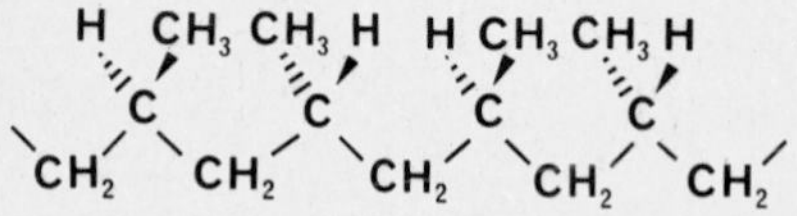

b.

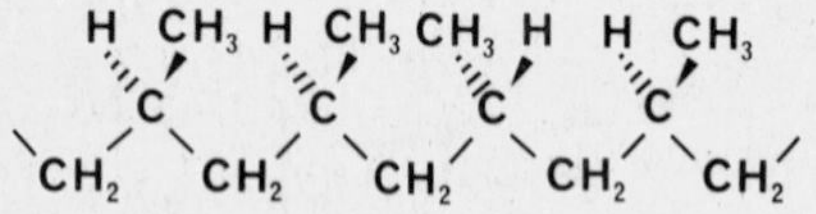

16.15 Each contains three monomer units, so the next product will contain six monomer units. Its structure is

$$HO_2C \wasp C \overset{O}{\overset{\|}{}} {-}0 \wasp 0 {-} C \wasp C \overset{O\quad O}{\overset{\|\quad\|}{}} {-}0 \wasp 0 {-} C \wasp C \overset{O\quad O}{\overset{\|\quad\|}{}} {-}0 \wasp OH$$

The monomer units are marked off by dashed lines.

If the diester-diol reacted only with monomeric diacid, or if the diester-diacid reacted only with monomeric diol, the product would contain only four monomer units; but by reacting with each other, the two compounds engage in step growth to six monomer units.

16.16 $HO_2C \wasp CO_2H$ + $HO \wasp OH \longrightarrow HO \wasp 0 {-} C \wasp C \overset{O\quad O}{\overset{\|\quad\|}{}} {-} 0 \wasp OH$

       (large excess)   diester-diol

$HO \wasp OH$ + $HO {-} C \wasp C \overset{O\quad O}{\overset{\|\quad\|}{}} {-} OH \longrightarrow$

       (large excess)

$$HO {-} C \wasp C \overset{O\quad O}{\overset{\|\quad\|}{}} {-} 0 \wasp 0 {-} C \wasp C \overset{O\quad O}{\overset{\|\quad\|}{}} {-} OH$$

                        diester-diacid

16.17 Phenol, in its reaction with formaldehyde, is potentially *tri-*functional; it can condense *ortho* or *para* to the hydroxyl group:

Three reactive sites; each can contribute a hydrogen to the small elimination product, H—OH

Formaldehyde is *di*functional, in the sense that it can link two phenol molecules via a —CH_2— (difunctional) group.

16.18 The two reactions are (1) acid-catalyzed hydrolysis of a nitrile to a carboxyl group (eq. 10.11), and (2) acid-catalyzed cleavage of an ether (eq. 8.10).

16.19 The Fischer projection for L-methyldopa as shown in eq. 16.32 is

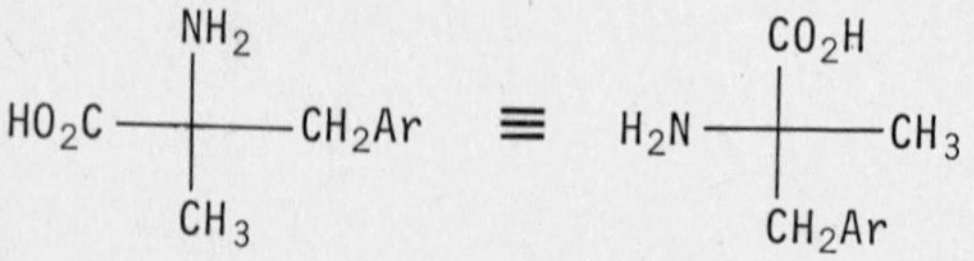

$$HO_2C \overset{NH_2}{\underset{CH_3}{\vert}} CH_2Ar \equiv H_2N \overset{CO_2H}{\underset{CH_2Ar}{\vert}} CH_3$$

$$Ar = -\overset{OH}{\underset{}{\bigcirc}}-OH$$

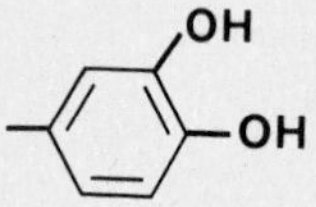

Two group interchanges give the Fischer projection shown above at the right. Replacement of —CH$_3$ with —H then gives

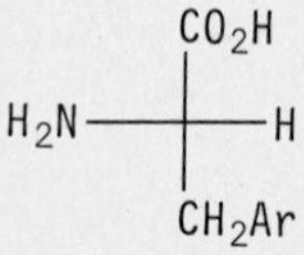

$$H_2N \overset{CO_2H}{\underset{CH_2Ar}{\vert}} H$$

which has the same configuration as the L-amino acids (Figure 14.1).

16.20

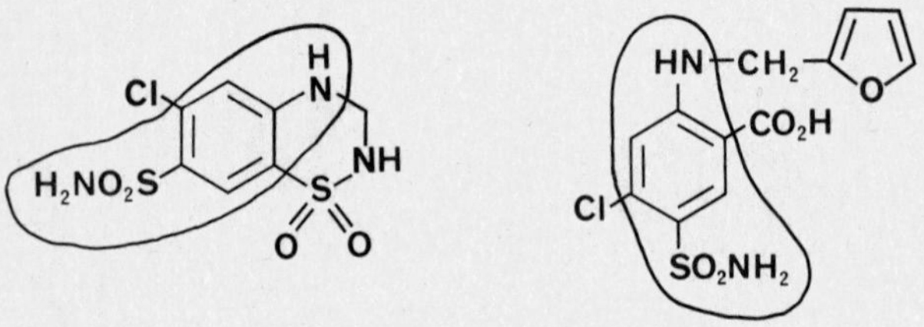

$$Cl \quad H_2NO_2S \overset{H}{\underset{}{N}} \overset{NH}{\underset{S}{\underset{O \quad O}{\parallel\parallel}}}$$

$$HN-CH_2-\bigcirc_O \quad Cl \overset{CO_2H}{\underset{SO_2NH_2}{}}$$

16.21

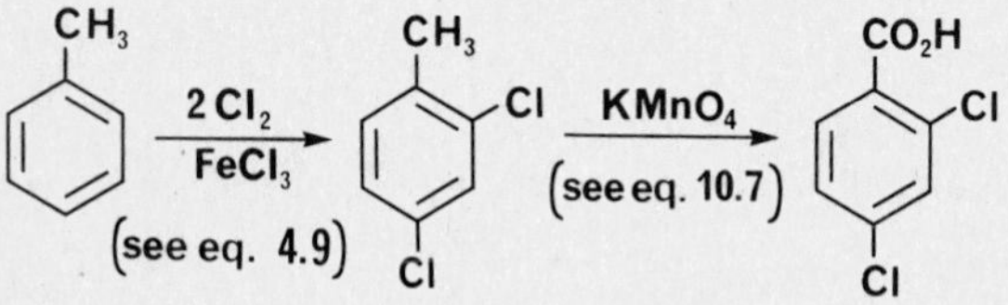

$$\underset{CH_3}{\bigcirc} \xrightarrow[\text{FeCl}_3]{2\,Cl_2} \underset{(see eq. 4.9)}{\overset{CH_3}{\underset{Cl}{\bigcirc}}Cl} \xrightarrow[\text{(see eq. 10.7)}]{\text{KMnO}_4} \overset{CO_2H}{\underset{Cl}{\bigcirc}}Cl$$

16.22

$$\underset{\bigcirc}{\overset{\bigcirc}{}}CHOCH_2CH_2\overset{H}{\overset{\oplus}{N}}(CH_3)_2 \qquad CH_3\text{—}N\underset{O}{\overset{O}{\underset{}{}}} \overset{N}{\underset{N}{\overset{\ominus}{}}}\text{—}Cl \quad \underset{CH_3}{}$$

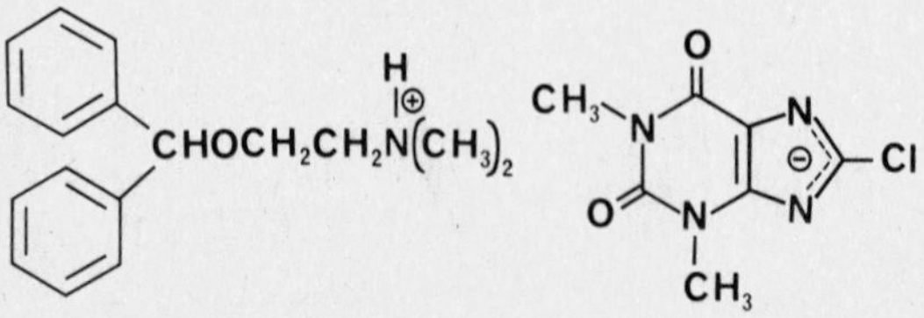

The hydrogen on the nitrogen in 8-chlorotheophylline is acidic and is donated to the basic nitrogen of diphenylhydramine. The 8-chlorotheophyllinate anion has the negative charge delocalized (resonance) over both nitrogens in the five-membered ring.

Additional Problems

16.23 Possible solutions are:

a. 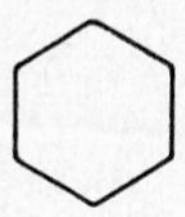 (cyclohexane hexagon)

b. $CH_3\!-\!\overset{\displaystyle Cl}{\underset{\displaystyle Cl}{C}}\!-\!CH_3$

c. $CH_3C\!\equiv\!CCH_3$

d. 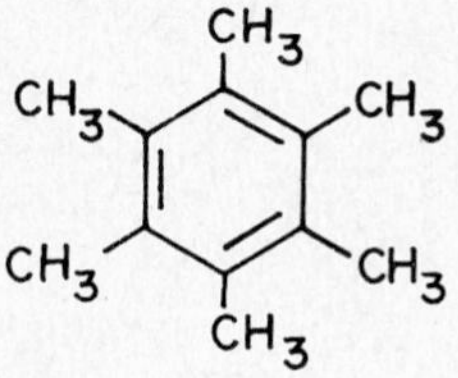 (benzene ring substituted with six CH_3 groups)

e. CH_3OCH_3

f. $CH_3\!-\!\overset{\displaystyle CH_3}{\underset{\displaystyle CH_3}{C}}\!-\!CH_3$

16.24 a. There are four different types of protons:

$CH_3\overset{a}{\diagdown}$ ... $\underset{CH_3}{\diagup}\,^{a}$ ... $CH\overset{b}{-}CH_2\overset{c}{-}CH_3\,^{d}$

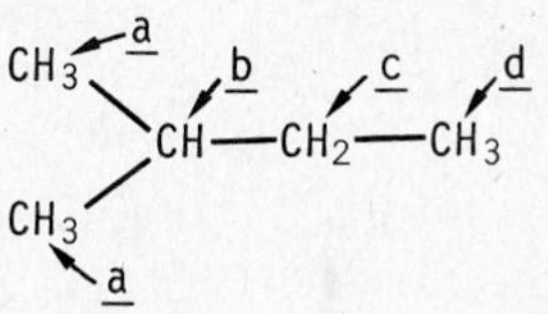

b. There are three different types of protons:

$CH_3\overset{a}{\diagup}$ ... $\underset{CH_3}{\diagdown}\,_{a}$ ... $N\overset{b}{-}CH_2\overset{c}{-}CH_3$

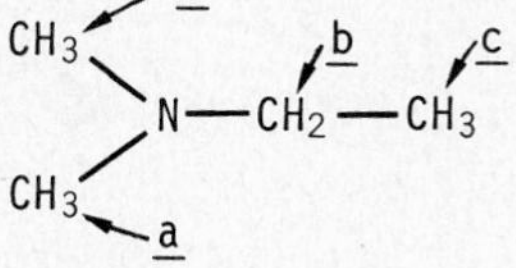

c. There are three different types of protons:

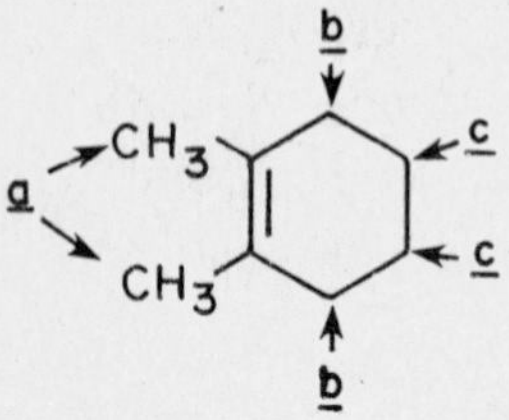

d. There are three different types of protons:

a b c

CH_3CH_2OH

16.25 The first compound must have nine equivalent hydrogens. The only possible structure is *t*-butyl bromide:

$$\begin{array}{c} CH_3 \\ | \\ CH_3-C-Br \\ | \\ CH_3 \end{array}$$

Its isomer has three different types of hydrogens, two of one kind, one unique, and six equivalent. The compound must be isobutyl bromide. The chemical shifts and spin-spin splitting pattern fit this structure:

δ1.9, complex, area 1

$(CH_3)_2CHCH_2Br$

δ 1.0, doublet, area 6

δ 2.7, doublet, area 2

16.26 a. CH_3CCl_3 All the protons will appear as a singlet.

$CH_2ClCHCl_2$ There will be two sets of peaks, with an area ratio of 2:1. The former will be a doublet, the latter a triplet.

b. $CH_3CH_2CH_2OH$

In addition to the O—H peak, there will be three sets of proton peaks, with area ratio 3:2:2.

(CH$_3$)$_2$CHOH

In addition to the O—H peak, there will be only two sets of proton peaks, with area ratio 6:1.

$$\underset{O}{\overset{\displaystyle O}{\|}}$$

c. CH$_3$CH$_2$C—OCH$_3$

The spectrum will show a singlet at about δ 3.5-3.8 for the O—CH$_3$ protons, and a quartet and triplet near δ 2.6 and 0.95, respectively.

CH$_3$C—OCH$_2$CH$_3$

The spectrum will show a singlet at about δ 2.1-2.6 for the

CH$_3$C— and a quartet and triplet near δ 3.8 and 0.95, respectively.

d.

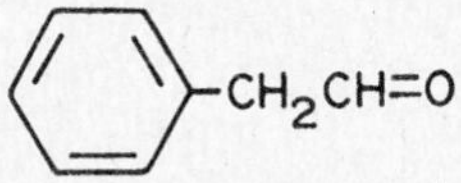

—CH$_2$CH=O

A one-proton aldehyde peak at δ 9.5-9.7 (a triplet) will easily distinguish this aldehyde from its ketone isomer.

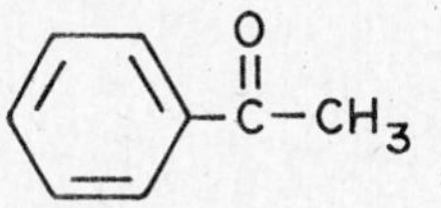

—C—CH$_3$

The aliphatic protons will give a sharp three-proton singlet at about δ 2.1-2.6.

16.27 At very low temperatures, the interconversion of one chair form to another can be frozen out. If this happens, there will be two sets of protons, equatorial and axial, with different chemical shifts. At room temperature, the interconversion is so fast that the spectrum shows only one signal, an average between the two types.

16.28 The peaks are assigned as follows:

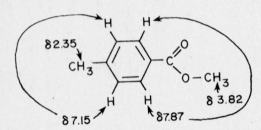

In general, aromatic protons show up at lower field strengths if they are adjacent to electron-withdrawing substituents; this is the basis for distinguishing between the two sets of aromatic protons.

16.29 a. CH_3CHO

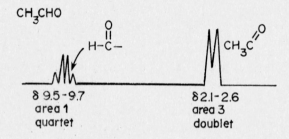

δ 9.5-9.7
area 1
quartet

δ 2.1-2.6
area 3
doublet

b. $(CH_3)_2CHOCH(CH_3)_2$

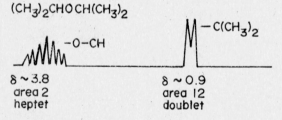

$\delta \sim 3.8$
area 2
heptet

$\delta \sim 0.9$
area 12
doublet

c. $CH_3CH{=}CCl_2$

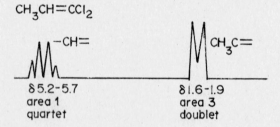

δ 5.2-5.7
area 1
quartet

δ 1.6-1.9
area 3
doublet

d.

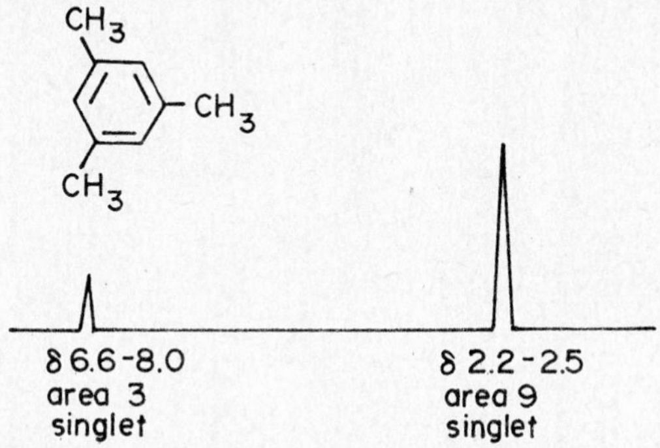

δ 6.6-8.0
area 3
singlet

δ 2.2-2.5
area 9
singlet

16.30 Replacement of the five chlorines with hydrogens gives C_3H_8. Thus there are no double bonds or rings. To split each other, the protons must be on adjacent carbon atoms. Two possibilities are:

$ClCH_2$—$CHCl$—CCl_3 and Cl_2CH—CH_2—CCl_3

However, neither of these alternatives fits the chemical shift data. For the first structure, the two-proton signal should be at δ 3.5 or a little lower, but not at δ 6.0. For the second structure, the CH_2 protons should be at higher field than the CH proton, whereas the opposite is true. We must therefore seek another possibility, and this one comes to mind:

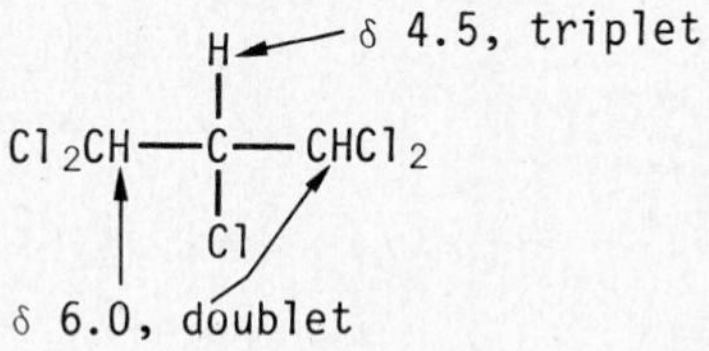

The two protons on C-1 and C-3 are equivalent and appear at low field because of the two chlorines.

16.31 There are three possibilities. The number of different types of carbons is shown under each structure.

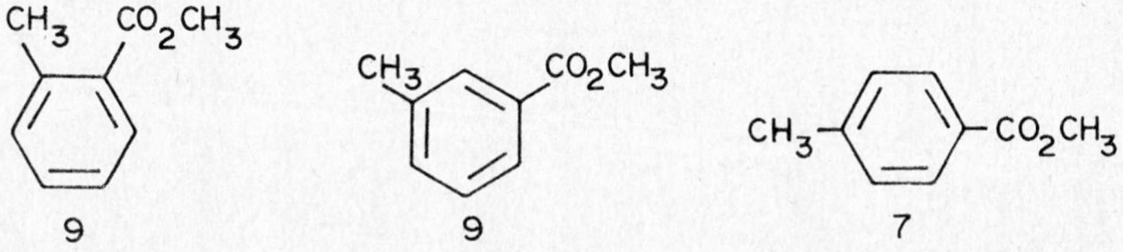

The compound must be the *para* isomer. For its proton NMR spectrum, see the answer to Problem 16.28.

16.32 The absence of a band at 3500 cm^{-1} indicates that there is no hydroxyl group. The absence of a band at 1720 cm^{-1} indicates that the compound is not an aldehyde or ketone. This suggests that the oxygen function is probably an ether. Possible structures are

CH$_3$OCH=CH$_2$

An NMR spectrum would readily distinguish between these possibilities.

16.33 These infrared data provide direct evidence for hydrogen bonding in alcohols. In dilute solution, the alcohol molecules are isolated, being surrounded by inert solvent molecules. The sharp band at 3580 cm^{-1} is due to the O—H stretching frequency in an isolated ethanol molecule. As the concentration of ethanol is increased, alcohol molecules can come in contact with one another and form hydrogen bonds. Hydrogen-bonded O—H is less "tight" than an isolated O—H and has a variable length (as the proton is transferred back and forth between oxygen atoms). Consequently, hydrogen-bonded O—H absorbs at a lower frequency and with a broader range (3250-3350 cm^{-1}) than the isolated O—H group.

16.34 Refer to Table 16.3, p. 453, when answering this problem.

   a. The first compound, a ketone, will have an intense band in the 1650-1780 cm^{-1} region. The ether will not absorb in this region.

   b. The distinguishing feature is the C=C bond in the second compound which should absorb in the 1600-1680 cm^{-1} region.

   c. The second compound will show an O—H stretching band near 3500 cm^{-1}.

   d. Both compounds will show C=O stretching bands, but only the acid will also show a broad O—H band in the 2500-3000 cm^{-1} region.

16.35 Both compounds will be similar in the functional group region of the spectrum with bands at 3500 cm^{-1} for the O—H stretch and 1700 cm^{-1} for the C=O stretch. But their fingerprint regions (1500-700 cm^{-1}) are expected to differ from one another.

16.36 The band at 1725 cm^{-1} in the infrared spectrum is due to a carbonyl group, probably a ketone. The quartet-triplet pattern in the NMR spectrum suggests an ethyl group. The compound is 3-pentanone:

$$\underset{\displaystyle \qquad\qquad\;\; \overset{\textstyle \delta\; 0.9,\; \text{triplet, area 3 (or 6)}}{}}{CH_3-CH_2-\overset{\displaystyle O}{\overset{\|}{C}}-CH_2-CH_3}$$

δ 2.7, quartet, area 2 (or 4)

δ 0.9, triplet, area 3 (or 6)

16.37 The quartet-triplet pattern suggests that the ten protons are present as two ethyl groups. This gives a partial structure of $(CH_3CH_2)_2CO_3$. The chemical shift of the CH_2 groups (δ 4.15) suggests that they are attached to the oxygen atoms. Finally, the infrared band at 1745 cm^{-1} suggests a carbonyl function. The structure is diethyl carbonate:

$$CH_3CH_2O\overset{\displaystyle O}{\overset{\|}{C}}OCH_2CH_3$$

16.38 Compounds a, c, and e have no unsaturation and will not absorb in the ultraviolet region of the spectrum.

16.39 See p. 456 in the text. As the number of double bonds increases, so does the extent of conjugation. Thus the absorption maximum moves to longer and longer wavelengths.

16.40 We must use Beer's law (eq. 16.1) to solve this problem.

$$A = \epsilon c l, \text{ or } c = \frac{A}{\epsilon l}$$

$$c = \frac{0.43}{215 \times 1} = 2 \times 10^{-3} \text{ mol/L}$$

To go further, 1 L of the cyclohexane will contain $2 \times 10^{-3} \times 78 = 0.156$ g of benzene as a contaminant. As you can see, ultraviolet spectroscopy can be a very sensitive tool for detecting impurities.

16.41 Follow the example in eq. 16.3.

$$[CH_3CH_2-\overset{..}{\underset{..}{O}}-H]^+$$

16.42 The molecular ion of 1-pentanol will be

$$[CH_3CH_2CH_2CH_2CH_2\overset{.}{\underset{..}{O}}H]^+$$

Fragmentation between C-1 and C-2 would give a daughter ion with $m/e = 31$ (compare with eq. 16.4). A possible mechanism for this cleavage is

$$\left[CH_3CH_2CH_2\overset{\displaystyle H}{\underset{\displaystyle H}{C}}\text{—}\overset{\displaystyle H}{\underset{\displaystyle H}{C}}\text{—}\ddot{O}\text{—}H \right]^+$$

$$\downarrow$$

$$CH_3CH_2CH_2CH_2\cdot \;+\; \left[\overset{\displaystyle H}{\underset{\displaystyle H}{\diagdown\diagup}}C\!\!=\!\!\ddot{O}\text{—}H \right]^+$$

$$m/e = 31$$

16.43 The formula $C_5H_{12}O = C_5H_{11}OH$ tells us that the alcohols are saturated. The peak at $m/e = 59$ cannot be due to a four-carbon fragment ($C_4 = 4 \times 12 = 48$; this would require 11 hydrogens, too many for four carbons). Thus the peak must contain one oxygen, leaving $59 - 16 = 43$ for carbon and hydrogen. A satisfactory composition for the peak at 59 is $C_3H_7O^+$ (or $C_3H_6OH^+$). Similarly, the $m/e = 45$ peak corresponds to $C_2H_5O^+$ (or $C_2H_4OH^+$). Fragmentation of alcohols often occurs between the hydroxyl-bearing carbon and an attached hydrogen or carbon (eq. 16.4).

A possible structure for the first alcohol is

$$CH_3CH_2CH\text{——}OH \qquad or \qquad CH_3\overset{\displaystyle CH_3}{\underset{\displaystyle CH_2CH_3}{\underset{\displaystyle |}{C}}}\text{—}OH$$

3-pentanol   2-methyl-2-butanol

A possible structure for its isomer is

$$CH_3CH\text{——}OH \qquad or \qquad CH_3CH\text{——}OH$$
$$\;\;\;\;CH_2CH_2CH_3 \qquad\qquad\;\;\;\; CH(CH_3)_2$$

2-pentanol   3-methyl-2-butanol

The correct structure for the first isomer could be deduced by NMR spectroscopy as follows:

δ 4, 1 H, quintet

$CH_3CH_2CHCH_2CH_3$

δ 1
6 H
triplet

OH

δ 1.3
4 H
quintet or
multiplet

three ^{13}C peaks

CH_3 OH
   C
CH_3 CH_2CH_3

δ 1
6 H
singlet

δ 1.3
2 H
quartet

δ 1
3 H
triplet

four ^{13}C peaks

The correct structure for the second isomer also could be deduced by NMR spectroscopy. However, both proton spectra are likely to be quite complex because of similar chemical shifts and a great deal of spin-spin splitting. The ^{13}C spectra are simpler and diagnostic:

$CH_3CHCH_2CH_2CH_3$
       |
       OH

six types of protons

five ^{13}C peaks

          CH_3
CH_3CHCH
       | CH_3
       OH

five types of protons

four ^{13}C peaks

16.44 The NMR peak at δ 7.4 with an area of 5 suggests that the compound may have a phenyl group, C_6H_5—. If so, this accounts for 77 of the 102 mass units. This leaves only 25 mass units, one of which must be a hydrogen (for the NMR peak at δ 3.08). The other 24 units must be two carbon atoms, since the compound is a hydrocarbon (no other elements present except C and H). Phenylacetylene fits all the data:

$C\equiv C-H$

16.45 For the definitions, see the indicated sections in the text.

a. Sec. 16.8   b. Sec. 16.8   c. Sec. 16.8
d. Sec. 16.8   e. Sec. 16.8   f. Sec. 16.8
g. Sec. 16.10   h. Sec. 16.10   i. Sec. 16.12
j. Sec. 16.9a

16.46 The molecular weight of the repeating unit is 54 (butadiene) + 104 (styrene) = 158. The degree of polymerization is

$$n = \frac{79,000}{158} = 500$$

16.47 Follow eqs. 16.6-16.8, with X = Cl.

$$R\cdot + CH_2\!\!=\!\!CHCl \longrightarrow RCH_2\overset{\cdot}{C}HCl$$

$$RCH_2\overset{\cdot}{C}HCl + CH_2\!\!=\!\!CHCl \longrightarrow RCH_2CH\underset{\underset{Cl}{|}}{CH_2}\overset{\cdot}{C}HCl, \text{ etc.}$$

16.48

$$-CH_2CH\underset{\underset{OH}{|}}{}\!\!\left(\!CH_2CH\underset{\underset{OH}{|}}{}\!\right)_{\!n}\!\!-CH_2CH\underset{\underset{OH}{|}}{}\ldots$$

The vinyl monomer presumably would be $CH_2\!\!=\!\!CHOH$, but this is the *enol* of acetaldehyde, which exists almost completely as $CH_3CH\!\!=\!\!O$ and thus cannot be a vinyl monomer. Polyvinyl acetate can, however, serve as the precursor to polyvinyl alcohol:

$$HC\!\!\equiv\!\!CH + H\!-\!\overset{\overset{\textstyle O}{\|}}{OCCH_3} \longrightarrow CH_2\!\!=\!\!CH\!-\!\overset{\overset{\textstyle O}{\|}}{OCCH_3}$$

vinyl acetate

$$CH_2\!\!=\!\!CH\!-\!\overset{\overset{\textstyle O}{\|}}{OCCH_3} \xrightarrow[\text{polymerization}]{\text{radical}} \left(\!CH_2\!-\!\underset{\underset{\underset{O}{\|}}{\underset{OCCH_3}{|}}}{CH}\!\right)_{\!n}$$

The polyester can then be saponified with base;

$$\left(\!CH_2CH\!\right)_{\!n}\underset{\underset{\underset{O}{\|}}{OCCH_3}}{} \xrightarrow[\text{H}_2\text{O}]{\text{NaOH}} -\!\left(\!CH_2CH\!\right)_{\!n}\underset{\underset{OH}{|}}{} + n\,CH_3CO_2H$$

16.49 The growing polymer chain can abstract a hydrogen atom from the methyl group of a propylene monomer:

$$\sim\!\!\sim\!\!CH_2\!-\!\overset{\cdot}{C}H\underset{\underset{CH_3}{|}}{} + CH_3CH\!\!=\!\!CH_2 \longrightarrow \sim\!\!\sim\!\!CH_2\!-\!CH_2\underset{\underset{CH_3}{|}}{} + \overset{\cdot}{C}H_2\!-\!CH\!\!=\!\!CH_2$$

This process terminates the chain growth. It is an energetically favorable process because the resulting radical is allylic and resonance-stabilized:

$$\left[\dot{C}H_2\text{---}CH{=\!=}CH_2 \longleftrightarrow CH_2{=\!=}CH\text{---}\dot{C}H_2 \right]$$

16.50 Chain termination by ion combination is not possible because the ions have like charges. In cationic polymerizations, termination by proton loss is possible:

$$\text{\textasciitilde\textasciitilde CH}_2\overset{H}{\underset{R}{\text{---}\overset{+}{C}\text{---}R}} \xrightarrow{\;-H^+\;} \text{\textasciitilde\textasciitilde CH}_2{=\!=}\overset{}{\underset{R}{C}}\text{---}R$$

In anionic polymerizations, termination by proton abstraction (carbanions are strong bases) is one possibility:

$$\text{\textasciitilde\textasciitilde CH}_2\overset{\ominus}{\underset{R}{\text{---}\overset{\cdot\cdot}{C}\text{---}R}} + R'H \longrightarrow \text{\textasciitilde\textasciitilde CH}_2\overset{H}{\underset{R}{\text{---}\overset{|}{C}\text{---}R}} + R'^{\ominus}$$

16.51 The organometallic reagent might simply add to the ester carbonyl group (Sec. 10.15):

$$CH_2{=\!=}\overset{}{\underset{CH_3}{C}}\text{---}\overset{O}{\overset{\|}{C}}\text{---}OCH_3 + CH_3CH_2CH_2CH_2^-Li^+ \longrightarrow$$

$$CH_2{=\!=}\overset{}{\underset{CH_3}{C}}\text{---}\overset{O^-Li^+}{\underset{OCH_3}{\overset{|}{C}}}\text{---}CH_2CH_2CH_2CH_3 \longrightarrow$$

$$CH_2{=\!=}\overset{}{\underset{CH_3}{C}}\text{---}\overset{O}{\overset{\|}{C}}\text{---}CH_2CH_2CH_2CH_3 + CH_3O^-Li^+$$

The reaction could continue further, to give a tertiary alcohol.

16.52 See Sec. 16.10 and Example 16.10.

$$\overset{H\;\;\;Ph\;\;H\;\;\;Ph\;\;H\;\;\;Ph}{\underset{CH_2\;\;\;\;CH_2\;\;\;\;CH_2\;\;\;\;CH_2}{C\quad\quad C\quad\quad C}}$$

$$\left(Ph = \text{---}\langle\!\bigcirc\!\rangle \right)$$

16.53 The polymer has the following structure:

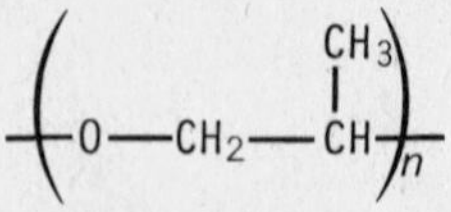

$$\left(\!\!-O-CH_2-\overset{\overset{\displaystyle CH_3}{|}}{CH}\!\!-\right)_{\!n}$$

It is formed by S_N2 displacements on the primary carbon of the epoxide (Sec. 8.9):

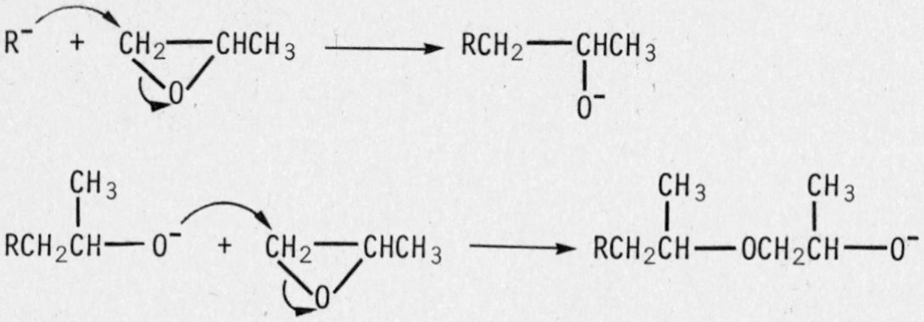

$$R^- \ + \ CH_2\!\!-\!\!\overset{\diagdown\!\!O\diagup}{CHCH_3} \ \longrightarrow \ RCH_2-\overset{\overset{\displaystyle }{|}}{\underset{\underset{\displaystyle O^-}{|}}{CHCH_3}}$$

$$RCH_2\overset{\overset{\displaystyle CH_3}{|}}{CH}-O^- \ + \ CH_2\!\!-\!\!\overset{\diagdown\!\!O\diagup}{CHCH_3} \ \longrightarrow \ RCH_2\overset{\overset{\displaystyle CH_3}{|}}{CH}-OCH_2\overset{\overset{\displaystyle CH_3}{|}}{CH}-O^-$$

etc.

16.54 Polyethylenes obtained by free-radical polymerization have highly branched structures as a consequence of chain-transfer reactions (see eq. 3.37 and the structure below it, on p. 90 of the text). Ziegler-Natta polyethylene is mainly linear: $-\!(CH_2CH_2\!)_n\!-$. It has a higher degree of crystallinity and a higher density than the polyethylene obtained by the free-radical process.

16.55 a. $$\left(\!\!-\overset{\overset{\displaystyle O}{\|}}{C}(CH_2)_6\overset{\overset{\displaystyle O}{\|}}{C}NH(CH_2)_6NH\!\!-\right)_{\!n}$$

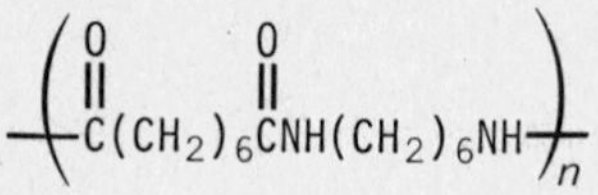

   b. $$\left(\!\!-\overset{\overset{\displaystyle O}{\|}}{C}NH-\!\!\left\langle\!\!\!\bigcirc\!\!\!\right\rangle\!\!-CH_2-\!\!\left\langle\!\!\!\bigcirc\!\!\!\right\rangle\!\!-NH\overset{\overset{\displaystyle O}{\|}}{C}OCH_2CH_2O\!\!-\right)_{\!n}$$

   c. $$\left(\!\!-\overset{\overset{\displaystyle O}{\|}}{C}(CH_2)_4\overset{\overset{\displaystyle O}{\|}}{C}OCH_2CH_2O\!\!-\right)_{\!n}$$

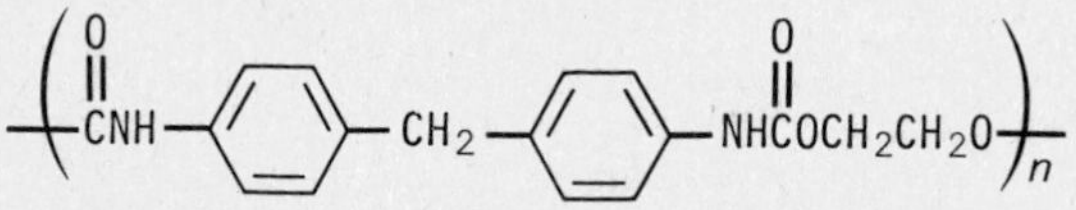

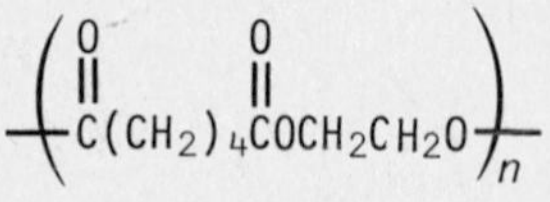

16.56 $$\left(\!\!-\overset{\overset{\displaystyle O}{\|}}{C}-O-\!\!\left\langle\!\!\!\bigcirc\!\!\!\right\rangle\!\!-\overset{\overset{\displaystyle CH_3}{|}}{\underset{\underset{\displaystyle CH_3}{|}}{C}}-\!\!\left\langle\!\!\!\bigcirc\!\!\!\right\rangle\!\!-O\!\!-\right)_{\!n}$$

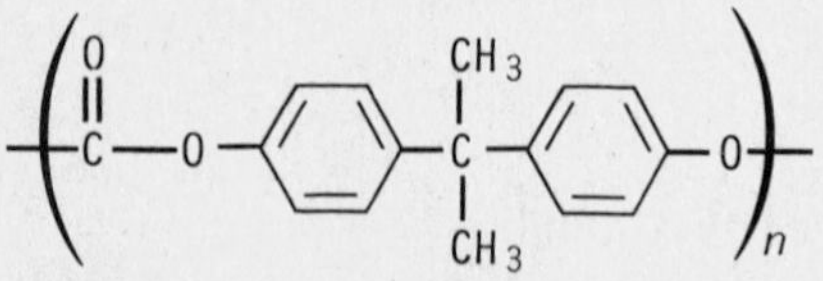

The other product of the polymerization is phenol.

16.57   The "unzipping" reaction, which gives formaldehyde, is expressed
        by the following curved arrows:

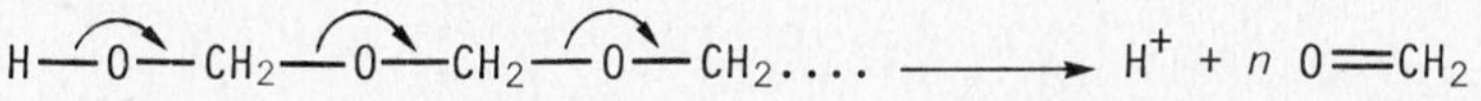

$$H-O-CH_2-O-CH_2-O-CH_2\ldots \longrightarrow H^+ + n\ O{=\!=}CH_2$$

This reaction is made possible by dissociation of a proton from a
terminal hydroxyl group (an alcohol; see eq. 7.9). If the polymer
is treated with acetic anhydride, the terminal hydroxyl groups are
esterified, and "unzipping" is no longer possible:

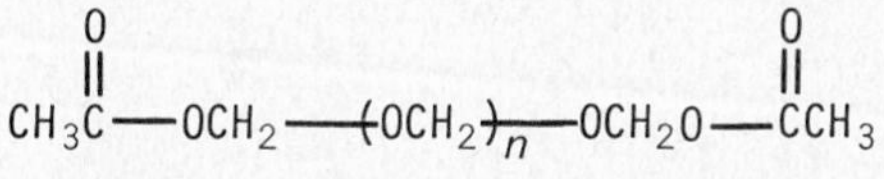

$$CH_3\overset{O}{\overset{\|}{C}}-OCH_2-\!\!\left(OCH_2\right)_{\!\!n}\!-OCH_2O-\overset{O}{\overset{\|}{C}}CH_3$$

16.58

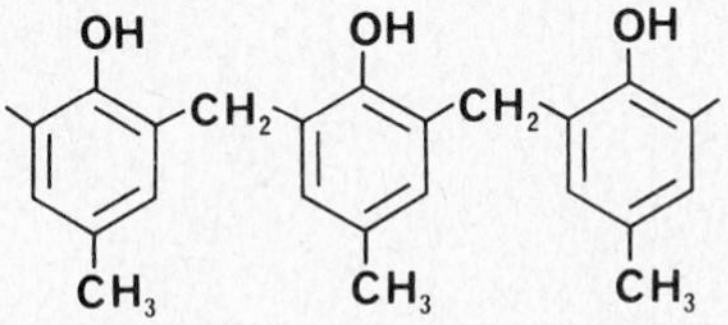

With the *para* position "blocked" by a methyl substituent, condensation can only occur *ortho* to the phenolic hydroxyl, leading to
a linear (*not* a cross-linked) polymer.

16.59

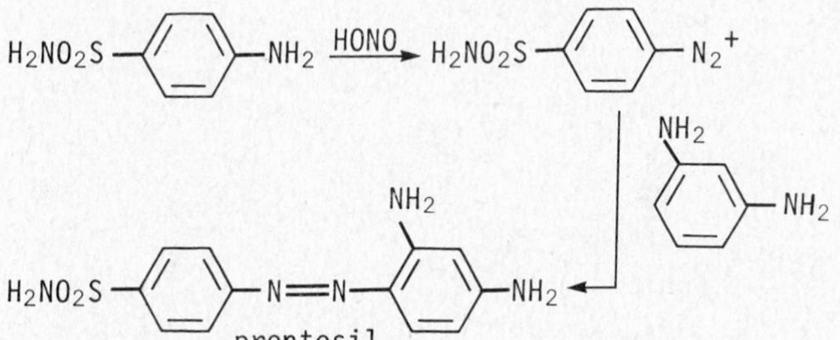

prontosil

Coupling occurs *ortho* to one amino group and *para* to the other
amino group in *m*-phenylenediamine (1,3-diaminobenzene).

16.60   The electrophile may be formed as follows:

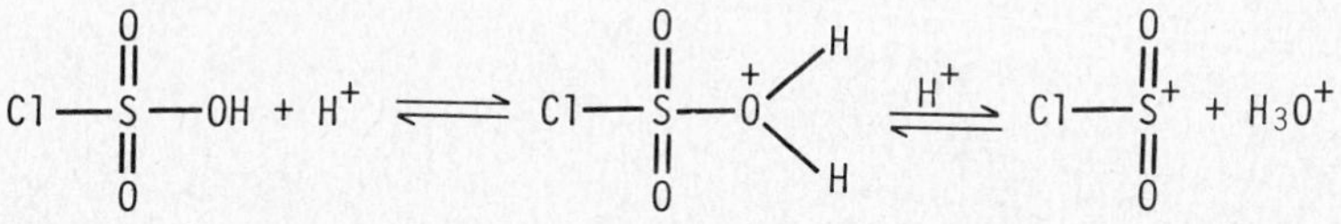

The $CH_3\overset{O}{\overset{||}{C}}—\overset{..}{N}H—$ substituent is *o,p*-directing:

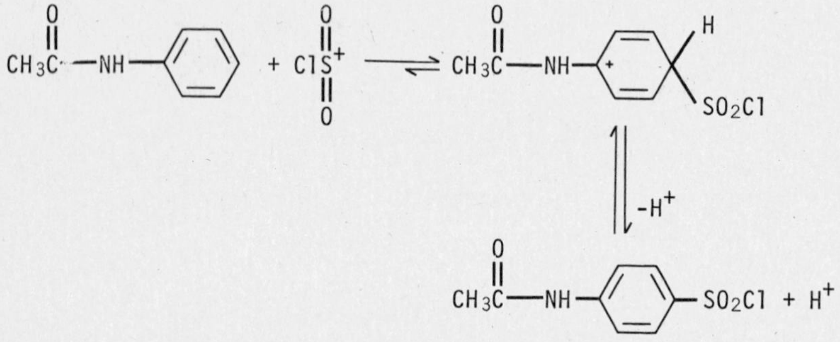

Alternatively, the sulfonating agent may be $^+SO_3H$, which will give a sulfonic acid that is then converted to the sulfonyl chloride by the excess chlorosulfonic acid.

16.61

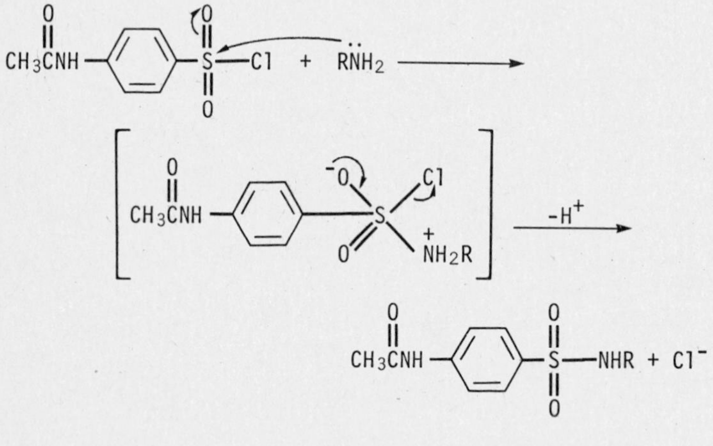

Sulfur, being a second-row element, can expand its valence shell beyond eight electrons.

16.62 The final step is the hydrolysis of an amide (eq. 10.31):

$$CH_3CO_2H + H_2N\text{---}\bigcirc\text{---}SO_2NHR$$

16.63

16.64 The three subunits of streptomycin are linked by acetal groups, which are easily hydrolyzed by dilute acid:

16.65 The compound has two chiral centers, marked with asterisks:

There are $2^2 = 4$ possible stereoisomers, represented by the following Fischer projections:

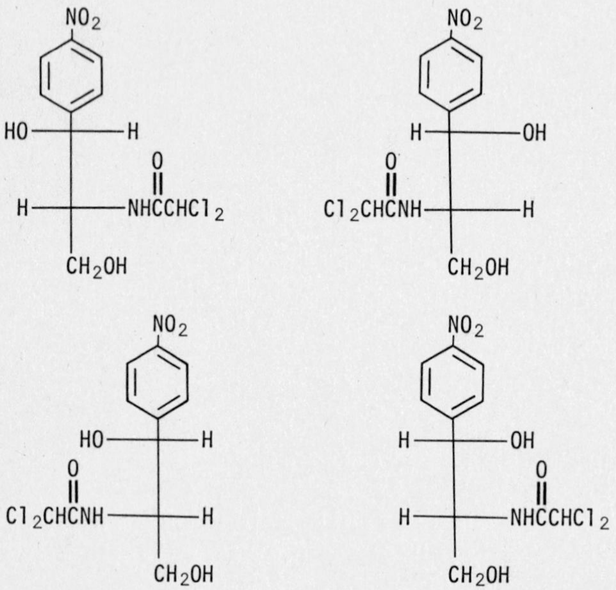

16.66

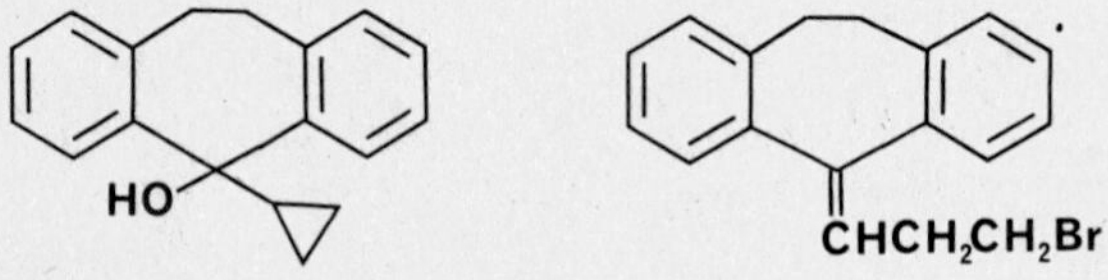

A   B

A is a tertiary alcohol formed from a ketone and a Grignard reagent in the usual way (Sec. 9.10). The formation of **B** from **A** involves a rearrangement:

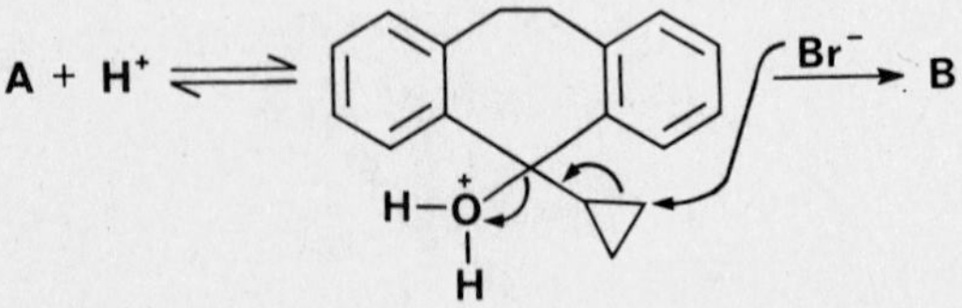

The final step is an alkylation of an amine by the primary alkyl bromide (Sec. 12.5).

16.67   The two chlorine substituents are *o,p*-directing, and —CO_2H is
        m-directing. The possible substitution positions are shown by
        arrows:

        Of these two, substitution at <u>a</u> gives the least sterically crowded
        product.

16.68   The first step is an S_N2 displacement by the 1-naphthoxide ion on
        the primary alkyl chloride (Sec. 6.5):

        The second step is another S_N2 displacement, this time on the
        primary carbon of the epoxide (Sec. 8.9):

16.69 The first step involves a free-radical chain mechanism (Secs. 2.15 and 2.16). The second step is an S_N1 reaction (Sec. 6.6).

$$\left(\bigcirc\!\!\!\bigcirc\right)_2\!\!-CHBr \rightleftharpoons \left(\bigcirc\!\!\!\bigcirc\right)_2\!\!-CH^+ + Br^-$$

$$\left(\bigcirc\!\!\!\bigcirc\right)_2\!\!-CH^+ + H\ddot{O}CH_2CH_2N(CH_3)_2 \longrightarrow$$

$$\left(\bigcirc\!\!\!\bigcirc\right)_2\!\!-CH-\overset{+}{\underset{H}{\ddot{O}}}-CH_2CH_2N(CH_3)_2 \rightleftharpoons$$

$$\left(\bigcirc\!\!\!\bigcirc\right)_2\!\!-CH-O-CH_2CH_2N(CH_3)_2 + H^+$$

The intermediate carbocation is stabilized by resonance. The positive charge can be delocalized to the *ortho* and *para* positions of both phenyl rings. Some of the contributors to the resonance hybrid are the following:

In the text, methods for preparing the particular classes of compound are presented in two ways. In some chapters, methods for synthesizing the class of compound with which the chapter deals are enumerated and discussed. More frequently, however, syntheses are presented less formally as a consequence of various reactions.

In this summary of the main synthetic methods for each important class of compound, a general equation for each reaction is given. Also, section numbers, given in parentheses, refer you to the place or places of the book where the reaction is described. You can easily find sections in the text by referring to the headings at the top of all the odd-numbered (right-hand) pages. References to "Word About" sections are abbreviated WA.

1. Alkanes and Cycloalkanes

    a. alkenes + H_2 (3.9; 4.8; 11.11)

$$\text{C}=\text{C} + H_2 \xrightarrow[\text{catalyst}]{\text{Ni or Pt}} -\underset{H}{\overset{|}{C}}-\underset{H}{\overset{|}{C}}-$$

    b. alkynes + H_2 (3.1)

$$-\text{C}\equiv\text{C}- + H_2 \xrightarrow[\text{catalyst}]{\text{Ni or Pt}} -\underset{H}{\overset{H}{C}}-\underset{H}{\overset{H}{C}}-$$

    c. cyclohexanes from aromatic compounds + H_2 (4.8; WA 5, p. 126)

$$\bigcirc + 3\ H_2 \xrightarrow[\text{catalyst}]{\text{Ni or Pt}} \bigcirc$$

    d. Grignard reagent + H_2O (or D_2O) (8.5)

$$RMgX + H\!-\!OH \longrightarrow R\!-\!H + Mg(OH)X$$

2. <u>Alkenes and Cycloalkenes</u>

    a. alkynes + H_2 (3.25)

$$-C\equiv C- \ + \ H_2 \ \xrightarrow[\text{catalyst}]{\text{Lindlar's}} \ \underset{H}{\overset{H}{>}}C=C< \qquad (cis\text{-alkene})$$

    b. dehydration of alcohols (7.9; 11.6)

$$-\underset{OH}{\overset{|}{C}}-\underset{H}{\overset{|}{C}}- \ \xrightarrow{H^+} \ >C=C< \ + \ H_2O$$

    c. elimination reaction; alkyl halide + strong base (6.8; 6.9)

$$-\underset{X}{\overset{|}{C}}-\underset{H}{\overset{|}{C}}- \ \xrightarrow{\text{base}} \ >C=C< \ + \ BH^+ \ + \ X^-$$

    d. *cis-trans* isomerism (3.5)

$$\underset{B}{\overset{A}{>}}C=C\underset{B}{\overset{A}{<}} \ \underset{\text{light}}{\overset{\text{heat or}}{\rightleftharpoons}} \ \underset{B}{\overset{A}{>}}C=C\underset{A}{\overset{B}{<}}$$

    e. elimination from alkyldiazonium ions (12.12)

$$RCH_2CH_2NH_2 \ \xrightarrow{HONO} \ RCH_2CH_2N_2^+ \ \xrightarrow[-H^+]{-N_2} \ RCH=CH_2$$

    f. Diels-Alder reaction (3.17)

    g. cracking of alkanes (WA 3, p. 97; WA 4, p. 102)

$$CH_3CH_3 \ \xrightarrow{700\text{-}900°C} \ CH_2=CH_2 \ + \ H_2$$

$$C_nH_{2n+2} \ \xrightarrow{\text{catalyst}} \ C_mH_{2m} \ + \ C_pH_{2p+2} \quad (m+p=n)$$

3. Alkynes

   a. from acetylides and alkyl halides (3.26; 6.3)

$$R—C \equiv C—H \xrightarrow[\text{NH}_3]{\text{NaNH}_2} R—C \equiv C^{\ominus} Na^{\oplus} \xrightarrow{R'X} R—C \equiv C—R'$$

                      (best for R' = primary)

   b. pyrolysis of methane (WA 3, p. 98)

$$2 \; CH_4 \xrightarrow{1500°C} HC \equiv CH + 3 \; H_2$$

4. Aromatic Compounds

   a. alkylbenzenes from Friedel-Crafts reactions (4.9; 4.10c; WA 5, p. 125; 11.13)

$$R—X + Ar—H \xrightarrow{AlCl_3} R—Ar + HX$$

$$\diagdown C = C \diagup + Ar—H \xrightarrow{H^+} Ar—\overset{|}{\underset{|}{C}}—\overset{|}{\underset{|}{C}}—H$$

   b. aromatic nitro compounds, by nitration (4.9; 4.10; 4.12; 4.13; 7.16; 12.17)

$$Ar—H + HONO_2 \xrightarrow{H^+} Ar—NO_2 + H_2O$$

   c. aromatic sulfonic acids, by sulfonation (4.9; 4.10b; 11.13)

$$Ar—H + HOSO_3H \longrightarrow Ar—OSO_3H + H_2O$$

   d. aromatic halogen compounds, by halogenation (4.2; 4.9; 4.10; 4.12; 4.13; 7.16)

$$Ar—H + X_2 \xrightarrow{FeX_3} Ar—X + HX \; (X = Cl, \; Br)$$

   e. aromatic halogen compounds from diazonium salts (12.13)

$$Ar—N_2^+ + HX \xrightarrow{Cu_2X_2} Ar—X + N_2 \; (X = Cl, \; Br)$$

$$Ar—N_2^+ + KI \longrightarrow Ar—I + N_2 + K^+$$

   f. alkylbenzenes from alkanes (WA 4, p. 103)

5. Alcohols

a. hydration of alkenes (3.10; 3.13)

$$\text{C=C} \quad + \quad H\text{—}OH \quad \xrightarrow{H^+} \quad \text{—}\overset{|}{\underset{OH}{C}}\text{—}\overset{|}{\underset{H}{C}}\text{—} \quad \text{(follow Markovnikov's rule)}$$

b. hydroboration-oxidation of alkenes (3.20)

$$RCH\text{=}CH_2 \quad \xrightarrow[\substack{2.\ H_2O_2, \\ OH^-}]{1.\ BH_3} \quad RCH_2CH_2OH$$

c. alkyl halides + aqueous base (6.2; 6.3; 6.6; 6.7)

$$R\text{—}X \; + \; OH^- \; \longrightarrow \; R\text{—}OH \; + \; X^- \quad \text{(best for R = primary)}$$

d. Grignard reagent + carbonyl compound (9.10)

$$\text{C=O} \; + \; RMgX \; \longrightarrow \; R\text{—}\overset{|}{\underset{|}{C}}\text{—}OMgX \; \xrightarrow[H^+]{H_2O} \; R\text{—}\overset{|}{\underset{|}{C}}\text{—}OH$$

For primary alcohols, use formaldehyde. For secondary alcohols, use other aldehydes. For tertiary alcohols, use a ketone.

e. reduction of aldehydes or ketones (9.13; 9.20; 13.10)

$$\text{C=O} \; \xrightarrow[\substack{or \\ LiAlH_4,\ NaBH_4}]{H_2,\ catalyst} \; H\text{—}\overset{|}{\underset{|}{C}}\text{—}OH$$

f. Grignard reagent + ethylene oxide (8.9)

$$RMgX \; + \; \underset{O}{CH_2\text{—}CH_2} \; \longrightarrow \; R\text{-}CH_2CH_2OMgX \; \xrightarrow[H^+]{H_2O} \; R\text{-}CH_2CH_2OH$$

Useful only for primary alcohols.

g. Grignard reagent (excess) + ester (10.15)

$$\overset{O}{\overset{\|}{RC}}\text{—}OR' \; + \; 2\ R''MgX \; \longrightarrow \; R\text{—}\overset{OMgX}{\underset{R''}{C}}\text{—}R'' \; \xrightarrow[H^+]{H_2O} \; R\text{—}\overset{OH}{\underset{R''}{C}}\text{—}R''$$

Useful for tertiary alcohols with at least two identical R groups.

h. saponification of esters (10.13)

$$R-\overset{\overset{\text{O}}{\|}}{C}-OR' \ + \ Na^+OH^- \ \longrightarrow \ R-\overset{\overset{\text{O}}{\|}}{C}-O^-Na^+ \ + \ R'OH$$

i. reduction of esters (10.16; 11.13)

$$R-\overset{\overset{\text{O}}{\|}}{C}-OR' \ \xrightarrow[\substack{\text{or} \\ \text{LiAlH}_4}]{\text{H}_2, \ \text{Ni}} \ RCH_2OH \ + \ R'OH$$

j. diazotization of primary amines (12.12)

$$RNH_2 \ + \ HONO \ \longrightarrow \ ROH \ + \ N_2 \ + \ H_2O$$

k. methanol from carbon monoxide and hydrogen (WA 11, p. 203)

$$CO \ + \ 2 \ H_2 \ \xrightarrow{\text{catalyst}} \ CH_3OH$$

6. Phenols

a. from diazonium salts and base (12.13)

$$ArN_2^+X^- \ + \ NaOH^- \ \xrightarrow[\text{warm}]{H_2O} \ Ar-OH \ + \ N_2 \ + \ Na^+X^-$$

b. from phenoxides and acid (7.7)

$$ArO^-Na^+ \ + \ H^+X^- \ \longrightarrow \ ArOH \ + \ Na^+X^-$$

c. phenol from isopropylbenzene (WA 5, p. 126)

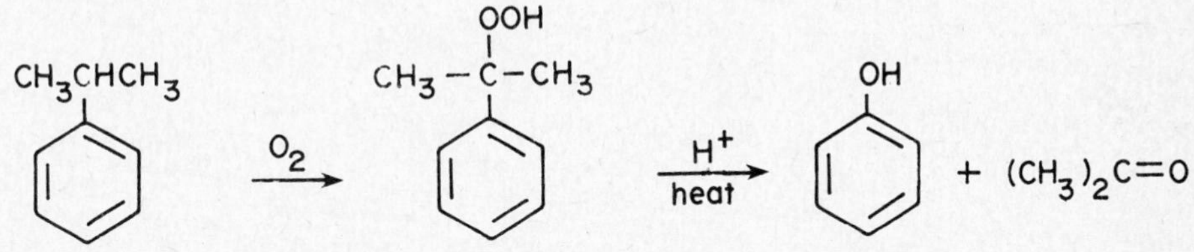

d. reduction of quinones (WA 18, p. 254)

7. Glycols

a. ring opening of epoxides (8.9)

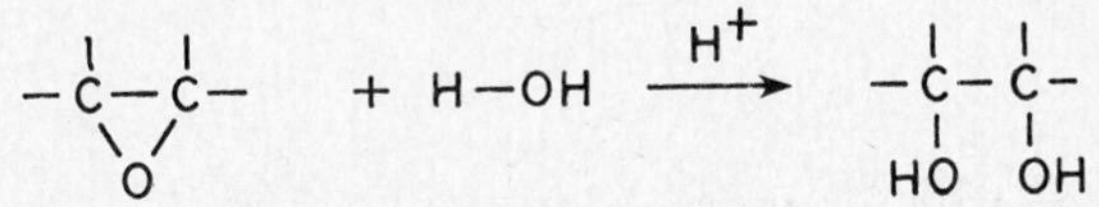

387

b. oxidation of alkenes (3.21)

$$\underset{}{\diagdown}C=C\underset{}{\diagup} \xrightarrow{KMnO_4} -\underset{|}{\underset{HO}{C}}-\underset{|}{\underset{OH}{C}}-$$

c. hydrolysis of a fat or oil to give glycerol (11.10; 11.12)

$$
\begin{array}{l}
CH_2O\overset{O}{\overset{\|}{C}}R \\
| \\
CHO\overset{O}{\overset{\|}{C}}R \\
| \\
CH_2O\overset{O}{\overset{\|}{C}}R
\end{array}
+ \; 3 \; NaOH \xrightarrow{heat}
\begin{array}{l}
CH_2OH \\
| \\
CHOH \\
| \\
CH_2OH
\end{array}
+ \; 2 \; R\overset{O}{\overset{\|}{C}}-O^-Na^+
$$

8. <u>Ethers and Epoxides</u>

a. from alkoxides and alkyl halides; Williamson synthesis (8.6; 13.11; 16.16)

$$RO^-Na^+ \; + \; R'X \longrightarrow R-OR' \; + \; Na^+X^-$$

$$ArO^-Na^+ \; + \; R'X \longrightarrow Ar-OR' \; + \; Na^+X^-$$

Best for R' = primary or secondary

b. dehydration of alcohols (8.6)

$$2 \; ROH \xrightarrow{H^+} ROR \; + \; H_2O$$

Most useful for symmetric ethers.

c. ethylene oxide from ethylene and air (8.8)

$$CH_2{=}CH_2 \; + \; O_2 \xrightarrow[catalyst]{Ag} CH_2\underset{\diagdown \; O \; \diagup}{-}CH_2$$

d. alkenes and peracids (8.8)

$$\underset{}{\diagdown}C{=}C\underset{}{\diagup} \; + \; R-\overset{O}{\overset{\|}{C}}-O-OH \longrightarrow -\underset{|}{C}-\underset{|}{\underset{\diagdown O \diagup}{C}}- \; + \; R-\overset{O}{\overset{\|}{C}}-OH$$

R is usually $CH_3{-}$, $C_6H_5{-}$, or m-$ClC_6H_4{-}$

e. intramolecular S_N2 (or Williamson) reaction (WA 16, p. 223)

$$-\overset{X}{\underset{OH}{\overset{|}{C}}}-\overset{|}{\underset{|}{C}}- \quad \xrightarrow{NaOH} \quad -\overset{|}{\underset{O^-}{\overset{X}{C}}}-\overset{X}{\underset{}{\overset{|}{C}}}- \quad \longrightarrow \quad -\overset{|}{\underset{\diagdown O}{C}}-\overset{|}{\underset{}{C}}- \quad + \; X^- \; (X = Cl, Br)$$

f. ring opening of epoxides with alcohols (8.9)

$$-\overset{|}{\underset{\diagdown O \diagup}{C}}-\overset{|}{\underset{}{C}}- \quad + \; ROH \quad \xrightarrow{H^+} \quad -\overset{|}{\underset{OH}{C}}-\overset{|}{\underset{}{C}}-OR$$

9. Alkyl Halides

a. halogenation of alkanes (2.15)

$$-\overset{|}{\underset{|}{C}}-H \quad + \; X_2 \quad \xrightarrow[\text{light}]{\text{heat or}} \quad -\overset{|}{\underset{|}{C}}-X \quad + \; H-X$$

b. alkenes (or dienes) + hydrogen halides (3.11; 3.16; 5.14)

$$\diagup C = C \diagdown \quad + \; H-X \quad \longrightarrow \quad -\overset{|}{\underset{H}{C}}-\overset{|}{\underset{X}{C}}-$$

c. vinyl halides from alkynes + hydrogen halides (3.25)

$$-C \equiv C- \quad + \; HX \quad \longrightarrow \quad \overset{H}{\diagdown} C = C \overset{\diagup}{\diagdown X}$$

d. alcohols + hydrogen halides (7.10)

$$R-OH \; + \; H-X \longrightarrow R-X \; + \; H_2O$$

(Catalysts such as ZnX_2 are required when R is primary.)

e. alcohols + thionyl chloride or phosphorus halides (7.11)

$$R-OH \; + \; SOCl_2 \longrightarrow R-Cl \; + \; SO_2 \; + \; HCl$$

$$3 \; R-OH \; + \; PX_3 \longrightarrow 3 \; R-X \; + \; H_3PO_3$$

f. cleavage of ethers with hydrogen halides (8.7)

$$R\!-\!O\!-\!R' \;+\; 2\;H\!-\!X \longrightarrow R\!-\!X \;+\; R'\!-\!X \;+\; H_2O$$

g. alkyl iodides from alkyl chlorides (6.3)

$$R\!-\!Cl \;+\; NaI \xrightarrow{\text{acetone}} R\!-\!I \;+\; NaCl$$

10. <u>Polyhalogen Compounds</u>

   a. halogenation of alkanes (2.15; 6.10)
   b. addition of halogen to alkenes, dienes, and alkynes (3.8; 3.25)

11. <u>Aldehydes and Ketones</u>

   a. oxidation of alcohols (7.13; 9.4)

   Primary alcohols give aldehydes. The best oxidant is pyridinium
   chlorochromate (PCC). Secondary alcohols give ketones; the reagent
   is usually CrO_3, H^+.

   b. hydration of alkynes (3.25; 9.4)

$$R\!-\!C\!\equiv\!C\!-\!H \;+\; H\!-\!OH \xrightarrow[\text{H}_2\text{SO}_4]{\text{HgSO}_4} R\!-\!\overset{\displaystyle O}{\overset{\|}{C}}\!-\!CH_3$$

   c. ozonolysis of alkenes (3.22; 9.4)

   d. decarboxylation of β-keto acids (11.8)

   e. hydrolysis of acetals and ketals (9.8)

f. acetaldehyde from ethylene (9.3)

$$2\ CH_2{=\!\!=}CH_2\ +\ O_2\ \xrightarrow[\text{100-130°C}]{\text{Pd-Cu}}\ 2\ CH_3CH{=\!\!=}O$$

g. formaldehyde from methanol (9.3)

$$CH_3OH\ \xrightarrow[\text{600-700°C}]{\text{Ag}}\ CH_2{=\!\!=}O\ +\ H_2$$

h. deuterated aldehydes or ketones (9.17)

$$\underset{\underset{|}{\overset{|}{C}}}{\overset{H}{\underset{}{}}}\overset{O}{\underset{}{\overset{\parallel}{C}}}\ \xrightarrow[\text{CH}_3\text{OD}]{\text{CH}_3\text{O}^-\text{Na}^+}\ \underset{\underset{|}{\overset{|}{C}}}{\overset{D}{\underset{}{}}}\overset{O}{\underset{}{\overset{\parallel}{C}}}$$

Only α-hydrogens exchange; acid catalysis is also used, especially with aldehydes.

i. β-hydroxy carbonyl compounds and α,β-unsaturated carbonyl compounds by way of the aldol condensation (9.18; 9.19; 9.20)

$$2\ \underset{H}{\overset{|}{C}}{\overset{O}{\overset{\parallel}{C}}}\ \xrightarrow{\ OH^-\ }\ \overset{OH}{\underset{|}{C}}\underset{H}{\overset{|}{C}}\overset{O}{\overset{\parallel}{C}}\ \xrightarrow[\ H^+\ ]{\text{heat or}}\ \diagdown C{=\!\!=}C\overset{O}{\overset{\parallel}{C}}$$

an aldol

12. <u>Carboxylic Acids</u>

a. hydrolysis of nitriles (cyanides) (10.8d)

$$R{-\!\!-}C{\equiv}N\ \xrightarrow[\text{H}^+\text{ or OH}^-]{\text{H}_2\text{O}}\ R{-\!\!-}C\overset{\displaystyle O}{\underset{\displaystyle OH}{\diagup}}\quad (+\ NH_3\ or\ NH_4{}^+)$$

b. Grignard reagents + carbon dioxide (10.8c)

$$R{-\!\!-}MgX\ +\ O{=\!\!=}C{=\!\!=}O\ \longrightarrow\ R{-\!\!-}\overset{O}{\overset{\parallel}{C}}{-\!\!-}OMgX\ \xrightarrow[\text{H}^+]{\text{H}_2\text{O}}\ R{-\!\!-}\overset{O}{\overset{\parallel}{C}}{-\!\!-}OH$$

c. oxidation of aromatic side chains (10.8b)

$$Ar{-\!\!-}CH_3\ \xrightarrow[\text{H}^+]{\text{CrO}_3}\ Ar{-\!\!-}\overset{O}{\overset{\parallel}{C}}{-\!\!-}OH$$

d. oxidation of aldehydes (9.14; 10.8a; 13.9)

$$R-CH=O \xrightarrow[\substack{\text{or} \\ \text{other} \\ \text{oxidant}}]{Ag^+} \underset{\displaystyle RC-OH}{\overset{\displaystyle O}{\|}}$$

e. oxidation of primary alcohols (10.8a; 13.9)

$$RCH_2OH \xrightarrow[H^+]{K_2Cr_2O_7} RCO_2H$$

f. saponification of esters (10.13)

$$\underset{\displaystyle R-C-OR'}{\overset{\displaystyle O}{\|}} + NaOH \longrightarrow R'OH + \underset{\displaystyle R-C-O^-Na^+}{\overset{\displaystyle O}{\|}} \xrightarrow{H^+} \underset{\displaystyle R-C-OH}{\overset{\displaystyle O}{\|}}$$

g. hydrolysis of acid derivatives (10.18; 10.19; 10.20)

$$\underset{\displaystyle R-C-Cl}{\overset{\displaystyle O}{\|}} + H_2O \longrightarrow \underset{\displaystyle R-C-OH}{\overset{\displaystyle O}{\|}} + HCl$$

$$\underset{\displaystyle R-C-O-C-R}{\overset{\displaystyle O \qquad O}{\| \qquad \|}} + H_2O \longrightarrow 2\ \underset{\displaystyle R-C-OH}{\overset{\displaystyle O}{\|}}$$

$$\underset{\displaystyle R-C-NH_2}{\overset{\displaystyle O}{\|}} + H_2O \xrightarrow[OH^-]{H^+ \text{ or}} \underset{\displaystyle R-C-OH}{\overset{\displaystyle O}{\|}} + NH_3$$

h. decarboxylation of malonic-type acids (11.3)

i. dicarboxylic acids from cyclic ketones (9.14)

$$+ HNO_3 \xrightarrow{V_2O_5} HOOC(CH_2)_4 COOH$$

j. phenolic acids from phenols and carbon dioxide (11.7)

$$\text{phenol} + CO_2 \xrightarrow[\text{heat, pressure}]{\text{base}} \text{salicylic acid}$$

k. keto acids from hydroxy acids (5.8)

$$\underset{\underset{OH}{|}}{RCHCOOH} \xrightarrow[\substack{\text{chemical}\\\text{oxidation}}]{\text{enzymatic or}} \underset{\underset{O}{\|}}{RCCOOH}$$

13. <u>Esters</u>

a. from an alcohol and an acid (10.11; 10.12; 14.5; 16.11)

$$\underset{\overset{\|}{O}}{R-C}-OH + R'OH \xrightarrow{H^+} \underset{\overset{\|}{O}}{R-C}-OR' + H_2O$$

b. from an alcohol and an acid derivative (10.18; 10.19; 10.21; 11.4; 11.5; 11.7; 13.11; 13.14b)

$$\underset{\overset{\|}{O}}{R-C}-Cl + R'OH \longrightarrow \underset{\overset{\|}{O}}{R-C}-OR' + HCl$$

$$\underset{\overset{\|}{O}}{R-C}-O-\underset{\overset{\|}{O}}{C}-R + R'OH \longrightarrow \underset{\overset{\|}{O}}{R-C}-OR' + RCO_2H$$

c. salt + alkyl halide (6.3; 14.11)

$$\underset{\overset{\|}{O}}{R-C}-O^-Na^+ + R'X \longrightarrow \underset{\overset{\|}{O}}{R-C}-OR' + Na^+X^-$$

d. lactones from hydroxy acids (11.6)

$$\underset{\underset{OH}{|}}{C-C-C-\overset{\overset{O}{\|}}{C}-OH} \xrightarrow[\text{or } H^+]{\text{heat}} \underset{C-C}{\overset{C-C}{}} \overset{O}{\underset{O}{\diagdown}} + H_2O$$

e. lactides from α-hydroxy acids (11.6)

$$2 \text{ RCHCOOH} \quad \xrightarrow{heat}$$
$$\overset{|}{\underset{OH}{}}$$

f. β-keto esters by way of the Claisen condensation (11.9)

$$2 \text{ RCH}_2\overset{O}{\overset{||}{C}}\text{—OR'} \quad \xrightarrow{base} \quad \text{RCH}_2\overset{O}{\overset{||}{C}}\text{—}\underset{R}{\overset{|}{CH}}\text{—}\overset{O}{\overset{||}{C}}\text{—OR'} \quad + \quad \text{R'OH}$$

14. Amides

a. acyl halides + ammonia (or primary or secondary amines) (10.18; 12.10; 14.5)

$$R\text{—}\overset{O}{\overset{||}{C}}\text{—Cl} \quad + \quad H\text{—}N\overset{R'}{\underset{R''}{}} \quad \longrightarrow \quad R\text{—}\overset{O}{\overset{||}{C}}\text{—}N\overset{R'}{\underset{R''}{}} \quad + \quad \text{HCl}$$

(R' and R" = H, alkyl, or aryl)

b. acid anhydrides and ammonia (or primary or secondary amines) (10.19; 12.10)

$$R\text{—}\overset{O}{\overset{||}{C}}\text{—O—}\overset{O}{\overset{||}{C}}\text{—R} \quad + \quad H\text{—}N\overset{R'}{\underset{R''}{}} \quad \longrightarrow \quad R\text{—}\overset{O}{\overset{||}{C}}\text{—}N\overset{R'}{\underset{R''}{}} \quad + \quad \text{RC}\overset{O}{\overset{||}{}}\text{—OH}$$

c. esters and ammonia (or primary or secondary amines) (10.14)

$$R\text{—}\overset{O}{\overset{||}{C}}\text{—OR'} \quad + \quad H\text{—}N\overset{R''}{\underset{R'''}{}} \quad \longrightarrow \quad R\text{—}\overset{O}{\overset{||}{C}}\text{—}N\overset{R''}{\underset{R'''}{}} \quad + \quad \text{R'OH}$$

d. from acids and ammonia (or primary or secondary amines) (10.20)

$$R\text{—}\overset{O}{\overset{||}{C}}\text{—OH} \quad + \quad \text{HNR}_2' \quad \xrightarrow{heat} \quad R\text{—}\overset{O}{\overset{||}{C}}\text{—NR}_2' \quad + \quad H_2O \quad (\text{R'} = \text{H or alkyl})$$

e. polyamides from diamines and dicarboxylic acids or their derivatives (12.15; WA 25, p. 340)

$$H_2N\sim\sim NH_2 \ + \ HOOC\sim\sim COOH \ \longrightarrow \ \sim\sim\underset{\text{C}}{\overset{\text{O}}{\|}}\text{-NH}\sim\sim\text{HN}-\overset{\text{O}}{\overset{\|}{\text{C}}}\sim\sim_n$$

f. barbiturates from malonic esters and urea (12.19)

$$\begin{matrix} \diagdown \\ \diagup \end{matrix}C\begin{matrix} \diagup\text{COOR} \\ \diagdown\text{COOR} \end{matrix} \ + \ \begin{matrix} H_2N \\ H_2N \end{matrix}C{=}O \ \longrightarrow \ \begin{matrix} \diagdown \\ \diagup \end{matrix}C\begin{matrix} \diagup\overset{\overset{\text{O}}{\|}}{\text{C}}{-}\text{NH} \\ \diagdown\underset{\underset{\text{O}}{\|}}{\text{C}}{-}\text{NH} \end{matrix}C{=}O \ + \ 2\ ROH$$

g. peptides (14.11)

15. <u>Other Carboxylic Acid Derivatives</u>

a. salts from acids and bases (10.7)

$$R-\overset{\overset{\text{O}}{\|}}{\text{C}}-OH \ + \ NaOH \ \longrightarrow \ R-\overset{\overset{\text{O}}{\|}}{\text{C}}-O^-Na^+ \ + \ H_2O$$

b. salts (soaps) by saponification of esters (10.13; 11.12)

$$R-\overset{\overset{\text{O}}{\|}}{\text{C}}-OR' \ + \ NaOH \ \xrightarrow{\text{heat}} \ R-\overset{\overset{\text{O}}{\|}}{\text{C}}-O^-Na^+ \ + \ R'OH$$

c. acyl halides from acids (10.18)

$$R-\overset{\overset{\text{O}}{\|}}{\text{C}}-OH \ \xrightarrow[\text{PCl}_3]{\text{SOCl}_2 \text{ or}} \ R-\overset{\overset{\text{O}}{\|}}{\text{C}}-Cl$$

d. anhydrides from diacids (11.3)

16. <u>Nitriles (Cyanides)</u>

    a. nitriles from alkyl halides and inorganic cyanides (6.3; 10.8d)

$$R\!-\!X \;+\; Na^{+}CN^{-} \longrightarrow R\!-\!CN \;+\; Na^{+}X^{-} \;\; (R \;=\; \text{primary or secondary})$$

    b. nitriles from diazonium ions (12.13)

$$Ar\!-\!N_2^{+}X^{-} \;+\; KCN \xrightarrow{\;Cu_2(CN)_2\;} Ar\!-\!CN \;+\; N_2 \;+\; Cu_2X_2$$

    c. dehydration of primary amides (10.20)

$$\underset{\displaystyle RCNH_2}{\overset{\displaystyle O\atop\displaystyle \|}{}} \xrightarrow{\;P_2O_5\;} RC\!\equiv\!N$$

17. <u>Amines and Related Compounds</u>

    a. alkylation of ammonia or amines (6.3; 12.5)

$$2 \;\;{\gt}N\!-\!H \;+\; R\!-\!X \longrightarrow {\gt}N\!-\!R \;+\; \overset{\oplus}{N}\!\!\begin{smallmatrix}H\\ \\H\end{smallmatrix} \;\; X^{\ominus}$$

    b. reduction of nitriles (12.6)

$$R\!-\!C\!\equiv\!N \;+\; 2\,H_2 \xrightarrow[\text{H}_2,\text{ Ni catalyst}]{\text{LiAlH}_4\text{ or}} RCH_2NH_2$$

    c. reduction of nitro compounds (12.6)

$$Ar\!-\!NO_2 \xrightarrow[\text{H}_2,\text{ catalyst}]{\text{SnCl}_2,\text{ HCl or}} Ar\!-\!NH_2 \;(+\;\; 2\,H_2O)$$

    d. reduction of amides (10.20; 12.6)

$$\underset{\displaystyle R\!-\!\overset{\displaystyle O\atop\displaystyle \|}{C}\!-\!N\!\!\begin{smallmatrix}R'\\ \\R''\end{smallmatrix}}{} \xrightarrow{\;\text{LiAlH}_4\;} RCH_2\!-\!N\!\!\begin{smallmatrix}R'\\ \\R''\end{smallmatrix}$$

    e. hydrolysis of amides (10.20)

$$R\!-\!\overset{\displaystyle O\atop\displaystyle \|}{C}\!-\!N\!\!\begin{smallmatrix}R'\\ \\R''\end{smallmatrix} \;+\; H_2O \xrightarrow{\;OH^-\;} R\!-\!\overset{\displaystyle O\atop\displaystyle \|}{C}\!-\!O^- \;+\; R'\!-\!\overset{\displaystyle H\atop\displaystyle |}{N}\!-\!R''$$

    (R' and R" = alkyl or aryl)

f. amine salts from amines (4.16; 12.9)

$$\underset{\diagup}{\overset{\diagdown}{-}}\!N\!: \;+\; HX \;\longrightarrow\; -\overset{\displaystyle |}{\underset{\displaystyle |}{N}}\!\!\overset{+}{}\!\!-H \;+\; X^-$$

g. quaternary ammonium salts (6.3; 12.11)

$$R\!-\!\overset{\displaystyle ..}{\underset{\displaystyle |}{\underset{\displaystyle R''}{N}}}\!\!-R' \;+\; R'''\!-\!X \;\longrightarrow\; R\!-\!\overset{\displaystyle R'''}{\underset{\displaystyle |}{\underset{\displaystyle |}{\underset{\displaystyle R''}{N^+}}}}\!\!-R' \;\;X^-$$

$$(R''' \;=\; \text{primary or secondary})$$

h. amino alcohols from epoxides (8.9)

$$-\overset{\displaystyle |}{\underset{\displaystyle |}{C}}\!-\!\overset{\displaystyle |}{\underset{\displaystyle |}{C}}\!- \;\;+\;\; H\!-\!N\!\!\underset{R'}{\overset{R}{\diagup}} \;\longrightarrow\; -\overset{\displaystyle |}{\underset{\displaystyle |}{\underset{\displaystyle HO}{C}}}\!-\!\overset{\displaystyle |}{\underset{\displaystyle |}{\underset{\displaystyle NRR'}{C}}}\!-$$

$$(R \text{ and } R' = H \text{ or organic groups})$$

18. Miscellaneous Nitrogen Compounds

a. oximes, hydrazones, and imines from carbonyl compounds and ammonia derivatives (9.12)

$$\overset{\diagdown}{\underset{\diagup}{}}C\!=\!O \;+\; H_2N\!- \;\longrightarrow\; \overset{\diagdown}{\underset{\diagup}{}}C\!=\!N\!\overset{\diagdown}{} \;+\; H_2O$$

b. cyanohydrins from carbonyl compounds and hydrogen cyanide (9.11)

$$\overset{\diagdown}{\underset{\diagup}{}}C\!=\!O \;+\; HCN \;\longrightarrow\; \overset{\diagdown}{\underset{\diagup}{}}C\!\!\overset{OH}{\underset{CN}{\diagup}}$$

c. diazonium compounds from primary aromatic amines and nitrous acid (12.13)

$$Ar\!-\!NH_2 \;+\; HONO \;+\; HX \;\longrightarrow\; ArN_2^+X^- \;+\; 2\,H_2O$$

d. azo compounds, via coupling reactions (12.14)

$$ArN_2^+X^- + \;\langle\!\!\bigcirc\!\!\rangle\!-OH\,(or\,NR_2) \;\longrightarrow\; Ar\!-\!N\!\!\equiv\!\!N\!-\!\langle\!\!\bigcirc\!\!\rangle\!-OH\,(or\,NR_2)$$

397

e. alkyl nitrates (7.12; 7.14; 13.14b)

$$ROH + HONO_2 \longrightarrow RONO_2 + H_2O$$

f. alkyl nitrites (7.12)

$$ROH + HONO \longrightarrow RONO + H_2O$$

g. ureas (WA 20, p. 280; WA 24, p.333; 16.12)

h. isocyanates (WA 24, p. 331)

i. carbamates (urethanes) (WA 24, p. 332; 16.12)

j. nitroso compounds and nitrosamines (12.12)

$$R_2NH + HONO \xrightarrow{H^+} R_2N-N=O + H_2O$$

19. _Organic Sulfur Compounds_

a. thiols from alkyl halides and sodium hydrosulfide (6.3; 7.18)

$$R-X + Na^+SH^- \longrightarrow R-SH + Na^+X^- \text{ (best when R is primary)}$$

b. thioethers from alkyl halides and sodium mercaptides (6.3; 8.11)

$$R-X + Na^{+-}SR' \longrightarrow R-S-R' + Na^+X^- \text{ (best when R is primary)}$$

c. disulfides from thiols (7.18; 14.8)

$$2\ R-SH \xrightarrow{H_2O_2} R-S-S-R$$

d. sulfoxides and sulfones, by oxidation of sulfides (8.11)

e. sulfonium salts (6.3; 8.11; WA 8, p. 177)

$$R\!-\!\underset{\cdot\cdot}{\overset{\cdot\cdot}{S}}\!-\!R' \;+\; R''\!-\!X \;\longrightarrow\; R\!-\!\overset{\overset{\displaystyle R''}{|}}{S^+}\!-\!R' \;\; X^-$$

f. alkyl hydrogen sulfates from alcohols or from alkenes (3.11; 7.12; 11.13)

$$R\!-\!OH \;+\; HOSO_3H \;\xrightarrow{\;cold\;}\; ROSO_3H \;+\; H_2O$$

$$\underset{/}{\overset{\backslash}{C}}\!\!=\!\!\underset{\backslash}{\overset{/}{C}} \;+\; HOSO_3H \;\xrightarrow{\;cold\;}\; -\!\overset{|}{\underset{|}{C}}\!-\!\overset{|}{\underset{|}{C}}\!-\;\;\; \underset{H\quad OSO_3H}{}$$

g. sulfonic acids by sulfonation (4.9; 4.10b; 11.13)

h. sulfonamides (WA 27, p. 371; 16.14; 16.16)

20. **Miscellaneous Classes of Compounds**

a. Grignard reagents (8.5)

$$R\!-\!X \;+\; Mg \;\xrightarrow{\;ether\;}\; RMgX$$

b. organolithium compounds (8.5)

$$R\!-\!X \;+\; 2\,Li \;\longrightarrow\; RLi \;+\; LiX$$

c. acetylides (3.26)

$$R\!-\!C\!\equiv\!C\!-\!H \;+\; NaNH_2 \;\xrightarrow{\;NH_3\;}\; R\!-\!C\!\equiv\!C^-Na^+ \;+\; NH_3$$

d. alkoxides and phenoxides (7.7)

$$2\,ROH \;+\; 2\,Na \;\longrightarrow\; 2\,RO^-Na^+ \;+\; H_2$$

$$ArOH \;+\; Na^+OH^- \;\longrightarrow\; ArO^-Na^+ \;+\; H_2O$$

e. hemiacetals and acetals (9.8; 13.5; 13.12)

$$\underset{/}{\overset{\backslash}{C}}\!\!=\!\!O \;+\; ROH \;\rightleftharpoons^{H^+}\; \underset{\backslash OR}{\overset{/OH}{C}} \;\xrightarrow[\;\;]{ROH,\,H^+}\; \underset{\backslash OR}{\overset{/OR}{C}} \;+\; H_2O$$

f. quinones (7.17)

g. vinyl polymers (3.18; 3.19; 6.10; 11.5; 16.9; 16.10)

SUMMARY OF REACTION MECHANISMS

Although a substantial number of reactions are described in the text, they belong to a relatively modest number of mechanistic types. The preparation of alkyl halides from alcohols and HX, the cleavage of ethers, the preparation of amines from alkyl halides and ammonia (and many other reactions) all, for example, occur by a nucleophilic substitution mechanism.

The following is a brief review of the main mechanistic pathways discussed in the text.

1. Substitution Reactions

   a. free-radical chain reaction (2.16)

      initiation step:

$$X_2 \xrightarrow[\text{heat}]{\text{light or}} 2 \ X\cdot \ (X = Cl, Br)$$

      propagation steps:

$$R\!-\!H + X\cdot \longrightarrow R\cdot + H\!-\!X$$

$$R\cdot + X_2 \longrightarrow R\!-\!X + X\cdot$$

      termination steps:

$$X\cdot + X\cdot \longrightarrow X_2$$

$$R\cdot + R\cdot \longrightarrow R\!-\!R$$

$$R\cdot + X\cdot \longrightarrow R\!-\!X$$

   The sum of the propagation steps gives the overall reaction.

   b. electrophilic aromatic substitution (4.10)

      Reaction occurs in two steps: addition of an electrophile to the aromatic π electron system, followed by loss of a ring proton.

401

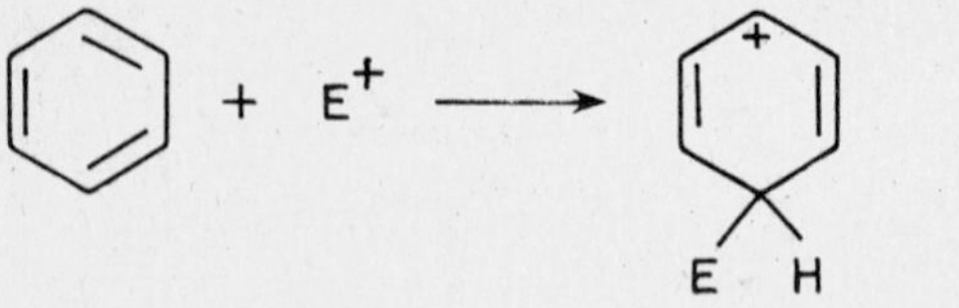

addition

benzenonium ion

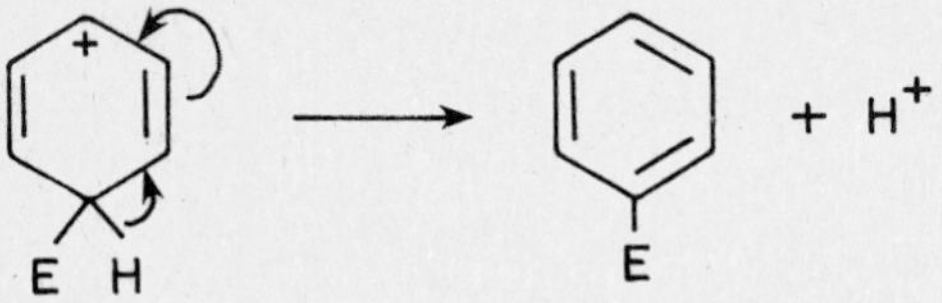

$+ H^+$ elimination

Substituents already on the aromatic ring affect the reaction rate and the orientation of subsequent substitutions (4.11; 4.12).

c. nucleophilic aliphatic substitution (6.5; 6.6; 6.7)

S_N2 (substitution, nucleophilic, bimolecular):

(1) rate depends on the concentration of both reactants, that is, the nucleophile and the substrate
(2) inversion of configuration at carbon
(3) reactivity order CH_3 > primary > secondary >> tertiary
(4) rate only mildly dependent on solvent polarity

S_N1 (substitution, nucleophilic, unimolecular):

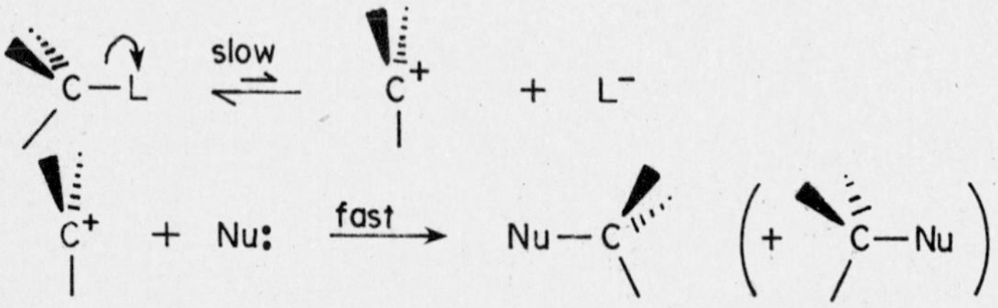

(1) rate depends on the concentration of the substrate, but *not* on the concentration of the nucleophile
(2) racemization at carbon
(3) reactivity order tertiary > secondary >> primary or CH_3
(4) reaction rate increased markedly by polar solvents

d. nucleophilic aromatic substitution (12.18; 14.9b)

Reaction occurs in two steps: addition of a nucleophile to the aromatic ring, followed by loss of a leaving group, often an anion.

The reaction is facilitated by electron-withdrawing substituents *ortho* or *para* to X, since they stabilize the intermediate carbanion.

2. Addition Reactions

a. electrophilic additions to C=C and C≡C (16.9c)

addition of acids (3.13):

Nu: = H—Ö—H, R—Ö—H, X⁻, ⁻OSO₃H, etc.

Addition proceeds via the most stable carbocation intermediate (Markovnikov's rule) (3.14).

addition of halogens (3.15):

Addition is *trans*, via a bromonium ion intermediate.

addition to alkynes (3.25):

1,4-addition to conjugated dienes (3.16):

(+ normal 1,2-addition)   **allylic carbocation**

cycloaddition (3.17):

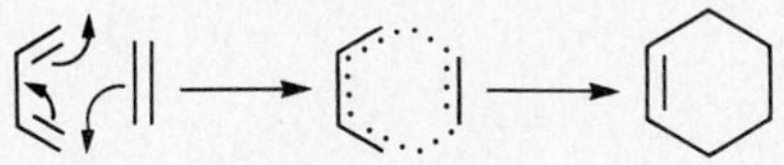

**transition
state**

The reaction involves six π electrons in a cyclic transition state. The dotted bonds in the transition state represent partially broken or partially formed bonds.

b.   free-radical addition (3.18; 16.9)

initiation step: ROOR $\xrightarrow{\text{heat}}$ 2 RO·

propagation steps: RO· + $\text{C}=\text{C}$ $\longrightarrow$ RO—C—C·

RO—C—C· + $\text{C}=\text{C}$ $\longrightarrow$ RO—C—C—C—C·

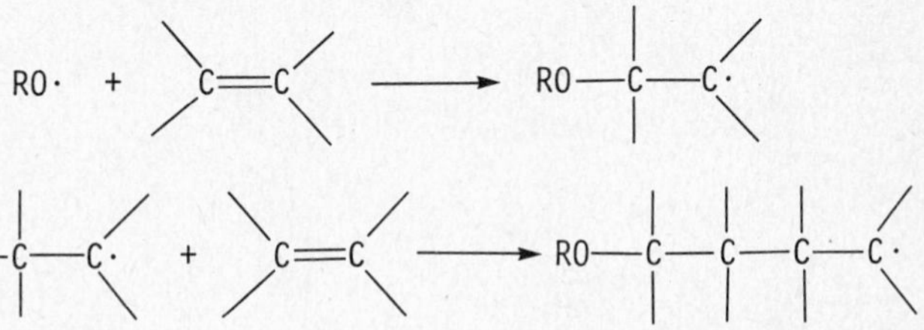

etc.

termination steps: radical combination
                                  or
                       radical disproportionation

Another reaction type associated with such reactions is chain-transfer (or hydrogen abstraction) (16.9).

R· + H—C— $\longrightarrow$ R—H + ·C—

Cationic and anionic chain-growth polymerizations occur by chain reactions similar to those for free-radical polymerizations but involving charged intermediates (16.9c; 16.9d).

c. hydroboration via a cyclic transition state (3.20)

The boron adds to the least crowded carbon; both the boron and hydrogen add to the same "face" of the double bond. Subsequent oxidation with H_2O_2 and OH^- places a hydroxyl group in the exact position previously occupied by boron:

d. nucleophilic addition to C=O (9.7; 10.12)

The reaction is often followed by protonation of the oxygen. The C=O group may be present in an aldehyde, ketone, ester, anhydride, acyl halide, amide, etc.

Acid catalysts may be necessary with weak nucleophiles. The acid protonates the oxygen, making the carbon more positive, and thus more electrophilic.

Nucleophilic *substitution* at a carbonyl group (10.12) takes place by an addition-elimination mechanism (10.17):

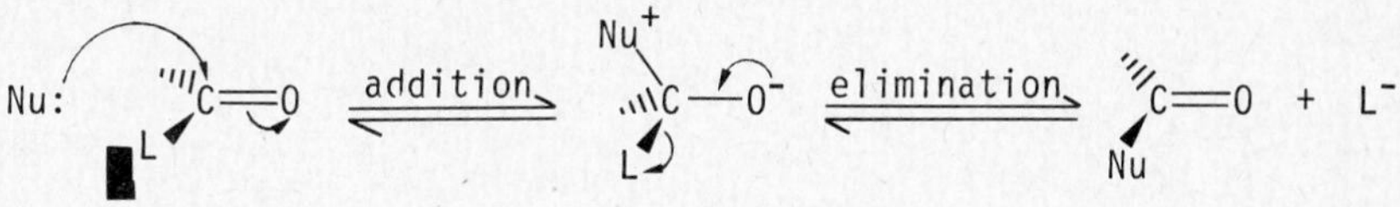

tetrahedral
intermediate

3. Elimination Reactions

a. the E2 mechanism (6.8)

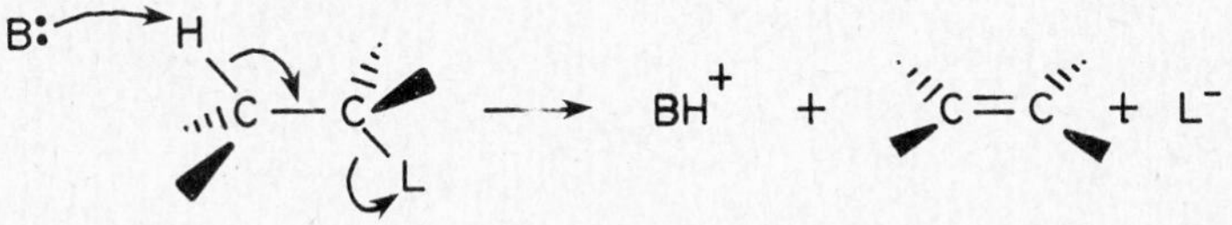

A planar, *anti* arrangement of the eliminated groups (H and L) is preferred. The reaction rate depends on the concentration of both the base (B) and the substrate. This mechanism can be important regardless of whether L is attached to a primary, secondary, or tertiary carbon.

b. the E1 mechanism (6.8)

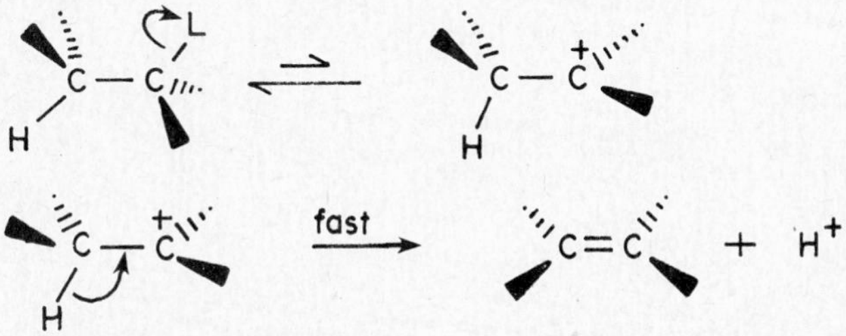

The first step is the same as the first step of the S_N1 mechanism. The reaction rate depends only on the substrate concentration. This mechanism is most important when L is attached to a tertiary carbon.

407

c. elimination via a cyclic transition state (11.3; 11.8)

R= alkyl or OH

REVIEW PROBLEMS ON SYNTHESIS

The following problems are designed to give you practice in multistep syntheses. They should be helpful in preparing for quizzes or examinations. To add to the challenge and make this preparation more realistic, no answers are given. If you do have difficulty with some of the problems, consult your textbook first, then your instructor.

The best technique for solving a multistep synthesis problem is to work backward from the final goal. Keep in mind the structure of the available starting material, however, so that eventually you can link the starting material with the product. The other constraint on synthesis problems is that you *must* use combinations of known reactions to achieve your ultimate goal. Although research chemists do try to discover and develop new reactions, you cannot afford that luxury until you have already mastered known reactions. All of the following problems can be solved using reactions in your text.

1. Show how each of the following can be prepared from propene:

   a. propane
   b. 2-bromopropane
   c. 1,2-dichloropropane
   d. 2-propanol
   e. 1-propanol

2. Each of the following conversions requires two reactions, in the proper sequence. Write equations for each conversion.

   a. *n*-propyl bromide to propene to 1,2-dibromopropane
   b. isopropyl alcohol to propene to 2-iodopropane
   c. 1-bromobutane to 2-chlorobutane
   d. 2-butanol to butane
   e. bromocyclopentane to 1,2-dibromocyclopentane
   f. *t*-butyl chloride to isobutyl alcohol

3. Starting with acetylene, write equations for the preparation of:

   a. ethane
   b. ethyl iodide
   c. 1-butyne
   d. 1,1-diiodoethane
   e. 2,2-dibromobutane
   f. 3-hexyne
   g. 1,1,2,2-tetrabromoethane
   h. *cis*-3-hexene

4. Write equations for each of the following conversions:

    a. 2-butene to 1,3-butadiene (two steps)
    b. 2-propanol to 1-propanol (three steps)
    c. 1-bromopropane to 1-aminobutane (two steps)
    d. 1,3-butadiene to 1,4-dibromobutane (two steps)

In solving problems of this type, carefully examine the structures of the starting material and the final product. Seek out similarities and differences. Note the types of bonds that must be made or broken in order to go from one structure to the other. Sometimes it is profitable to work backward from the product and forward from the starting material simultaneously, with the goal of arriving at a common intermediate.

5. Using benzene or toluene as the only organic starting material, devise a synthesis for each of the following:

    a. *m*-chlorobenzenesulfonic acid   b. 2,4,6-tribromotoluene

    c.

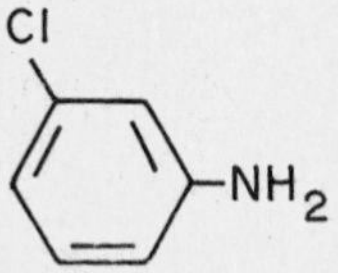

    d.

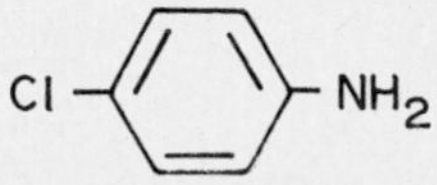

    e.

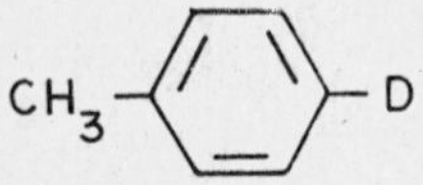

    f.

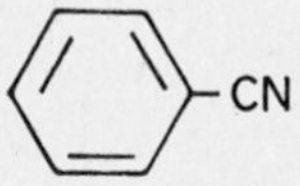

6. Write equations for the preparation of:

    a. 1-phenylethanol from styrene
    b. 2-phenylethanol from styrene
    c. 1-butanol from 1-bromobutane
    d. sodium 2-butoxide from 1-butene
    e. 1-butanethiol from 1-butanol
    f. 2,4,6-tribromobenzoic acid from toluene
    g. ethyl cyclohexyl ether from ethanol and phenol
    h. di-*n*-butyl ether from 1-butanol

7. Write equations that show how 2-propanol can be converted to each of the following:

    a. isopropyl chloride   b. 1,2-bromopropane
    c. 2-methoxypropane   d. isopropylbenzene

8. 2-Bromobutane can be obtained in one step from each of the following precursors: butane, 2-butanol, 1-butene, and 2-butene. Write an equation for each method. Describe the advantages or disadvantages of each.

9. Starting with an unsaturated hydrocarbon, show how each of the following can be prepared:

    a. 1,2-dibromobutane
    c. 1,2,3,4-tetrabromobutane
    e. 1,4-dibromo-2-butene
    g. 1-bromo-1-phenylethane

    b. 1,1-dichloroethane
    d. cyclohexyl iodide
    f. 1,1,2,2-tetrachloropropane
    h. 1,2,5,6-tetrabromocyclooctane

10. Give equations for the preparation of the following carbonyl compounds:

    a. 2-pentanone from an alcohol
    c. cyclohexanone from phenol (two steps)

    b. pentanal from an alcohol
    d. acetone from propyne

11. Complete each of the following equations, giving the structures and names of the main organic products:

    a. benzoic acid + ethylene glycol + H^+
    c. n-propylamine + acetic anhydride
    e. n-propylbenzene + $K_2Cr_2O_7$ + H^+
    g. pentanedioic acid + thionyl chloride
    i. methyl 3-butenoate + $LiAlH_4$

    b. $C_6H_5CH_2MgBr$ + CO_2, followed by H_3O^+
    d. p-hydroxybenzoic acid + acetic anhydride
    f. phthalic anhydride + methanol + H^+
    h. cyclopropanecarboxylic acid + NH_4OH, then heat
    j. ethyl propanoate + NaOH, heat, then acidify

12. Show how each of the following conversions can be accomplished:

    a. n-butyryl chloride to methyl n-butyrate
    c. butanoic acid to 1-butanol
    e. propionyl bromide to N-ethyl propionamide
    g. urea to ammonia and CO_2

    b. propionic anhydride to propionamide
    d. 1-pentanol to pentanoic acid
    f. oxalic acid to diethyl oxalate
    h. benzoyl chloride to N-methylbenzamide

13. Show how each of the following compounds can be prepared from the appropriate acid:

    a. propionyl bromide
    c. n-butanamide
    e. calcium oxalate
    g. isopropyl benzoate

    b. ethyl pentanoate
    d. phthalic anhydride
    f. phenylacetamide
    h. m-nitrobenzoyl chloride

411

14. Write equations to describe how each of the following conversions might be accomplished:

    a. *n*-butyl chloride to *n*-butyltrimethylammonium chloride
    b. *o*-toluidine to *o*-toluic acid
    c. *o*-toluidine to *o*-bromobenzoic acid
    d. 1-butene to 2-methyl-1-aminobutane

15. Starting with benzene, toluene, or any alcohol with four carbon atoms or fewer, and any essential inorganic reagents, outline steps for the synthesis of the following compounds:

    a. *n*-butylamine   b. *p*-toluidine
    c. 1-aminopentane   d. *N*-ethylaniline
    e. *m*-aminobenzoic acid   f. tri-*n*-butylamine
    g. 1,4-diaminobutane   h. ethyl cyclohexanecarboxylate

16. Show how each of the following compounds can be prepared using a Diels-Alder reaction:

    a.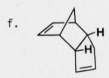

    b.

    c.

    d.

    e.

    f.

17. Plan a synthesis of each of the following compounds, using starting materials with five or fewer carbon atoms:

    a. 1-hexanol
    b. 2-hexanol
    c. 3-hexanol
    d. 1-heptanol
    e. ethylcyclopentane
    f. 1-ethynylcyclopentene
    g. *cis*-3-hexene
    h. 2-ethyl-2-hexenal

18. Show how you could accomplish the following conversions:

    a. 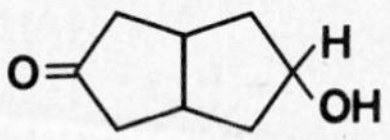 to

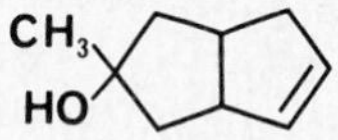

    b. 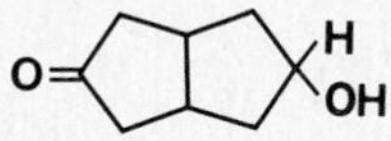 to

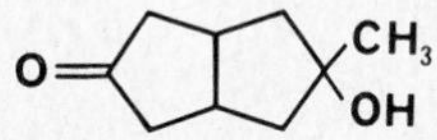

19. Show how each of the following compounds could be made starting with benzene:

    a.

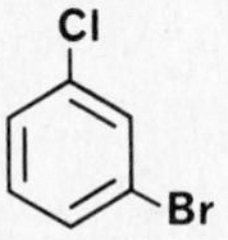

    b.

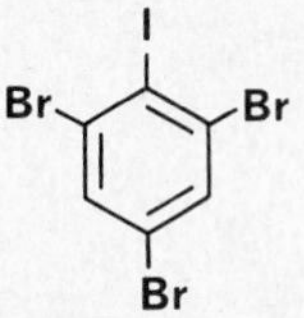

    c.

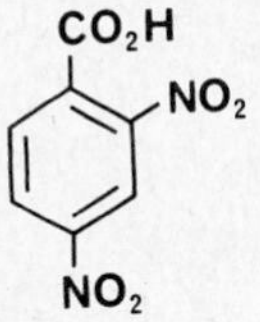

    d.

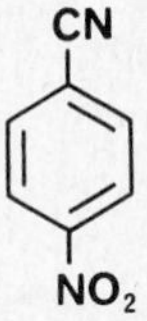

20. Give the structure of the expected product when each of the following is treated with strong aqueous base:

    a. 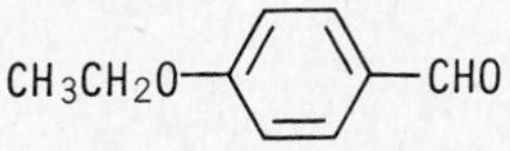 CH_3CH_2O—⟨phenyl⟩—CHO + CH_3CHO

    b. 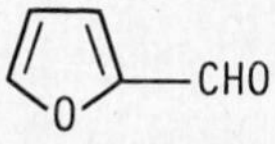 ⟨furyl⟩—CHO + CH_3CH_2CHO

413

c. + $(CH_3)_3CCHO$

d. CH_2O (excess) + CH_3CHO

e.

These questions, taken from mid-term and final examinations, are presented here for you to practice on. Although they do not cover all of the
subject matter in the text, these questions provide a fair review of the
material and may give you some idea of what to expect if (because of large
classes) your instructor uses multiple-choice exams.

Do not just guess at the correct answer. Take the time to write out the
structures or equations before you make a choice.

1. Which of the following compounds may exist as *cis-trans* isomers?

   a. 1-butene   b. 2-butene   c. 2-pentene

   1. a   2. b   3. a and b   4. b and c   5. a and c

2. What is the molecular formula of the following compound?

   1. $C_7H_{17}O$ 2. $C_8H_{15}O$ 3. $C_8H_{16}O$ 4. $C_8H_{17}O$ 5. $C_9H_{16}O$

3. How many σ (sigma) bonds are there in $CH_2{=}CH{-}CH{=}CH_2$?

   1. 3   2. 9   3. 10   4. 11   5. 12

4. The preferred conformation of butane is:

   5. none of these

5. The preferred conformation of *trans*-1,4-dimethylcyclohexane is:

1.

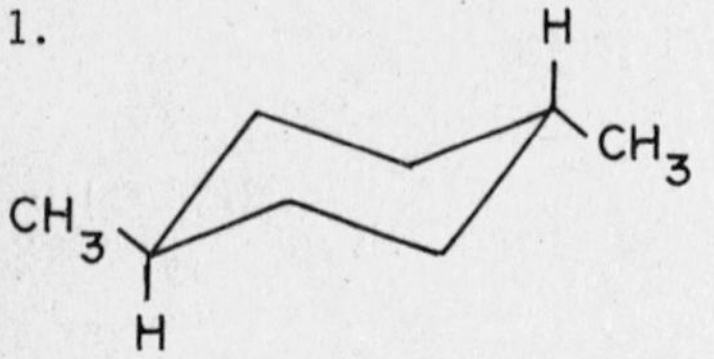

H

CH₃

CH₃

H

2.

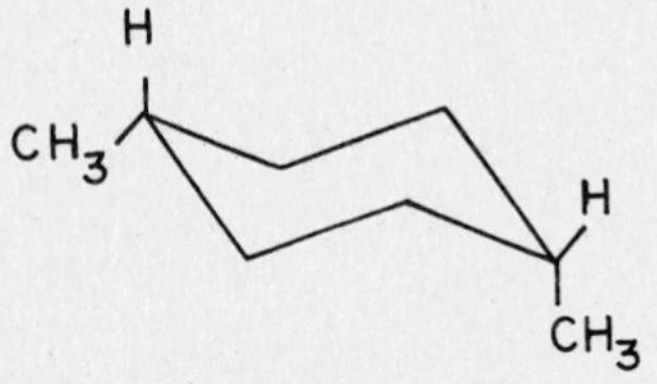

H

CH₃

H

CH₃

3.

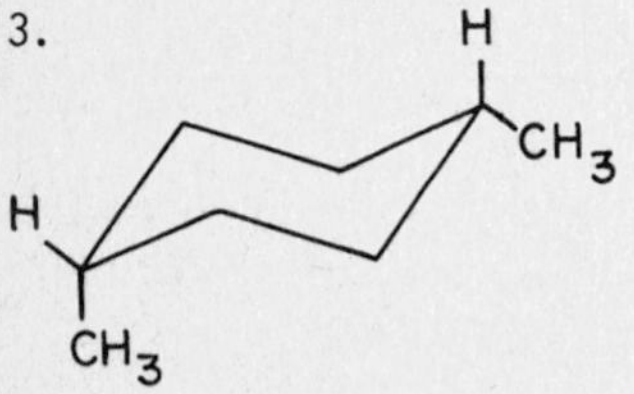

H

CH₃

H

CH₃

4.

CH₃

H

H

CH₃

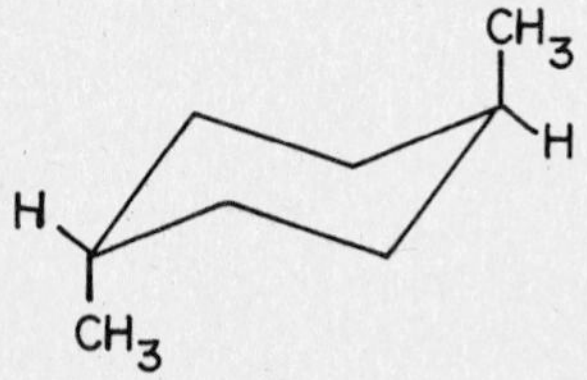

5. none of these

6. The major reaction product of

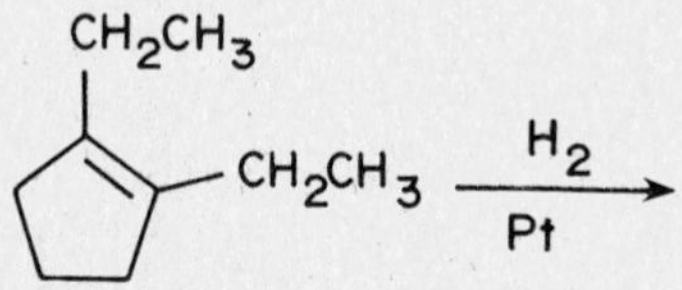

CH_2CH_3

CH_2CH_3

$\xrightarrow[Pt]{H_2}$ is:

1.

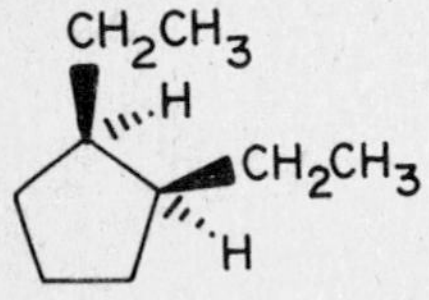

CH_2CH_3

...H

CH_2CH_3

...H

2.

CH_2CH_3

...H₂

CH_2CH_3

...H₂

3.

CH_2CH_3

...H

...CH_2CH_3

H

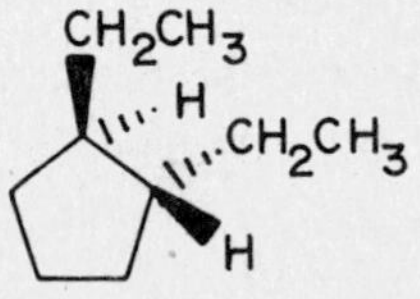

4.

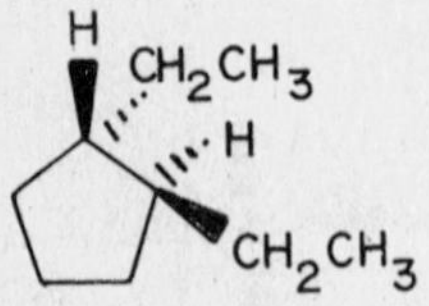

H

CH_2CH_3

...H

CH_2CH_3

5.

CH_2CH_3

..Pt

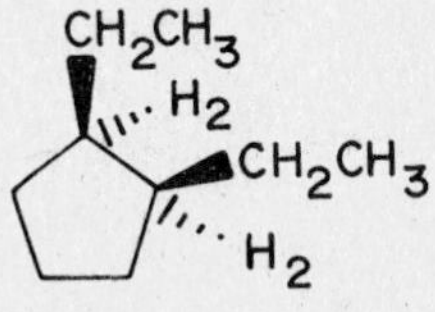

CH_2CH_3

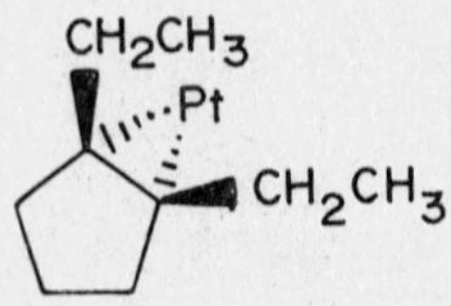

7. The structure that corresponds to a reaction intermediate in the reaction $CH_3\!\!-\!\!CH_3$ + Br_2 $\xrightarrow{\text{heat}}$ is:

   1. $CH_3CH_2^+$ 2. $CH_3CH_2\cdot$ 3. $CH_3CH_2^-$ 4. $CH_2CH_3^+$ 5. $\cdot CH_2\dot{C}H_2$

8. The best way to prepare $BrCH_2CH_2Br$ is to start with:

   1. $HC\!\equiv\!CH$ + Br_2 (excess)   2. $CH_2\!=\!CH_2$ + HBr

   3. $HC\!\equiv\!CH$ + HBr (excess)   4. $ClCH_2CH_2Cl$ + Br_2

   5. $CH_2\!=\!CH_2$ + Br_2

9. The IUPAC name (E)-2-methyl-3-hexene corresponds to:

10. The major product of the reaction

is:

11. The most likely structure for an unknown that yields only $CH_3CH_2CH\!=\!0$ on ozonolysis is:

4.   5.

12. Considering the hybridization of carbon orbitals, which of the following structures is *least* likely to exist?

1.   2.   3.   4.   5.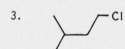

13. What is the correct name for the following structure?

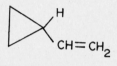

1. cyclopentene   2. cyclopropylpropene   3. vinylcyclopropane
4. vinylcyclobutane   5. 1-cyclopropyl-2-ene

14. H^+ often reacts as a(an):

1. electrophile   2. nucleophile   3. radical
4. electron   5. carbocation

15. The major reaction product of 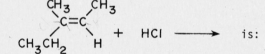 + HCl ⟶ is:

1.   2.   3.

4. Cl   5.

16. Which of the following classes of compounds is unreactive toward sulfuric acid?

    a. alkanes   b. cycloalkanes   c. alkenes   d. alkynes

    1. only a   2. only b   3. a and b   4. only d   5. a, b, and d

17. Which of the following structures represents a *trans*-dibromo compound?

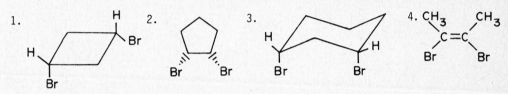

    5. none of these

18. How many *mono* chlorination products are possible in the reaction of 2,2-dimethylbutane with Cl_2 and light?

    1. 2   2. 3   3. 4   4. 5   5. 6

19. Teflon is represented by the structure $-CF_2CF_2(CF_2CF_2)_nCF_2CF_2-$. Which of the following monomers is used to make Teflon?

    1. $CF_2-\underset{\underset{F}{|}}{C}=F$ 2. $CF_2=CF-CF=CF_2$ 3. $CF_2=CFCF_3$

    4. CHF_2-CHF_2 5. $CF_2=CF_2$

20. Which of the following compounds contain one or more polar covalent bonds?

    a. CH_3CH_2OH b. $CH_3CH_2CH_3$ c. $CHCl=CHCl$ d. Cl_2

    1. a and b   2. a and c   3. b and c   4. only d   5. a and d

21. What is the relationship between [structure 1] and [structure 2] ?

1. structural isomers
2. geometric isomers
3. conformational isomers
4. identical
5. none of these

22. In which of the following structures does nitrogen have a formal charge of +1?

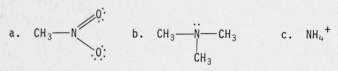

a. CH₃—N (with =O and O⁻)   b. CH₃—N—CH₃ with CH₃   c. NH₄⁺

1. a and c   2. a and b   3. b and c   4. a, b, and c   5. only a

23. Which of the following statements is(are) true of $CH_3CH_2CH_3$?

a. all bond angles are about 109.5°
b. each carbon is sp^3-hybridized
c. the compound is combustible

1. only a   2. only b   3. a and b   4. a and c   5. a, b, and c

24. Which reagent can accomplish this conversion?

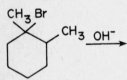

1. $KMnO_4$ 2. H_2O, H^+ 3. O_3, then Zn, H^+ 4. OH^- 5. H_2O, heat

25. How many elimination products are possible for the following reaction?

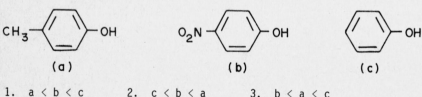

1. 0   2. 1   3. 2   4. 3   5. 4

26. Arrange the following compounds in order of increasing acidity:

CH₃—⟨ ⟩—OH   O₂N—⟨ ⟩—OH   ⟨ ⟩—OH

(a)   (b)   (c)

1. a < b < c   2. c < b < a   3. b < a < c
4. a < c < b   5. b < c < a

27. Which structure corresponds to *R*-1-chloroethylbenzene?

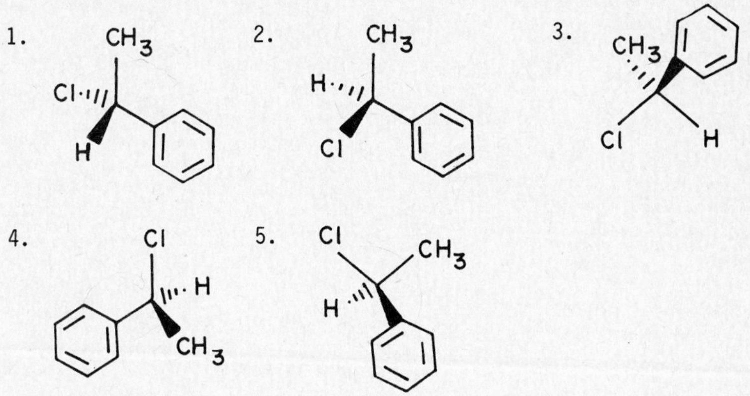

1.

CH₃

Cl····// H

2.

CH₃

H···// Cl

3.

CH₃

Cl   H

4.

Cl

H

CH₃

5.

Cl   CH₃

H

28. Which of the following structures represents a *meso* compound?

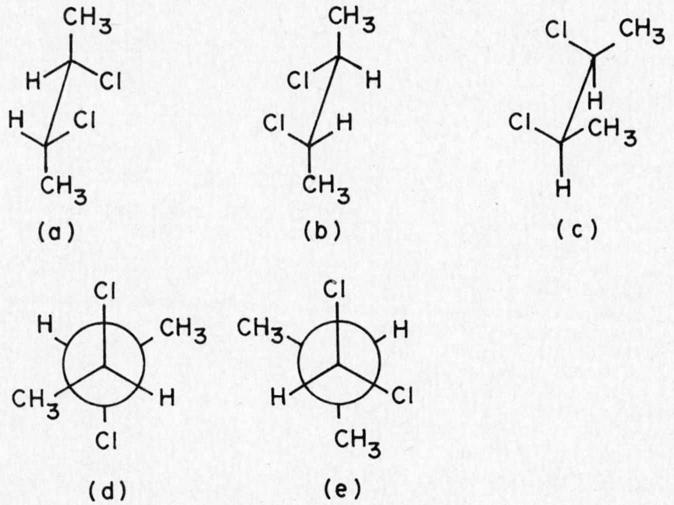

CH₃

H   Cl

H   Cl

CH₃

(a)

CH₃

Cl   H

Cl   H

CH₃

(b)

Cl   CH₃

H

Cl   CH₃

H

(c)

Cl

H   CH₃

CH₃   H

Cl

(d)

Cl

CH₃   H

H   Cl

CH₃

(e)

1. only e   2. a and b   3. c and d   4. d and e   5. a, b, and c

421

29. What is the correct name for this compound?

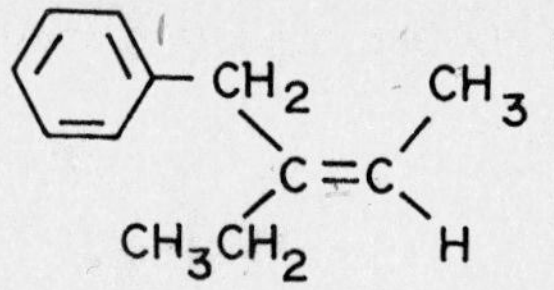

1. *cis* -3-phenyl-2-pentene
3. (Z)-3-benzyl-2-pentene
5. (Z)-3-benzyl-3-pentene

2. *trans* -3-phenyl-2-pentene
4. (E)-3-benzyl-2-pentene

30. Which of the following pairs of compounds can be separated by physical methods (for example, crystallization or distillation)?

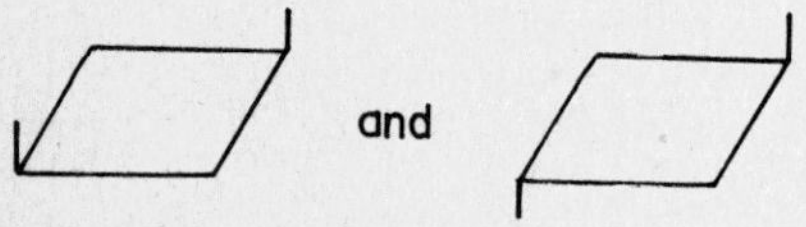

(a) and

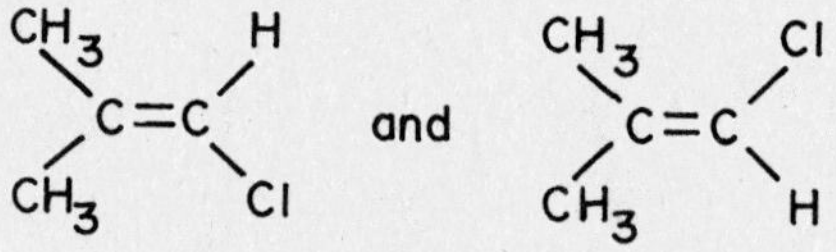

(b)

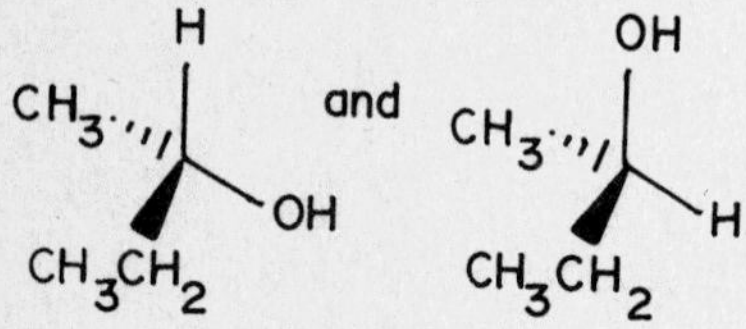

(c) and

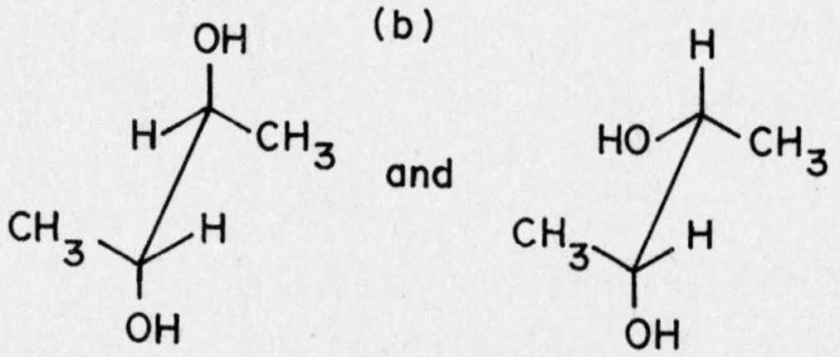

(d) and

1. all of them   2. a and d   3. b and c   4. only a   5. only c

31. To which Fischer projection formula does the following drawing correspond?

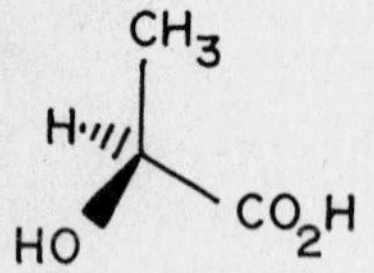

1. 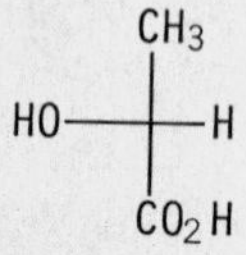 $HO \rule[0.5ex]{2em}{0.4pt} H$ (CH$_3$ top, CO$_2$H bottom)

2. $CH_3 \rule[0.5ex]{2em}{0.4pt} CO_2H$ (OH top, H bottom)

3. $CH_3 \rule[0.5ex]{2em}{0.4pt} CO_2H$ (H top, OH bottom)

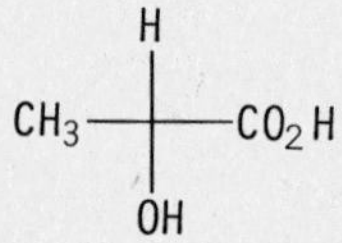

4. 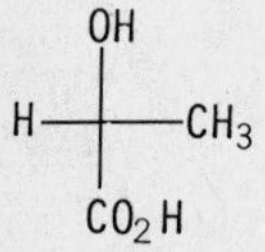 $H \rule[0.5ex]{2em}{0.4pt} CH_3$ (OH top, CO$_2$H bottom)

5. 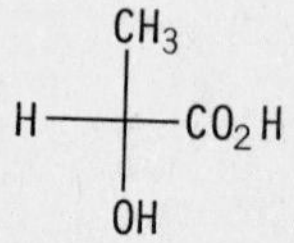 $H \rule[0.5ex]{2em}{0.4pt} CO_2H$ (CH$_3$ top, OH bottom)

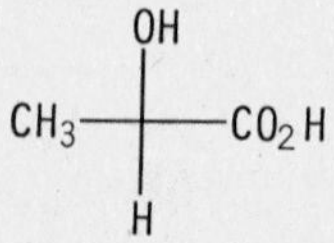

32. What is the correct name for the following compound?

1. (2R,3S)-2-chloro-3-fluorobutane
2. (2R,3S)-2-chloro-3-fluorobutane
3. (2S,3R)-2-chloro-3-fluorobutane
4. (2S,3S)-2-chloro-3-fluorobutane
5. none of the above

33. Which of the following compounds reacts least rapidly in electrophilic substitutions?

1.   2.   3.   4.   5.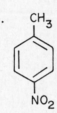

34. Which of the following compounds gives off the *least* amount of heat during hydrogenation?

1. 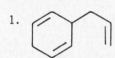 2. 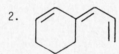 3.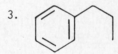

4. 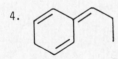 5.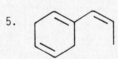

35. The resonance contributor that is most important in the intermediate for *para* electrophilic substitution in toluene is:

1. 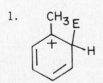 2. 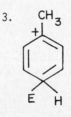 3. 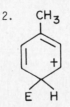 4. 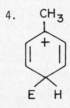 5.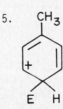

36. How are the following structures related?

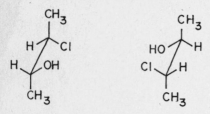

1. enantiomers
4. identical
2. diastereomers
5. achiral compounds
3. *meso* compounds

37. Which reagent would be most useful for carrying out the following transformation?

1. LiAlH₄   2. conc. H₂SO₄   3. H₂/Pd   4. CrO₃   5. NaOH

38. By which of the indicated mechanisms would the following reaction proceed?

1. S_N1   2. S_N2   3. E1   4. E2   5. free-radical

39. What is the most likely product of the following reaction?

1. 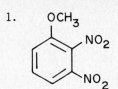 OCH₃ NO₂ NO₂

2. OCH₃ NO₂ OCH₃

3. OCH₃ NO₂

4. OCH₃ O₂N NO₂

5. 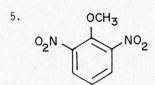 O₂N OCH₃ NO₂

40. Which of the following compound(s) is(are) not aromatic?

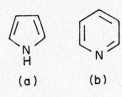

(a)   (b)   (c)   (d)   (e)

1. only d   2. only e   3. d and e   4. a and c   5. b and d

41. What are the appropriate reagents to accomplish the following transformations?

$$CH_3CH_2OCH_2CH_3 \xrightarrow{\text{A}} CH_3CH_2Br \xrightarrow{\text{B}} CH_3CH_3$$

1. Br₂   2. HBr   3. Na   4. conc. H₂SO₄   5. Mg, then H₂O
6. CrO₃

A = _____   B = _____

42. Which of the following compounds would react most rapidly in an S_N2 reaction?

1. — I   2. CH₃CH₂I   3. CH₂=CHI   4. (CH₃)₂CHI

5. (CH₃)₃CI

43. Which of the following compounds is a secondary alcohol?

    1. 3-hexanol   2. 1-hexanol   3. 2-methyl-2-hexanol
    4. 1-ethyl-1-cyclohexanol   5. cyclohexylmethanol

44. The conversion $CH_3C(=O)-OCH_3 \longrightarrow CH_3CH_2OH$ can be called:

    1. oxidation   2. reduction   3. hydrolysis
    4. saponification   5. dehydration

45. Which of the following compounds can be used as a detergent?

    1. NaOH

    2. Br

    3. (long chain) $O-C(=O)-CH_3$

    4. (long chain) $\overset{+}{N}(CH_3)_3 \ Cl^-$

    5. (long chain) CO_2H

46. Which of the following compounds is a tautomer of phenol?

    1.   2. 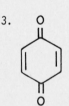   3.   4.   5.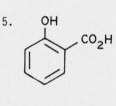

47. Which compound is the major product of the following reaction?

$$CH_3CH_2CH{=}O \ + \ CH_3MgBr \xrightarrow[\ H^+\ ]{\ H_2O\ }$$

426

1. $CH_3CH_2\overset{\overset{\displaystyle OH}{|}}{\underset{\underset{\displaystyle CH_3}{|}}{C}}$—$CH_3$ 2. $CH_3CH_2CH_2OH$ 3. $CH_3CH_2\overset{\overset{\displaystyle H}{|}}{\underset{\underset{\displaystyle CH_3}{|}}{C}}$—$CH_3$

4. $CH_3CH_2\overset{\overset{\displaystyle O}{||}}{C}$—$OCH_3$ 5. $CH_3CH_2\overset{\overset{\displaystyle OH}{|}}{C}HCH_3$

48. The compound [structure] is a:

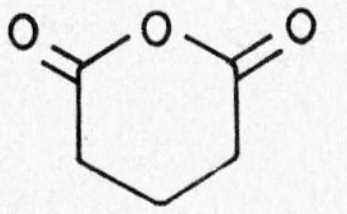

1. diketone   2. keto ether   3. lactone
4. anhydride   5. ketal

49. Nylon has the structure $-\left[\overset{\overset{\displaystyle O}{||}}{C}(CH_2)_4\overset{\overset{\displaystyle O}{||}}{C}NH(CH_2)_6NH\right]_n$. It can be called a:

1. polyamide   2. polyamine   3. polyester
4. vinyl polymer 5. polyurethane

50. The reaction $CH_3CH_2CH_2\overset{\overset{\displaystyle O}{||}}{C}$—$OCH_2CH_3 \xrightarrow[H_2O]{NaOH} CH_3CH_2CH_2\overset{\overset{\displaystyle O}{||}}{C}$—$O^-Na^+ + CH_3CH_2OH$
can be called:

1. enolization   2. elimination   3. condensation
4. esterification 5. saponification

51. The enol of 2-butanone is:

a. $CH_3\overset{\overset{\displaystyle OH}{|}}{C}HCH\text{=}CH_2$ b. $CH_3\overset{\overset{\displaystyle OH}{|}}{C}\text{=}CHCH_3$ c. $CH_2\text{=}\overset{\overset{\displaystyle OH}{|}}{C}CH_2CH_3$

d. $CH_2\text{=}\overset{\overset{\displaystyle OH}{|}}{C}$—$CH\text{=}CH_2$

1. only a   2. a and b   3. b and c   4. a and d   5. a, b, and c

52. What is the major product of the following reaction?

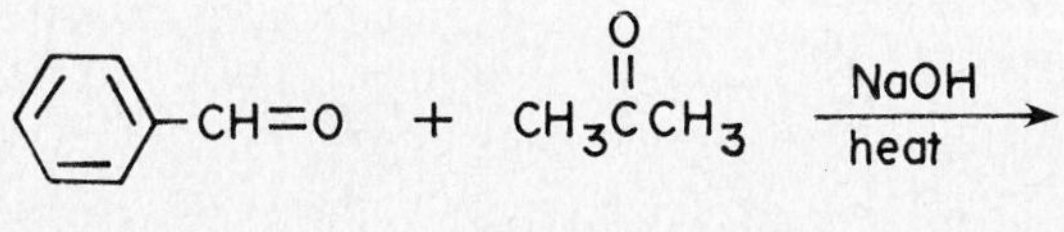

[structure] —CH=O + $CH_3\overset{\overset{\displaystyle O}{||}}{C}CH_3 \xrightarrow[\text{heat}]{NaOH}$

1. 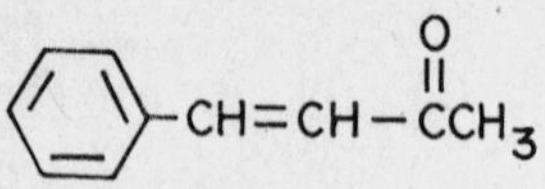
2. 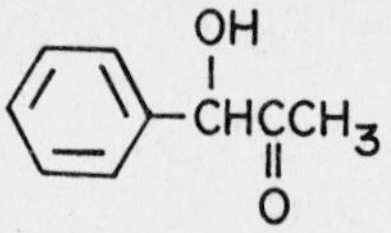
3. 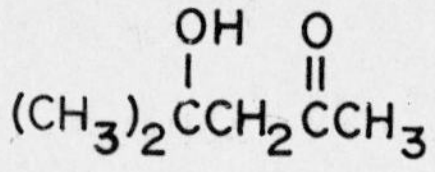

4. 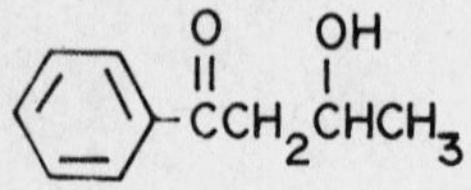
5. 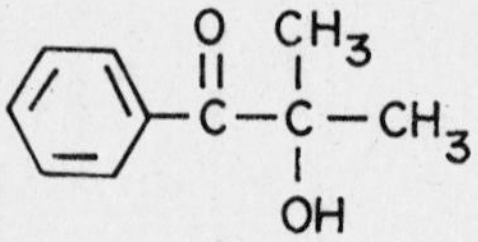

53. What is the major product of the following reaction?

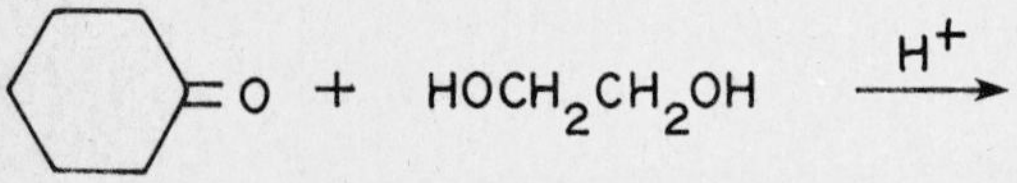

1. 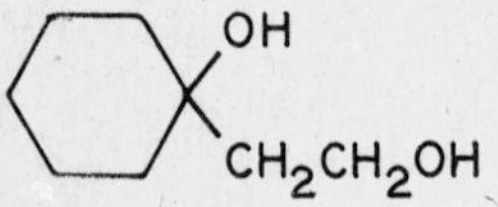
2. a polyester
3. 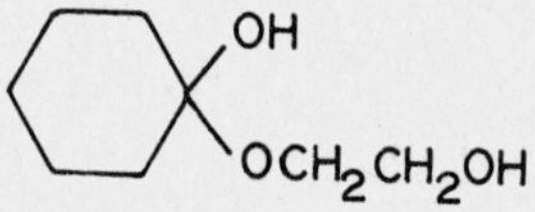

4. 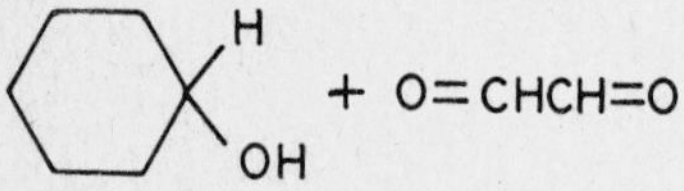
5. 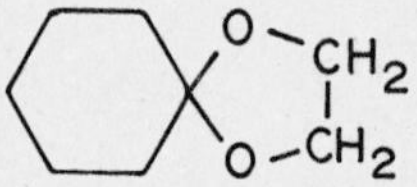

54. The structure of malonic acid is:

1. 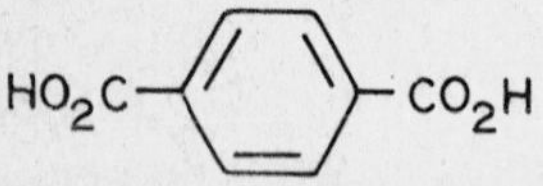
2. 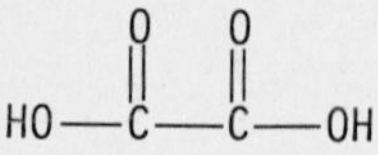
3. 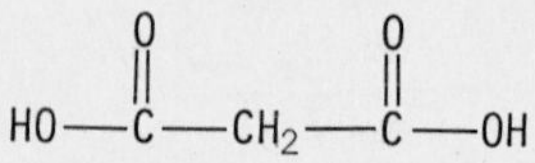

4. 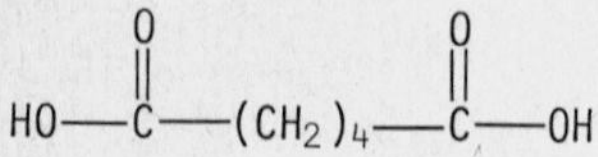
5. 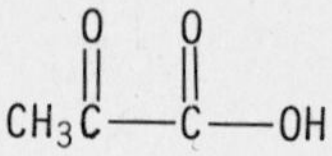

55. Which hybrid structure represents the most stable enolate anion formed in the following reaction?

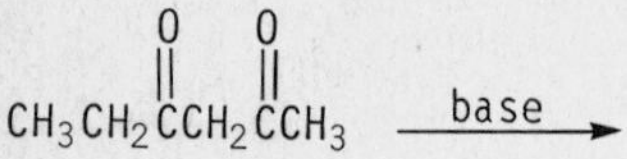

1. 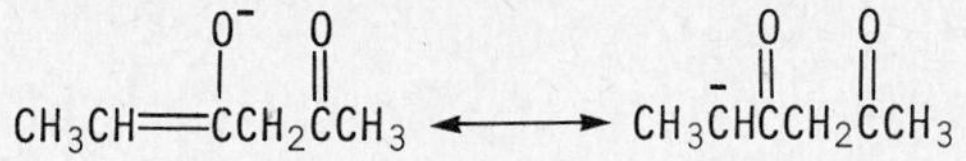

2. 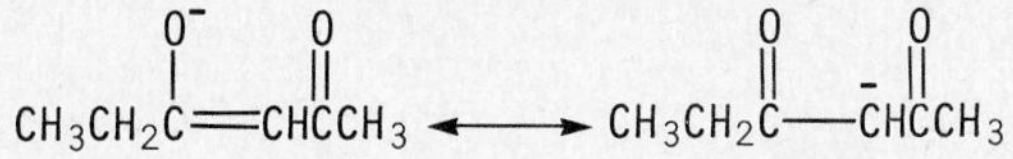

3. 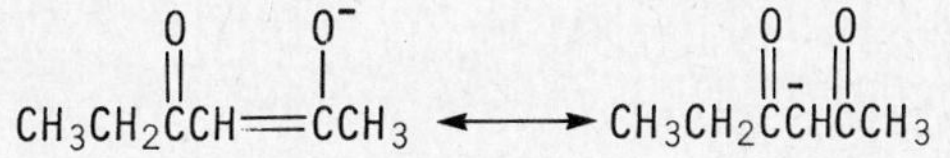

4. 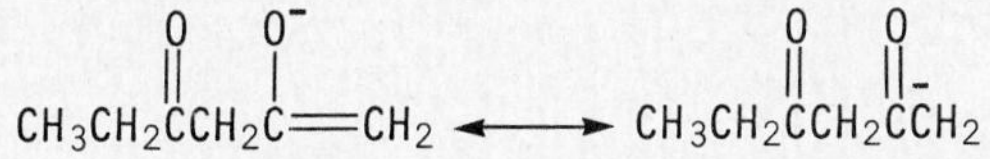

5. 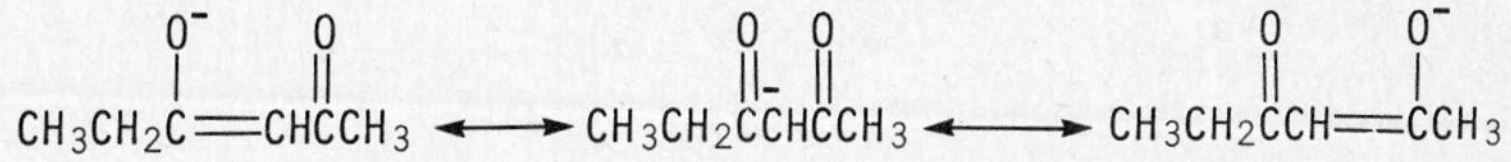

56. Arrange the following compounds in order of decreasing basicity:

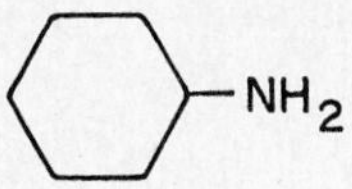

(a)

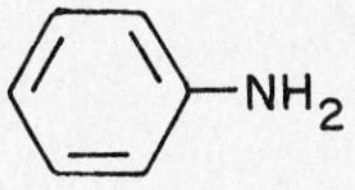

(b)

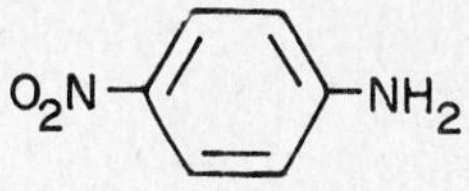

(c)

1. a > b > c   2. c > b > a   3. b > a > c
4. c > a > b   5. a > c > b

57. What is the major product of the following reaction?

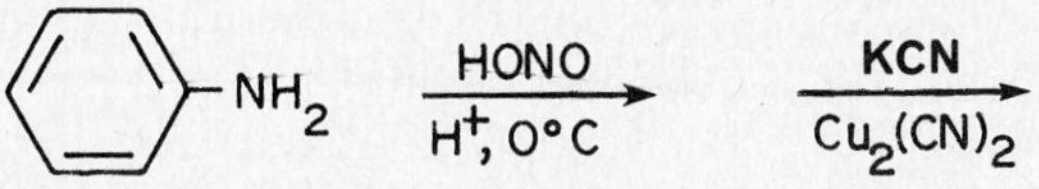

1. 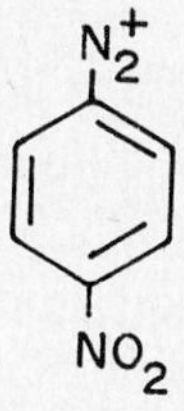

2. 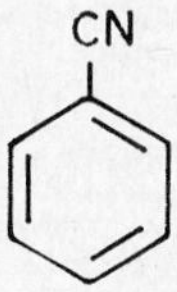

3. 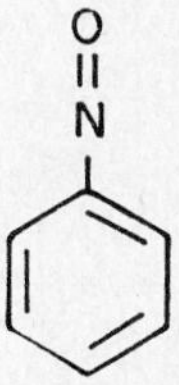

4. 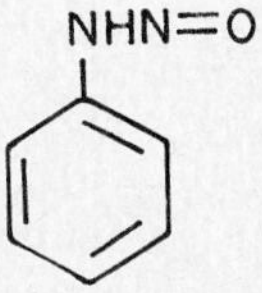

5. 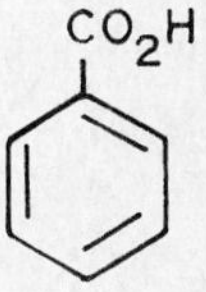

58. What is the major product of the following reaction?

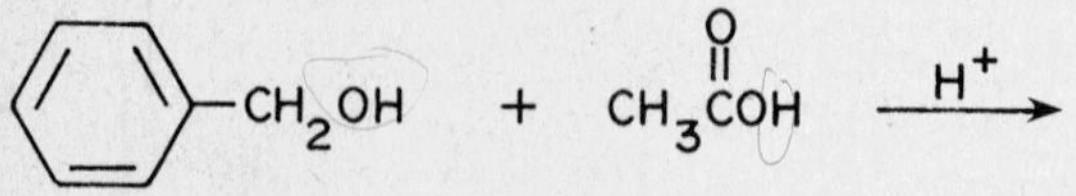

1.

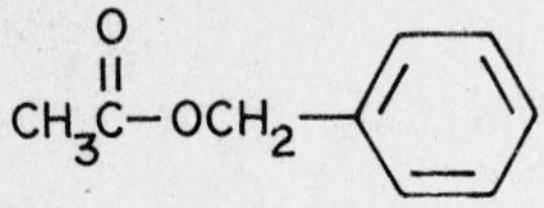

2.

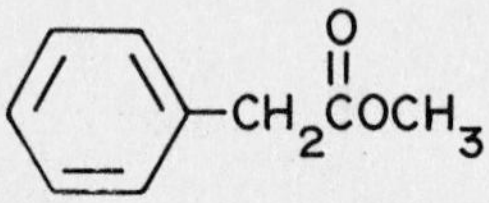

3.

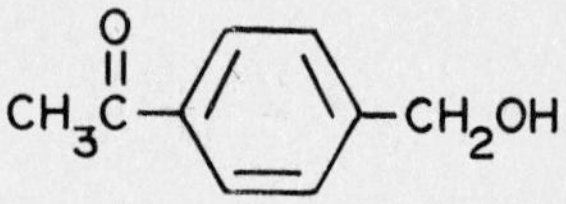

4.

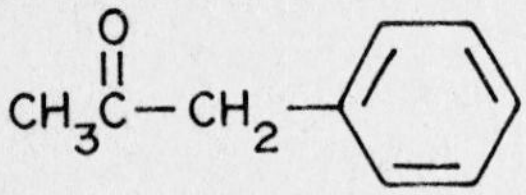

5.

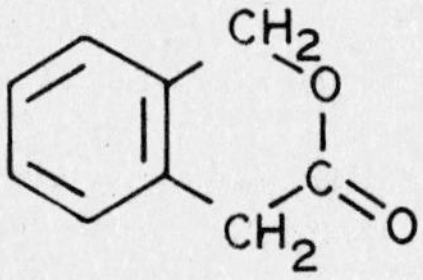

59. What is the major product of the following reaction?

$$CH_3CH_2COCH_2CH_3 \xrightarrow[CH_3CH_2OH]{CH_3CH_2O^-Na^+}$$

1. CH_3CH_2C—O—CCH_2CH_3

2. CH_3CH_2C—$CH_2CH_2COCH_2CH_3$

3. $CH_3CH_2CHCHCOCH_2CH_3$ (OH, CH_3)

4. $CH_3CH_2CCHCCH_2CH_3$ (CH_3)

5. CH_3CH_2C—$CHCOCH_2CH_3$ (CH_3)

60. Which of the following compounds is a secondary amine?

1. CH_3CHCH_3 (NH_2)

2. $CH_3CH_2NHCH_3$

3. CH_3C—N (CH_3, CH_3)

4. $H_2NCH_2CH_2NH_2$

5. $(CH_3)_3N$

430

61. Which statement concerning $(CH_3)_2N—N{=}O$ is incorrect?

    1. It is a nitrosamine.   2. It is carcinogenic.
    3. It is made from $(CH_3)_2NH$ and HONO.   4. It has polar bonds.
    5. It readily loses N_2 on heating.

62. What is the correct IUPAC name for $BrCH_2CH_2CH_2CH{=}O$?

    1. γ-bromobutyraldehyde   2. 1-bromo-3-carbonylpropane
    3. 1-bromo-4-butanal   4. 4-bromo-1-butanone
    5. 4-bromobutanal

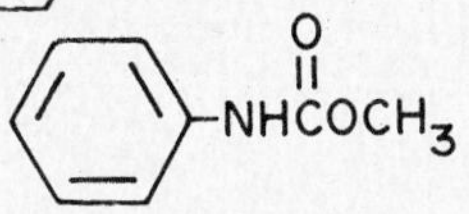

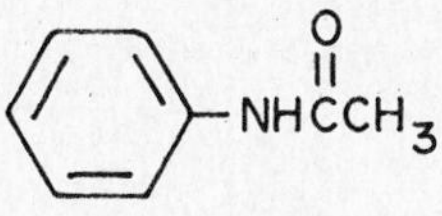

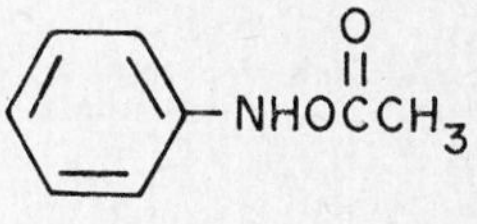

63. What is the product of the following reaction?

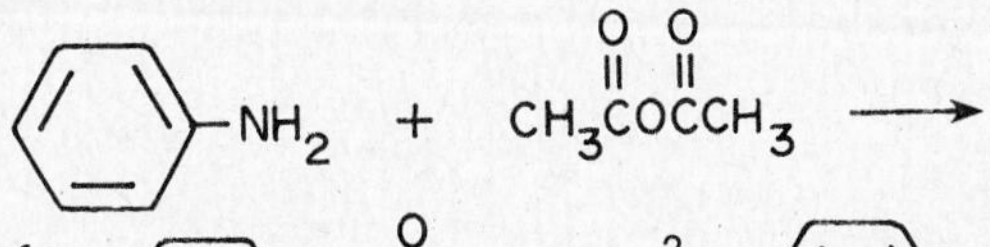

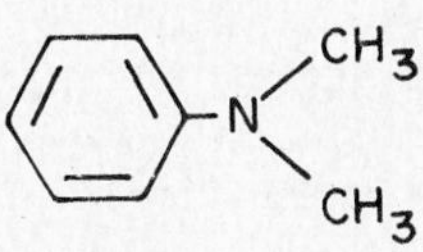

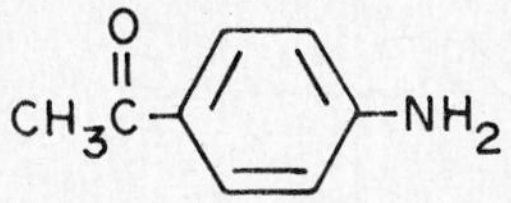

64. Which reaction is most appropriate for the preparation of $(CH_3)_2CHCH_2CH_2CO_2H$?

65. Arrange the following compounds in order of decreasing acidity:

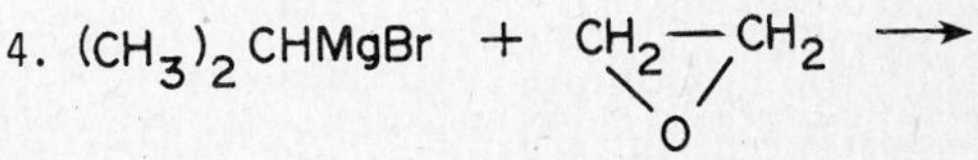

    a. $BrCH_2CH_2COOH$ b. $CH_3CHCOOH$ (with Br substituent)   c. $CH_3CHCOOH$ (with F substituent)

    1. a > b > c   2. c > b > a   3. c > a > b   4. b > a > c   5. b > c > a

431

66. Which structure corresponds to *cis* -3-pentenoic acid?

1.

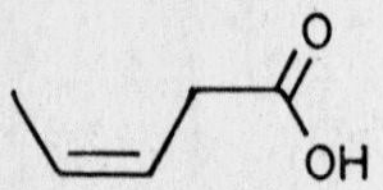

2.

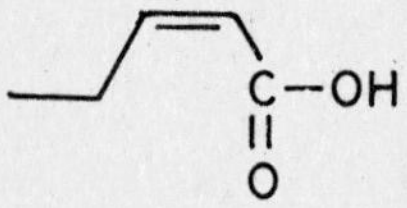

3.

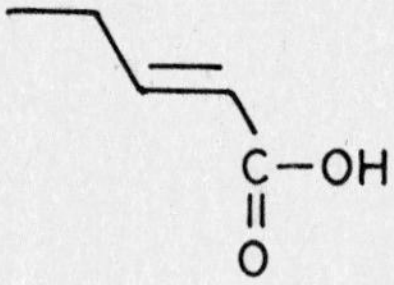

4.

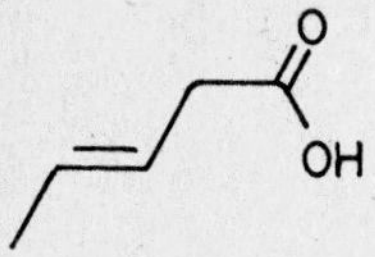

5.

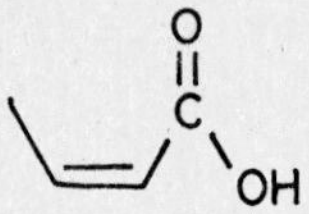

67. What is(are) the product(s) of the following reaction?

$$CH_3CH_2CH \begin{matrix} OCH_2CH_3 \\ \\ OCH_2CH_3 \end{matrix} \xrightarrow[\text{H}_2\text{O}]{\text{H}^+}$$

a. CH_3CH_2OH b. $CH_3CH_2\overset{O}{\overset{\|}{C}}OH$ c. $CH_3CH_2\overset{OH}{\underset{OH}{\overset{|}{\underset{|}{C}}}}OCH_2CH_3$ d. $CH_3CH_2CH{=}0$

1. a and b   2. a and d   3. a and c   4. only c   5. b and d

68. Cyclic ethers with a three-membered ring $\left[\begin{matrix} & O & \\ \diagdown & \diagup \diagdown & \diagup \\ C & - & C \\ \diagup & & \diagdown \end{matrix} \right]$ are called:

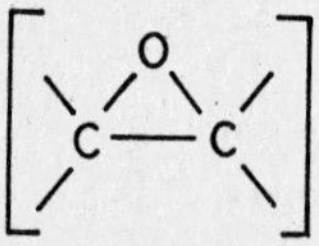

1. epoxy resins   2. lactones   3. oxiranes
4. lactams   5. alkoxides

69. The appearance of a red precipitate in a Fehling's test indicates the presence of:

1. sucrose   2. glucose   3. cellulose   4. starch   5. glyceride

70. The name benzyl α-bromopropionate refers to:

1.

2.

3.

4. $BrCH_2CH_2\overset{\overset{\displaystyle O}{\|}}{C}$—$OCH_2$—

5. $CH_3\overset{}{\underset{\underset{\displaystyle Br}{|}}{C}}H\overset{\overset{\displaystyle O}{\|}}{C}OCH_2$—

71. Compound X gives the following test results:

$$X + Ag(NH_3)_2^+ \longrightarrow \text{no silver mirror}$$

$$\underline{X} + H_2NNH—\text{(ring)} \longrightarrow \text{orange solid}$$

(with O_2N and NO_2 substituents on the ring)

$$X \xrightarrow[CH_3OD]{CH_3O^-Na^+} \text{exchanges } four \text{ hydrogens for deuterium}$$

Which of the following compounds is a possible structure for X?

1. $CH_3\overset{\overset{\displaystyle O}{\|}}{C}OC(CH_3)_3$

2. $CH_3CH_2CH_2\overset{\overset{\displaystyle O}{\|}}{C}CH_2CH_3$

3. $CH_3\overset{\overset{\displaystyle O}{\|}}{C}CH_2$—

4. O=$CHCH_2CH_2CH$=O

5.

72. Which statements regarding the following structure are true?

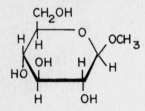

    a. It is a hemiacetal.   c. It can be hydrolyzed by H_3O^+.
    b. It is an acetal.   d. It gives a positive Tollens' test.

    1. b and d   2. a and d   3. only b   4. b and c   5. b, c, and d

73. The monosaccharide obtained from the hydrolysis of starch is:

    1. maltose   2. D-ribose   3. 2-deoxy-D-ribose
    4. D-galactose   5. D-glucose

74. The products of the complete hydrolysis of DNA are:

    1. D-ribose, phosphoric acid, and four heterocyclic bases
    2. nucleosides
    3. nucleotides
    4. glucose and fructose
    5. 2-deoxy-D-ribose, phosphoric acid, and four heterocyclic bases

75. Uracil is:

    1. a constituent of urine   2. a pyrimidine   3.   a purine
    4. a heterocyclic base present in DNA 5. a potent carcinogen

76. The primary structure of a protein refers to:

    1. the amino acid sequence in the polypeptide chain
    2. the presence or absence of an α helix
    3. the orientation of the amino acid side chains in space
    4. interchain cross-links with disulfide bonds
    5. whether the protein is fibrous or globular

77. Which of the following compounds is a fat?

                  O
                  ‖
    1. $CH_3(CH_2)_{16}CO(CH_2)_{15}CH_3$

2. 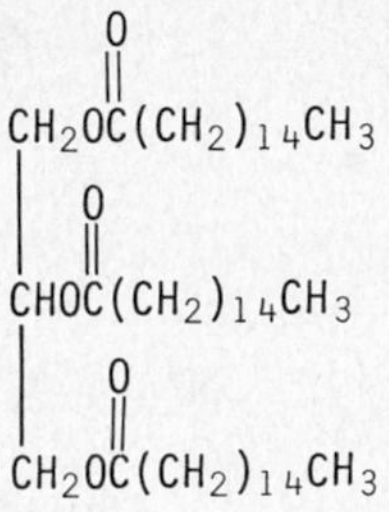
$$CH_2OC(CH_2)_{14}CH_3$$
|
$$CHOC(CH_2)_{14}CH_3$$
|
$$CH_2OC(CH_2)_{14}CH_3$$

3.
$$CH_2—C—O(CH_2)_{14}CH_3$$
|
$$CH—C—O(CH_2)_{14}CH_3$$
|
$$CH_2—C—O(CH_2)_{14}CH_3$$

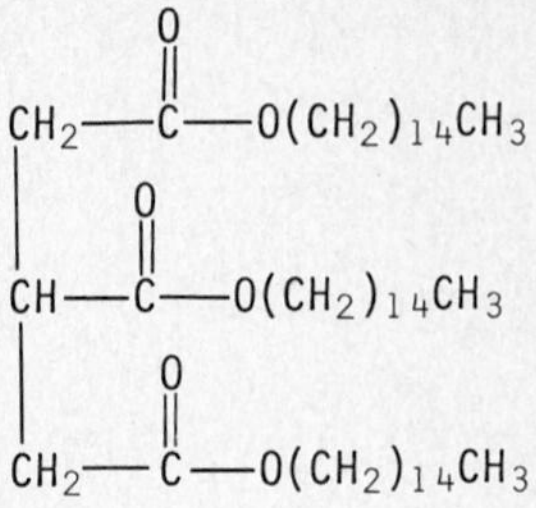

4.

HO

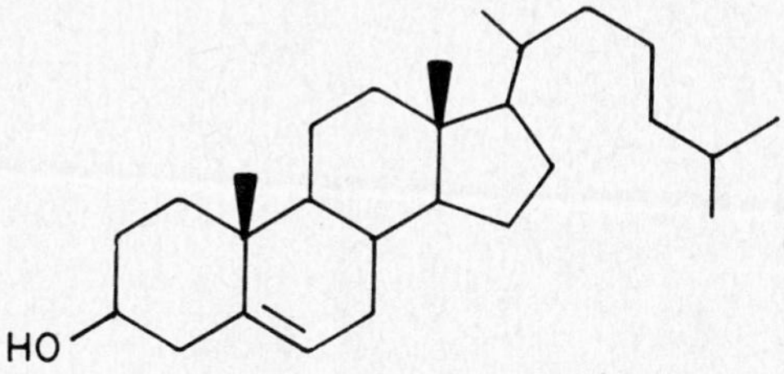

5.

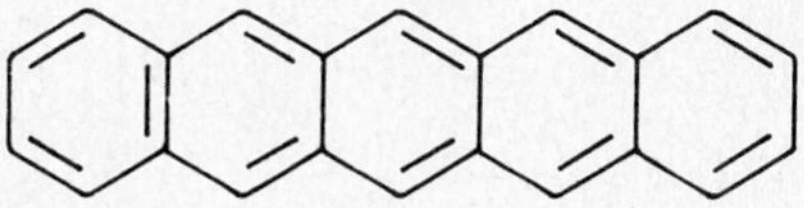

78. Which of the following structures is a dipolar ion?

1. $^-O—CCH_2CH_2CO^-$ 2. Ca^{2+} 3. $H_3\overset{+}{N}CH_2CO^-$

4. $CH_3(CH_2)_{16}CO^-Na^+$ 5. 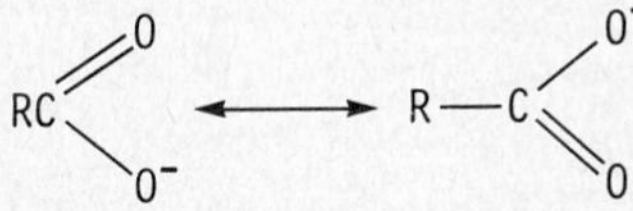 $RC \overset{O}{\underset{O^-}{}} \longleftrightarrow R—C \overset{O^-}{\underset{O}{}}$

79. Which of the following compounds would show only a single peak in a proton NMR spectrum?

a. $(CH_3)_4Si$ b. CCl_4 c. CH_2Cl_2 d. $CH_3CH_2OCH_2CH_3$

1. only b 2. a and c 3. only d 4. b and c 5. a, b, and c

80. Which of the following spectroscopic techniques is used to determine molecular weight?

1. proton NMR 2. ^{13}C NMR 3. infrared spectroscopy
4. ultraviolet-visible spectroscopy 5. mass spectrometry

81. The compound 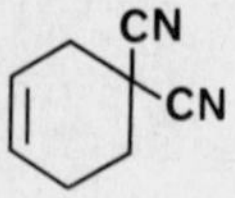 can best be prepared from:

1. 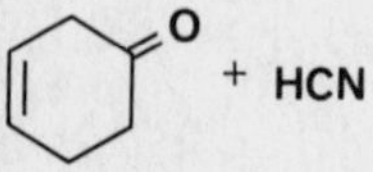

2. 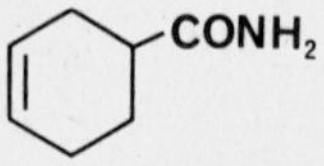

3. $CH_2{=}CH{-}CH{=}CH_2$ + $CH_2{=}C(CN)_2$

4. 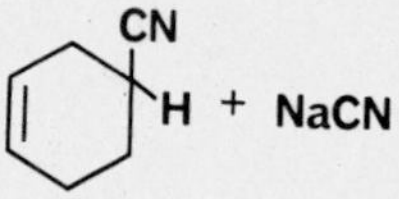

5. $CH_2{=}CH{-}CH{=}CH_2$ + $NC{-}C{\equiv}C{-}CN$

82. Which of the following terms can be used to describe this pair of compounds?

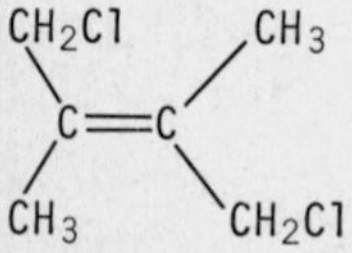 and 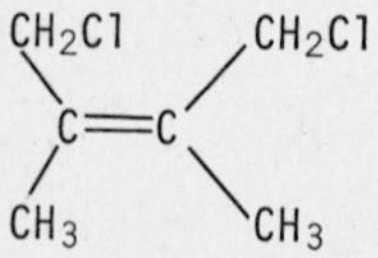

a. enantiomers b. diastereomers c. constitutional isomers
d. stereoisomers e. chiral f. achiral

1. a, b, and d 2. a, c, and e 3. b, d, and e 4. a, d, and e
5. b, d, and f

83. Which of the following best describes $C_6H_9Br_3$?

1. no rings, no double bonds
2. one ring, no double bonds
3. two double bonds
4. one triple bond
5. one ring and one double bond
6. two rings

84. The polymer 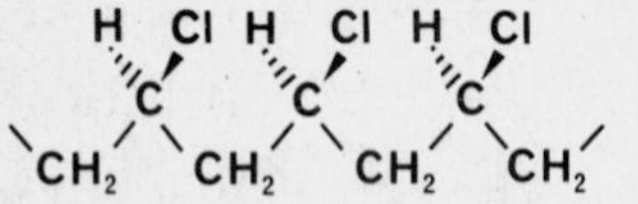 is:

1. syndiotactic
2. amorphous
3. atactic
4. block
5. glassy
6. isotactic

85. Which of the following is a step-growth polymer?

1. polystyrene
2. nylon
3. Teflon
4. polyvinyl chloride
5. synthetic rubber

86. How many chiral centers are present in androsterone?

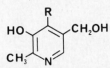

androsterone

1. 0   2. 3   3. 5   4. 7   5. 9

87.

$$CHO \quad\quad H$$

H————OH and CH_3————CHO

$$CH_3 \quad\quad OH$$

The above Fischer projection formulas represent:

1. the same compound   2. enantiomers   3. diastereomers
4. *meso* forms   5. rotamers

88. The parent heterocyclic ring present in vitamin B_6 is:

HO— ... R ... —CH_2OH

CH_3 — N

vitamin B_6

1. pyridine   2. pyrrole   3. pyrimidine
4. purine   5. imidazole

89. Oxidation of ethanol can give:

1. $CH_2{=}O$ 2. $HOCH_2CH_2OH$ and $CH_3CH(OH)_2$ 3. $CO_2 + H_2$
4. $CH_3CHO + CH_3CO_2H$ 5. $CH_3OCH_3 + CH_3CO_2H$

90. The structure of purine is . The least basic nitrogen present is:

1. N-1   2. N-3   3. N-7   4. N-9
5. They all are approximately equally basic.

437

91. The formulas for the amino acids glycine and alanine are as follows:

$$CH_2CO_2^-$$
$$|$$
$$^+NH_3$$
glycine

$$CH_3CHCO_2^-$$
$$|$$
$$^+NH_3$$
alanine

The correct formula for alanylglycine is:

1. $H_3\overset{+}{N}CH\overset{O}{\overset{\|}{C}}\text{—}\overset{+}{N}H_2CH_2CO_2^-$
   $|$
   CH_3

2. $H_3\overset{+}{N}CH_2\overset{O}{\overset{\|}{C}}\text{—}NHCHCO_2^-$
   $|$
   CH_3

3. $H_3\overset{+}{N}CH\overset{O}{\overset{\|}{C}}\text{—}CHCO_2^-$
   $|$ $|$
   CH_3 $^+NH_3$

4. $H_3\overset{+}{N}CH\overset{O}{\overset{\|}{C}}\text{—}NHCH_2CO_2^-$
   $|$
   CH_3

5. $H_3\overset{+}{N}CH\overset{O}{\overset{\|}{C}}\text{—}ONHCH_2CO_2^-$
   $|$
   CH_3

92. Which of the following can act as a nucleophile?

a. H_2O b. OH^- c. H^+ d. CN^- e. Na^+

1. a, b, and d   2. a and b   3. c and e   4. only d   5. b and d

93. Penicillin is a β-lactam. Its formula is:

1.

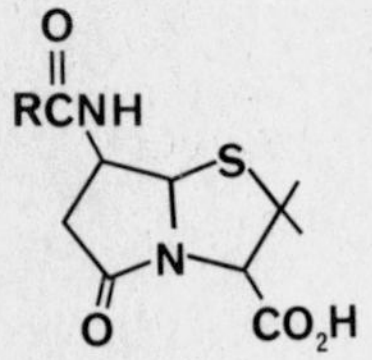

2.

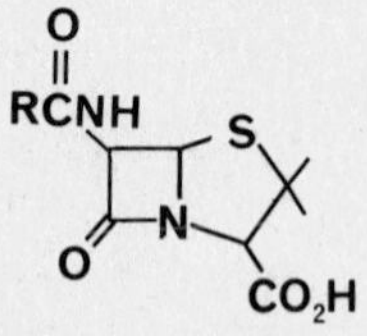

3.

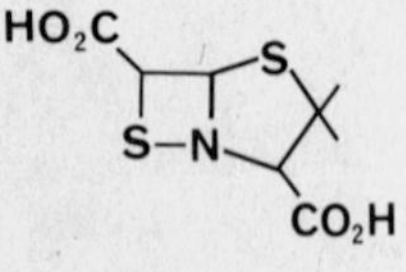

4.

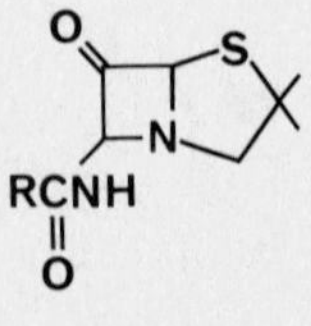

5.

94. The antihistamine diphenhydramine (Benadryl) has the formula

$\left(\text{phenyl}\right)_2$—CHOCH$_2CH_2$N(CH$_3$)$_2$. On heating with aqueous acid, it gives:

1. $\left(\text{phenyl}\right)_2$—CHOH + HOCH$_2CH_2$N(CH$_3$)$_2$

2. $\left(\text{phenyl}\right)_2$—CHOH + CH$_3CH_2$N(CH$_3$)$_2$

3. $\left(\text{phenyl}\right)_2$—CH$_2$ + HOCH$_2$CH$_2$N(CH$_3$)$_2$

4. $\left(\text{phenyl}\right)_2$—CHOCH$_2CH_2NH_2$ + CH$_3$—CH$_3$

5. $\left(\text{phenyl}\right)_2$—CHOCH$_2CH_2\overset{+}{N}$H(CH$_3$)$_2$

95. How many peaks appear in the ^{1}H NMR spectrum of the following compound?

1. 1   2. 2   3. 3   4. 6   5. 9

96. How many peaks appear in the ^{13}C NMR spectrum of the compound in Question 95?

1. 1   2. 2   3. 3   4. 6   5. 9

97. The ^1H NMR spectrum of one isomer of $C_3H_3Cl_5$ consists of a triplet at δ 4.5 and a doublet at δ 6.0, with relative areas 1:2. The structure of the isomer is

1. $CH_2ClCHClCCl_3$ 2. $CH_2ClCCl_2CHCl_2$ 3. $CHCl_2CH_2CCl_3$
4. $CHCl_2CHClCHCl_2$ 5. $CH_3CCl_2CCl_3$

98. Cocaine is an alkaloid. Its correct structure is

1.

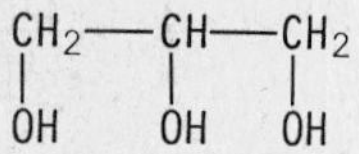

2.

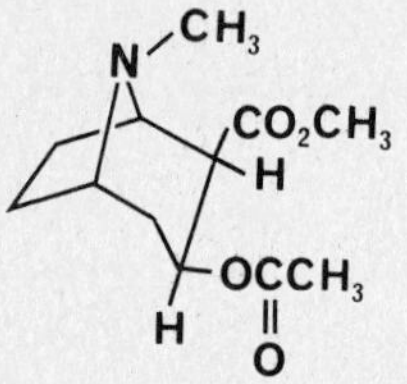

3.

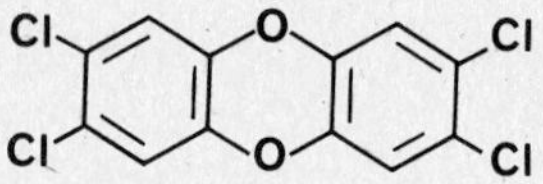

4.

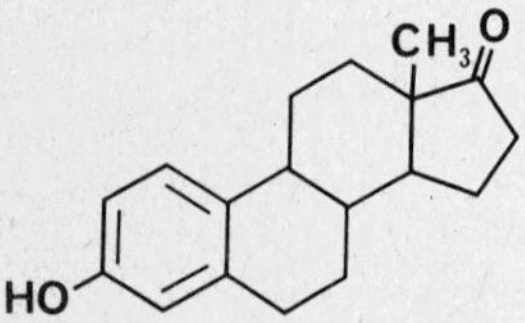

5.

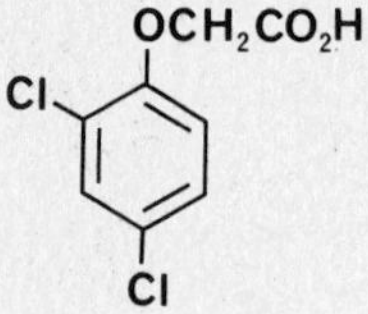

99. Which of the following statements is *not* true?

1. Natural rubber is a hydrocarbon.
2. Natural rubber is made of isoprene units.
3. Natural rubber is a polymer of 1,3-butadiene.
4. Natural rubber has *cis* double bonds.
5. Natural rubber can be vulcanized.

100. What is(are) the product(s) of the following reaction?

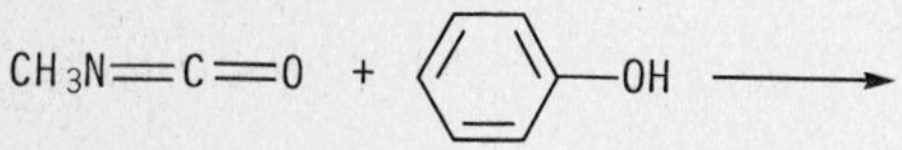

$$CH_3N{=}C{=}O \; + \; \text{(phenol)} {-}OH \longrightarrow$$

1.

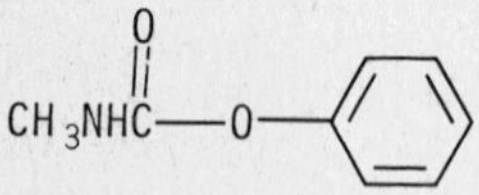

2.

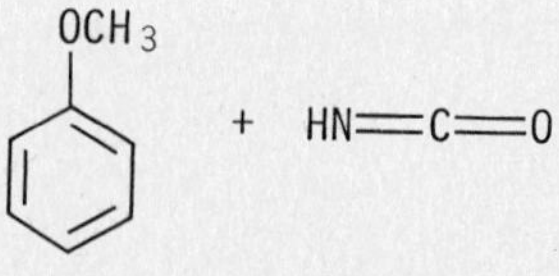

3. 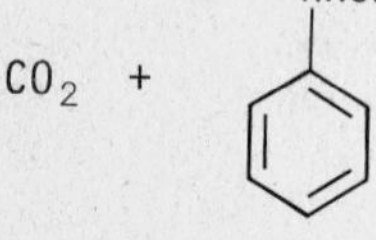 CO_2 +

4. a polyurethane

5.

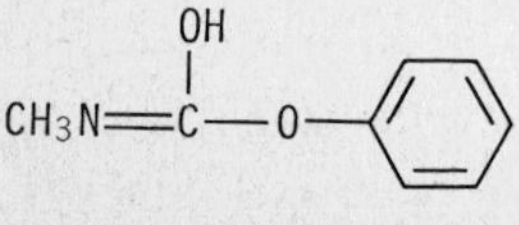